Ecological Studies, Vol. 183

Analysis and Synthesis

Ecological Studies

Volumes published since 2001 are listed at the end of this book.

E.A. Vaganov M.K. Hughes A.V. Shashkin

Growth Dynamics of Conifer Tree Rings

Images of Past and Future Environments

With 178 Figures and 22 Tables

Springer

Prof. Dr. Eugene A. Vaganov
V.N. Sukachev Institute of Forest
Russian Academy of Sciences
Siberian Branch
Academgorodok, Krasnoyarsk
660036, Russia

Prof. Dr. Malcolm K. Hughes
Laboratory of Tree-Ring Research
University of Arizona
W. Stadium #105
Tucson, AZ 85721, USA

Dr. Alexander V. Shashkin
V.N. Sukachev Institute of Forest
Russian Academy of Sciences
Siberian Branch
Academgorodok, Krasnoyarsk
66036, Russia

ISSN 0070-8356
ISBN-10 3-540-26086-2 Springer Berlin Heidelberg New York
ISBN-13 978-3-540-26086-8 Springer Berlin Heidelberg New York

Springer is a part of Springer Science+Business Media
springeronline.com

Printed in Germany

Editor: Dr. Dieter Czeschlik, Heidelberg, Germany
Desk editor: Dr. Andrea Schlitzberger, Heidelberg, Germany
Cover design: *design & production* GmbH, Heidelberg, Germany
Typesetting and production: Friedmut Kröner, Heidelberg, Germany

31/3152 YK – 5 4 3 2 1 0 – Printed on acid free paper

Preface

This is an expanded, enhanced, and updated version of "*Growth and Structure of Tree Rings of Conifers*" by E.A. Vaganov and A. Shashkin, a book that was published in Russian. Since the publication in 1976 of Harold Fritts' seminal work "*Tree Rings and Climate*", there has been a massive growth in the application of the techniques of dendrochronology in a wide range of scientific fields. Work flowing directly from that of Fritts has recently been extremely prominent in the study of the climate of recent centuries on a global scale; and other applications of dendrochronology have made significant impacts on, for example, disturbance ecology. In several recent cases, dendrochronologists have detected, and sought to explain, marked changes in the climate control of tree-ring growth in recent decades on very large spatial scales. These changes have important consequences for the projection of the role of forests in the global carbon cycle in coming years.

Although these applications have, from the beginning, been conducted on a rigorously quantitative basis and informed by an appreciation of the biology and ecology of the trees involved, they have been limited until the last few years by the almost complete lack of relevant process-based models. In this book, the reader is introduced to an attempt to fill this gap. This is not a text on tree biology, nor is it intended as a review of recent developments concerning the vascular cambium. Readers wishing to pursue these topics will find that a number of volumes in the "Springer Series in Wood Science" are of interest (Carlquist 2001; Larson 1994; Schweingruber 1993; Roberts et al. 1988).

Basic ideas concerning the nature and environmental control of the process of tree-ring growth are introduced here, exclusively for conifers. This is, however, done only to the extent necessary to explain the development, testing, and application of the simulation models described in this book. The reader is introduced to one family of process-based models of the environmental control of tree-ring growth, whose most distinguishing features are an emphasis on cambial dynamics and the modeling of intra- and inter-annual variability in tree-ring growth.

We have been greatly aided by numerous colleagues and institutions, both in the writing of the original Russian volume and in the development of this expanded and modified English language version. E.A. Vaganov is grateful for the generous support of the Alexander Von Humboldt Foundation, from which he received a research award in 2003, and to the Max-Planck-Gesellschafts Institute for Biogeochemistry in Jena, Germany, where he held the award. He also acknowledges the generous support of the Ministry of Science and Education of the Russian Federation (grant 2108.2003.4, supporting scientific schools). M.K. Hughes acknowledges the support of the Laboratory of Tree-Ring Research, University of Arizona. Erena Mikhina, Olga Sidorova, and Anastasiya Zelenova worked long and hard on the preparation of the manuscript and its many figures, tables, and references; and we are greatly indebted to them. The members of the Spring 2004 Graduate Class in "Dynamics of Tree-Ring Formation" at the Laboratory of Tree-Ring Research, University of Arizona, were a gracious and constructive "test audience". Kevin Anchukaitis, Franco Biondi, David Meko, Thomas Swetnam, and Frank Telewski made a number of valuable suggestions. We have been sustained in the task of preparing this volume by the support and encouragement of Ernst-Detlef Schulze, Director of the MPI for Biogeochemistry in Jena, and Galina Vaganova, Maria and Mark D'Alarcao, and Rachel Hughes.

Krasnoyarsk and Tucson
July 2005

E.A. Vaganov
M.K. Hughes
A.V. Shashkin

Contents

Table of Symbols

Symbols describing environmental factors

T	Temperature
W	Soil water content (absolute and relative)
P	Precipitation
E	Solar irradiation
S_{THAW}	Depth of soil thawing layer

Symbols describing the anatomy of tree rings

N	Total number of cells in annual file of tracheids; or annual xylem increment in the number of tracheids
L_T	Width of tree rings measured parallel to tracheid file direction
D_R	Radial size of tracheid (radial tracheid dimension)
D_T	Tangential size of tracheids
D_L	Size of lumen
V	Rate of radial enlargement (dD_R/dt)
L_W	Cell wall thickness
A_W	Cell wall area (cross-sectional area occupied by cell wall)
A_C	Cell cross-sectional area
A_L	Area of lumen (inner area surrounded by cell wall)
M_W	Mass of cell wall
ρ_W	Density of all wall material
ρ_X	Density (specific gravity) of wood

Symbols describing the processes of wood formation

t	Time
τ	Time interval
N_c	Number of cells in the cambial zone
N_e	Number of cells in the zone of radial enlargement
N_t	Number of cells in the zone of cell wall thickening
N_m	Number of cells in the zone of tracheids that have completed the differentiation process (mature cells)
j	Distance (position) of a xylem mother cell (dividing) in the cambial zone from the initial
M_I	Mitotic index (ratio of number of cells in mitosis to total number of cells in the dividing population)
M	Length of time for mitosis, process ending in the production of two new daughter cells from one mother cell in the cambial zone
C	Cell cycle (sometimes the term mitotic cycle is used): time between completed cell divisions In the case of all potentially existing dividing cells (the real meristem), the relationship between M_I, M and C is described as: $M_I = (\ln_2 M)/C$ (Ivanov 1974)
R_{SPL}	Specific growth rate (linear scale) within cambial zone, inversely related to C ($R_{SPL} = 1/C$)
R_{SPN}	Specific growth rate of cell production [$R_{SPN} = (1/N_c)/(dN/dt)$] For the meristem, R_{SPL} and R_{SPN} are the same if the size of xylem mother cells before division is constant throughout the cambial zone
$G(t)$	Calculated growth rate
g_T	Partial growth rate due to temperature
g_W	Partial growth rate due to soil water content
g_E	Partial growth rate due to solar irradiation

Terms describing the three major processes of cell differentiation

t_{GS}	Length of growing season (includes t_S, t_P, t_E; therefore $t_{GS} > t_P, t_E$)
t_S	Period of swelling in the cambial zone
t_P	Period of new cell production by the cambial zone
t_E	Period of enlargement from the first to the last tracheids formed

1 Introduction and Factors Influencing the Seasonal Growth of Trees

1.1 Introduction

1.1.1 Perspective

Our aim in this book is to introduce the way of thinking about the environmental control of tree-ring variability in conifers that is expressed clearly in its title: "*Growth dynamics of conifer tree rings: images of past and future environments*". In particular, each ring contains an image of the time when the ring formed, projected onto the ring's size, structure and composition. The lens through which this projection occurs is the vascular cambium, the site of development of each year's ring. We focus on its dynamics. Our particular perspective comes from our chosen task – the extraction of an image of past environments, especially climate variability, from the incomparable natural archives that tree rings offer. The emphasis on variability leads naturally to a dynamic rather than a static view of climate/tree-ring interactions.

In order to best use our understanding of the environmental control of tree-ring variability, simulation models have been developed. The aim is to capture those features of the system under investigation that are essential for the description of the behavior of interest, no more and no less. Our specific objective is to simulate the interannual and decadal variability of conifer tree rings as it is driven by climate variability. We do this by focusing strongly, and uniquely, on the direct environmental control of cambial activity, without any explicit treatment of photosynthesis, respiration, and transpiration. This may seem a radical, perhaps even extreme, approach to the readers of this series, who are especially aware of the complexities of ecological systems and the many physical, biophysical, and biological processes that may be linked to any particular phenomenon in myriad modalities. We hope to convince the reader that, even so, this strategy has merit.

Process-based simulation modeling should be viewed as an addition to the dendroclimatologists' already versatile toolkit. The mainly inductive empirical–statistical approach used in this field has helped change the way we think about environmental variability on multi-year timescales and large spatial scales, up to global. It can be complemented and enhanced by the application of understanding derived from experimental–deductive work more focused on daily to seasonal timescales at cellular and even biochemical levels. We recognize that we risk double jeopardy, both for trying to combine these ways of studying nature, and for perceived inadequacies from the point of view of either group of specialists. This will be a price worth paying, if those focused only on the history of large-scale climate variability gain some appreciation for the nature of the natural archive they depend on and its meaning for their interpretations, and those focused on the precise mechanisms of environmental control of xylogenesis are introduced to the many phenomena at larger scales that dendroclimatology reveals.

1.1.2 The Structure of This Book

The main factors affecting the seasonal growth of trees are introduced in this chapter. In Chap. 2, the reader is introduced to the idea of the tree ring's size and internal structure as an image of the growth conditions that existed at the time it was formed. The processes of ring development are set in a kinetic, seasonal context in Chaps 3, 4, and 5, which deal with the production, expansion, and maturation of xylem cells, respectively. A conceptual scheme for the environmental control of xylem differentiation is described in Chap. 6. This leads, in Chap. 7, to a discussion of modeling of external influence on tree-ring structure and, in particular to a description of the Vaganov–Shashkin (VS) model. The VS model is characterized by a strong emphasis on cambial activity, as if directly influenced by the external environment. Examples of the application of the VS model designed to illustrate its potential in tackling some pressing scientific questions are given in Chap. 8. An expansion of this modeling approach designed to explicitly include the influence of the canopy and the forest stand is described in Chap. 9. The Epilogue (Chap. 10) contains some parting thoughts.

1.2 The Environment and Tree-Ring Formation

The formation of annual rings in woody plants, as well as the formation of annual and growth layers in "recording structures" of other organisms (Mina and Klevezal 1970), is an outcome of the seasonal periodicity of growth processes. The growth of a woody plant represents an increase in the weight and volume of the whole plant or its parts as the result of the formation of new cells and the increase in their size (Reimers 1991). The seasonal periodicity or rhythm of biological processes in a woody plant is determined by regular environmental fluctuations associated with the annual cycle. The expression of these seasonal rhythms depends on the species of woody plant and the local conditions where the plant is growing.

Zones of formation of new cells in plants are called meristems. Apical meristems (meristems of shoots and roots) provide growth in height and length of the underground parts. Tree stems and roots increase in thickness as the result of the production of wood and bark cells by lateral meristems. There is a genetic relationship between apical and lateral meristems because the lateral meristem or cambium is a product of the apical meristem and consists of cells retaining the potential to proliferate (Zimmermann and Brown 1971). There is also a functional relationship, as demonstrated by the coordination of the processes of growth of different parts of a woody plant in the seasonal rhythm (Sinnot 1963; Kramer and Kozlowski 1983; Kozlowski and Pallardy 1997). Growth hormones are the mediators of this coordination. Their balance, together with assimilates and other nutritious substances, is important for the dynamics of growth processes and the formation of the cell structures of the tissues and organs of the plant (Kozlowski 1968; Philipson et al. 1971; Savidge 1996).

1.3 Internal Factors

We will limit our discussion to those internal factors whose influence on the dynamics of seasonal growth in woody plants has been established convincingly by experiment and observation. The genetic nature of the plant is one of the major factors. Investigating seasonal growth of 18 tree species during one growing season in Denmark, Ladefoged (1952) noted precise differences in phenology, the timing of bud opening, and the initiation of meristem activities in the shoots, stems and roots between different species (Fig. 1.1). He found a difference of three weeks in the date of initiation of new wood cell production amongst the conifer species he observed. The first divisions of lateral meristem cells in spruce (*Picea sitchensis*) stems are observed in the third ten-day period of April, in stems of a Scots pine (*Pinus sylvestris*) right at the beginning of May, and in stems of European larch (*Larix decidua*) only in the middle of May. The data by Henhappl (1965), who studied seasonal growth of seven species of conifers and 11 deciduous species near Freiburg, Germany, during three years, show that initiation of cell division in conifers ranged over 40 days. The earliest was Scots pine (*P. sylvestris*), the latest white pine (*P. strobus*). The length of the growth season varied from 104 days in European larch (*L. decidua*) to 137 days in fir (*Abies alba*).

Cambial initiation begins below the expanding/enlongating buds and the wave of activation propagates basipetally, so that the cambium at the base of

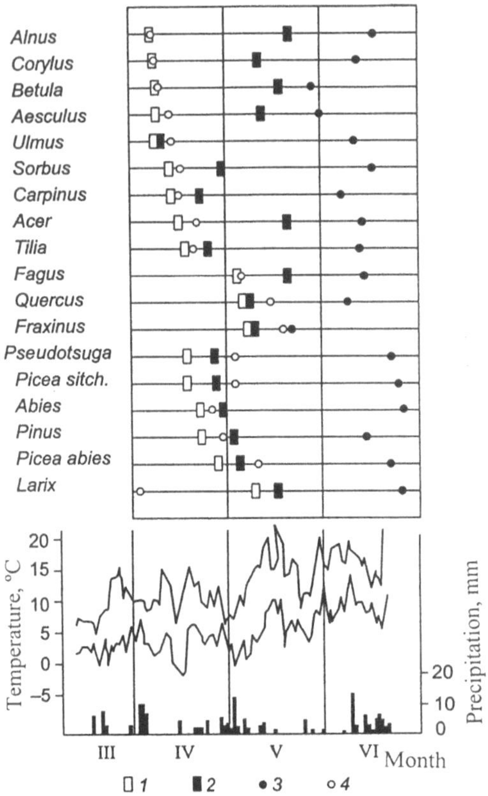

Fig. 1.1. Timing of onset of cambial activity in conifers and angiosperms in relationship to temperature and precipitation in Denmark. *1* At the bud base, *2* at breast height, *3* roots 10 cm deep and 1.5 m from the stem, *4* bud opening. Notice the time-lag between cambial activity in the branch and the root. This is possibly due to regulation through growth hormones (Ladefoged 1952)

the tree is the last to divide and differentiate (Zimmermann and Brown 1971; Kramer and Kozlowski 1983; Kozlowski and Pallardy 1997). Then, in the fall, the reverse process begins when the oldest cambial tissues (at the base of the tree) become dormant first and the process propagates acropetally back to the base of the apical meristems of the crown.

Differences between species are found not only in the timing of the initiation and termination of cell divisions in meristems, but also in the seasonal dynamics of growth. Ladefoged (1952) divided tree species into three groups on the basis of qualitative analysis of growth-rate curves: (1) with a growth rate maximum in the first third of the season, (2) with a more symmetrical curve of growth rate, (3) with uniform growth rate during the growing season. Similarly, in the Moscow region, Scots pine (*P. sylvestris*), European fir (*A. alba*), and European larch (*L. decidua*) demonstrate a distinct growth maximum in the first half of the season, while birch (*Betula pendula*) shows a uniform distribution of growth rate throughout the season (Smirnov 1964; Vaganov et al. 1975). The distribution through the season of the rate of growth is related to the characteristics of the species. For example, analysis of the sea-

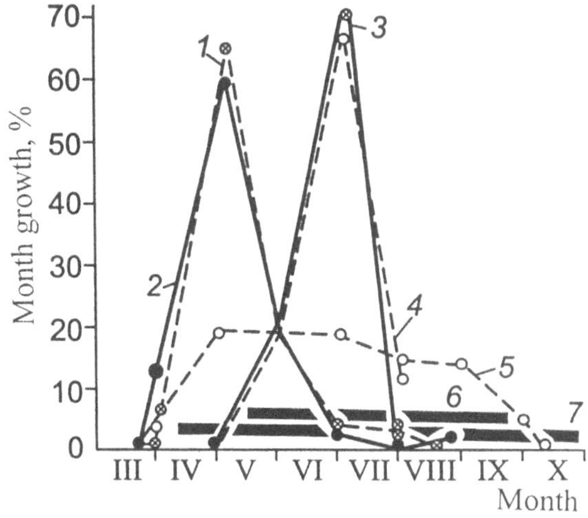

Fig. 1.2. Types of variation of height growth during the vegetation period in the northern and southern pine ecotypes. Two northern pines have preformed shoots and one shoot-forming period in a year. The southern pines have periodic shoot growth. *1 Pinus strobus* L. (North Carolina), *2 P. resinosa* Ait. (North Carolina), *3 P. resinosa* Ait. (New Hampshire), *4 P. strobus* L. (New Hampshire), *5* average for all southern pines, *6, 7* frost-free period in New Hampshire and North Carolina, accordingly (Kramer and Kozlowski 1983)

sonal dynamics of growth of different species of pines (*P. strobus* and *P. resinosa*) in the forests of the south- and northeastern United States reveals that northern ecotypes have an "explosive" character of seasonal growth rate change, whereas southern ecotypes have a more uniform distribution of growth rate during the season (Fig. 1.2; Kramer and Kozlowski 1983). This difference exists because the northern ecotypes have preformed shoots (limited growth) and only one period of shoot formation per year, whereas continuing periodic growth occurs in the southern ecotypes. Spurr and Barnes (1980) report differences in seasonal dynamics of growth rate ("explosive" or uniform) between *P. rigida*, *P. densiflora*, and *P. banksiana*.

The seasonal growth of shoots and needles (foliage), through the hormonal control of cambial activity, is of central importance to the character and distribution of the seasonal growth rate of stem wood (see reviews in Kramer and Kozlowski 1983; Zimmermann 1964; Zimmermann and Brown 1971; Barnett 1981; Savidge 1996). It is through this means that the processes of development of various tissues and organs are coordinated within the annual cycle, in a manner that varies from species to species. This is illustrated vividly by species-specific patterns of coordination of growth of aboveground (shoots) and underground (roots) parts of woody plants (Fig. 1.3; Lyr et al. 1974). Not only do the maximum increments vary between species, but so do the beginning, ending, and duration of the linear growth of shoots and roots. The coordination of the seasonal dynamics of growth of the various parts of woody plants results in allometric relations between them during the growth and development of the woody plant and the stand (Utkin 1982; Terskov and Vaganov 1978; Vaganov 1981; Utkin et al. 1996).

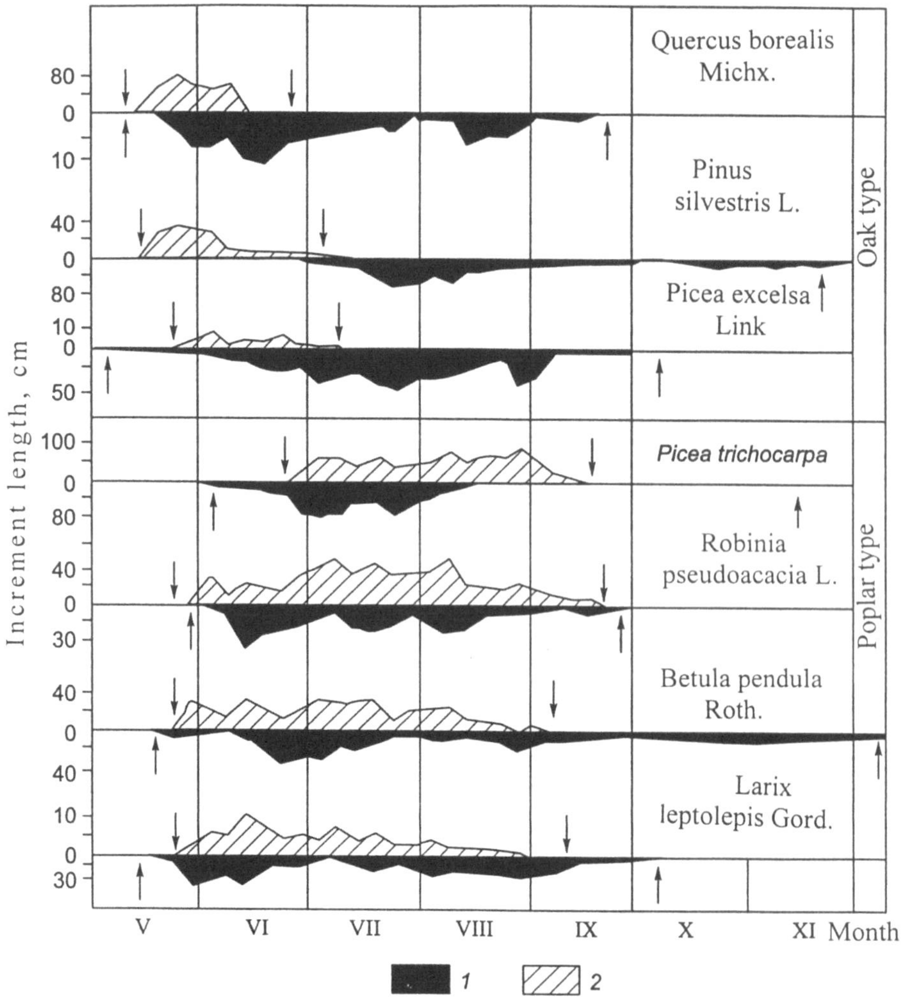

Fig. 1.3. Schematic diagram of annual root growth (*1*) and growth of the aboveground part (*2*) in some tree species in 1963, Eberswald, Germany. The *arrows* show the beginning and the end of growth (Lyr et al. 1974)

Age is also a major internal factor connected with the genotype and influencing the seasonal dynamics of growth processes in woody plants. There is a very wide range (30-fold or more) of maximum longevity in tree stems according to species (Table 1.1), across several major taxa of conifers. The association between longevity and adversity may be seen in situations where conditions are generally limiting to growth, so that the longest-lived species in semi-arid regions may typically be found in the most extreme locations within such regions (Schulman 1958). This is seen most dramatically in the five-needle pines of western North America (e.g. *P. longaeva, P. aristata, P. bal-*

Table 1.1 Maximum ages of conifer species used in dendrochronology. (Modified from Brown 1994). All ages are from cross-dated samples reported at http://www.rmtrr.org/oldlist.htm, with the exception of *Sequoia sempervirens*, which is derived from a ring-count. and the following species, which are from cross-dated samples reported elsewhere: *Austrocedrus chilensis* and *Larix decidua* (International Tree-Ring Data Bank), *Pinus sylvestris* (http://www.botanik.uni-bonn.de/conifers/pi/pin/sylvestris.htm), *Juniperus phoenicea* (R. Touchan, personal communication), *Larix cajanderi* (M. Naurzbaev, personal communication)

Species	Age (years)	Location
>4,000 years		
Pinus longaeva	4,844	Nevada, USA
>3,000 years		
Fitzroya cupressoides	3,622	Chile
Sequoiadendron giganteum	3,266	California, USA
>2,000 years		
Juniperus occidentalis	2,675	California, USA
P. aristata	2,435	Colorado, USA
Sequoia sempervirens	2,200	California, USA
P. balfouriana	2,110	California, USA
>1,000 years		
J. scopulorum	1,889	New Mexico, USA
P. flexilis	1,670	New Mexico, USA
Thuja occidentalis	1,653	Ontario, Canada
Taxodium distichum	1,622	N. Carolina, USA
P. albicaulis	1,267	Idaho, USA
Pseudotsuga menziesii	1,275	New Mexico, USA
Larix cajanderi	1,216	Yakutia, Russia
Lagarostrobus franklinii	1,089	Tasmania, Australia
>500 years		
Pinus edulis	973	Utah, USA
Larix decidua	932	France
Picea engelmannii	911	Colorado, USA
Austrocedrus chilensis	850	Chile
Pinus ponderosa	843	Utah, USA
Araucarua araucana	834	Argentina
L. siberica	750	Mongolia
L. lyalli	728	Alberta, Canada
P. sylvestris	711	Sweden
Abies magnifica var. *shastensis*	665	California, USA
P. sibirica	629	Mongolia
P. strobiformis	599	New Mexico, USA
J. phoenicea	526	Jordan
Picea glauca	522	Alaska, USA
>300 years		
Picea abies	468	Germany
P. rubens	405	New Hampshire, USA

fouriana, P. flexilis, P. albicaulis) and in other conifers, such as species of *Juniperus* and *Sabina* in central Asia and species of *Austrocedrus* and *Araucaria* in South America. These species are able to survive under conditions where they have few if any competitors and have characteristics of special value for surviving harsh and extremely variable environments, such as anatomical adaptations for an enhanced role of needles as storage organs (Connor and Lanner 1990) and the retention of needles for as long as 30 years (LaMarche 1974a). This has important consequences for the effect of the annual cycle and interannual variability on the formation of wood. The interannual variability of ring widths is greatest near the lower elevational limits of these species and is smallest at the upper limit, where there are very strong correlations between succeeding ring widths (LaMarche 1974b; Hughes and Funkhouser 2003). The maximum ages of all of these species are greater than 1,000 years, and in the case of *P. longaeva* approach 5,000 years. Similarly, the longest-lived larch species in the forest-tundra of Siberia, *Larix cajanderi*, is found at the northern timberline in the particularly extreme subarctic climate of northern Yakutia. Arno and Hammerly (1984) have described the larches of northern and elevational timberlines as trees that are able to grow in very extreme environments as a result of their deciduous habit, which lessens the effects of severe winters on the plant. This too has important consequences for the dynamics of wood formation in relation to the annual cycle and results in ring-width series where there is almost no persistence, that is, no correlation between one year's ring width and the next.

There are, however, several very long-lived (multimillennial) tree species that are not found at the extreme limits of tree growth. These are members of the families Taxodiaceae or Cupressaceae, such as *Taxodium distichum*, *Sequoia sempervirens*, and *Sequoiadendron giganteum* in North America, *Fitzroya cupressoides* in South America, and *Cupressus* species in eastern Asia. These species share several characteristics. Several of them first establish as pioneers, requiring mineral soil exposed by fire (*S. giganteum*) or neotectonic activity (*F. cupressoides*), and then reach great heights within a few centuries, obviating competition for light. All show a very strong taper in stem width near the base and usually a very parallel form of the stem above this zone, resulting in characteristic vertical patterns of ring width. As in the case of the long-lived pines of North America, they all produce wood that is extremely resistant to pests and pathogens and so may better survive repeated insult, such as lightning or wind damage. These species demonstrate a variety of "strategies" for longevity that are intimately associated with the interaction of the environment, including the annual cycle and the seasonal dynamics of wood formation. All of these long-lived species, however, provide excellent opportunities for future research on the relationship between the age of the cambium and the precise seasonal dynamics of wood formation, since it is possible to collect materials of a vast range of cambial ages from their stems and branches.

In the analysis of age changes, it is necessary to distinguish the age of a tissue or organ from that of the tree as a whole. For example, a tree at the age of 500 years has cambium aged about 500 years at the bottom of a trunk and aged 20–50 years in the upper part of the crown. The scheme in Fig. 1.4 shows that the actual age of cambium differently integrates the influences of internal and external factors, and this in turn is reflected in anatomical changes in the wood and tissue cells formed. As the actual age of the cambium increases from pith to bark, tracheid length, cross-sectional cell diameter, and production of resin ducts increase; and the width of annual rings and the percentage of late wood decrease (Vysotskaya and Vaganov 1989; Wimmer 1994). The wood cells made by young cambium have smaller radial sizes, a smaller cell-wall thickness, and a lower wood density, as well as more earlywood and less latewood (Telewski and Lynch 1991). The contrast between earlywood and latewood is also not as large, especially in wood density (Savva et al. 2002a,b). Similar changes are found in all woody plants and are associated with the phenomenon of "juvenile wood" (Fig. 1.5; Zobel and Jett 1995). Young cambium is characterized by high activity, resulting in the production of more

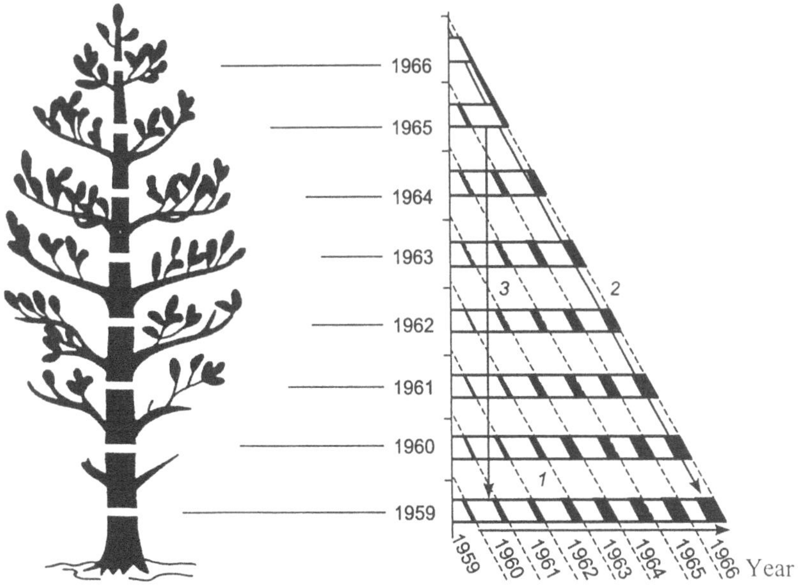

Fig. 1.4. In a series of cross-sections from various heights in the stem, a distinction can be made between endogenous and exogenous influencing factors. *Type 1* From the pith to the stem periphery. An ageing cambium experiences a variety of ecological conditions. *Type 2* Parallel to the youngest tree ring. A cambium with physiological varying ages (*above* young, *below* old) experiences the same ecological conditions. *Type 3* Parallel to the pith. Physiologically young cambium experiences a variety of ecological conditions (Duff and Nolan 1953, Smith and Wilsie 1961)

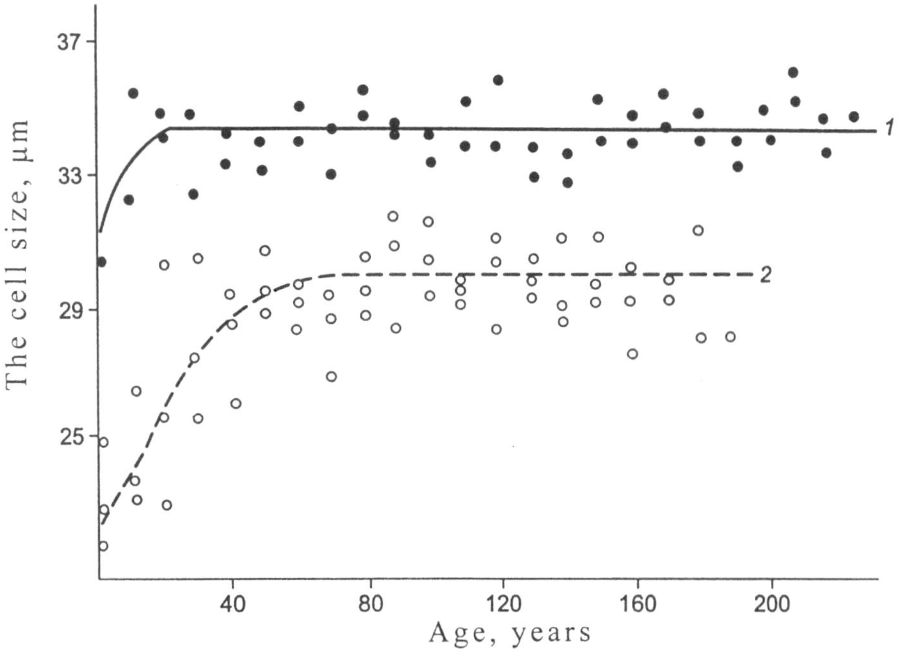

Fig. 1.5. Dependence of radial cell sizes on age of tree rings of pine growing in moist (*1*) and dry (*2*) sites (Vysotskaya and Vaganov 1989). This corresponds to series type 1 in Fig. 1.4

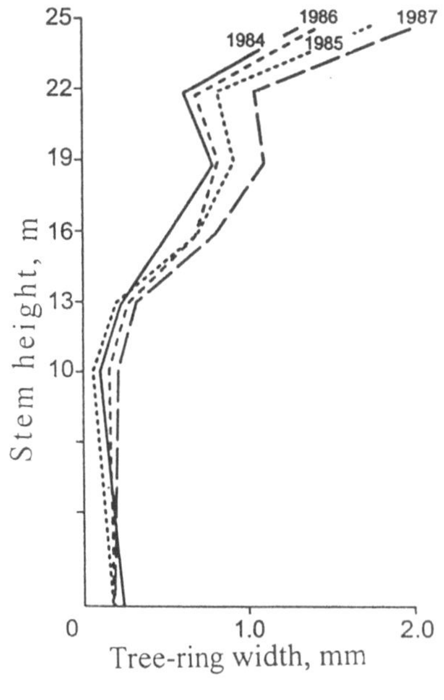

Fig. 1.6. Tree-ring width profiles for the years 1981–1987 along a spruce stem in northern Germany. Narrow rings are common in parts of the tree which grow the least (stem base, <10 m), and wide rings from each year in parts which grow the most (crown, >20 m). These series corresponds to type 2 in Fig. 1.4. (Krause and Eckstein 1992)

new wood cells and wider annual rings than those produced by older cambium (Fig. 1.6).

A typical example of the age changes in the main characteristics of wood is shown in Fig. 1.7. The dynamics of annual ring width and the width of the zones of early and late wood correspond to the so-called "curve of biological growth" or "curve of the grand growth period" (Shiyatov 1970, 1973a, 1986). Smaller relative changes are typical of tree-ring density (Braeker 1981). Carrer and Urbinati (2004) raise the interesting possibility that at least some of these changes and the strength of each tree's climate signal, may vary according to the capacity to support greater height growth by producing tracheids with larger lumina.

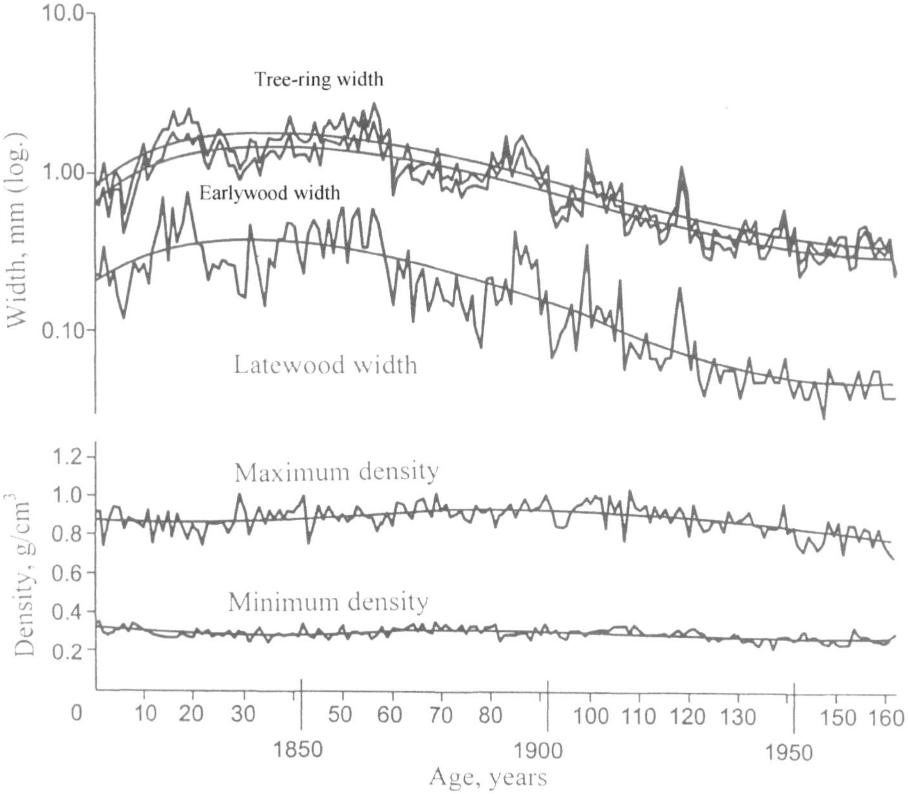

Fig. 1.7. Age changes in tree-ring width, early- and latewood width, maximum and minimum density in a *Larix sibirica* Ldb. tree

1.4 Physical–Geographical Factors

Among these factors, the most essential are regional climatology and relief. For example, there are strong gradients of average summer, winter, and annual temperature and precipitation along the Yenisei meridian in central Siberia (Fig. 1.8). As a result of these macro-climatic patterns, the duration of the growth season of trees extends from 35–40 days at the polar limit of forest (about 71° North) up to 95–105 days in the forest-steppe zone (51–52° North; Vaganov et al. 1999). An even longer period of seasonal tree growth is observed in the subtropical zone – up to 150–160 days (Vaganov and Park 1995). When tropical conditions exist with very small seasonal fluctuations of temperature, in an optimal zone for growth and with sufficient humidity, growth continues throughout the year.

Within a particular region, the timing of the beginning of the growth of woody plants and its duration varies according to elevation (Lobzhanidze 1961). This may be seen most clearly in mountain regions. In the case of European larch (*L. decidua*) at three different altitudes in the mountains of Tyrol, Austria (Tranquillini 1979), the greatest distinctions in growth are observed

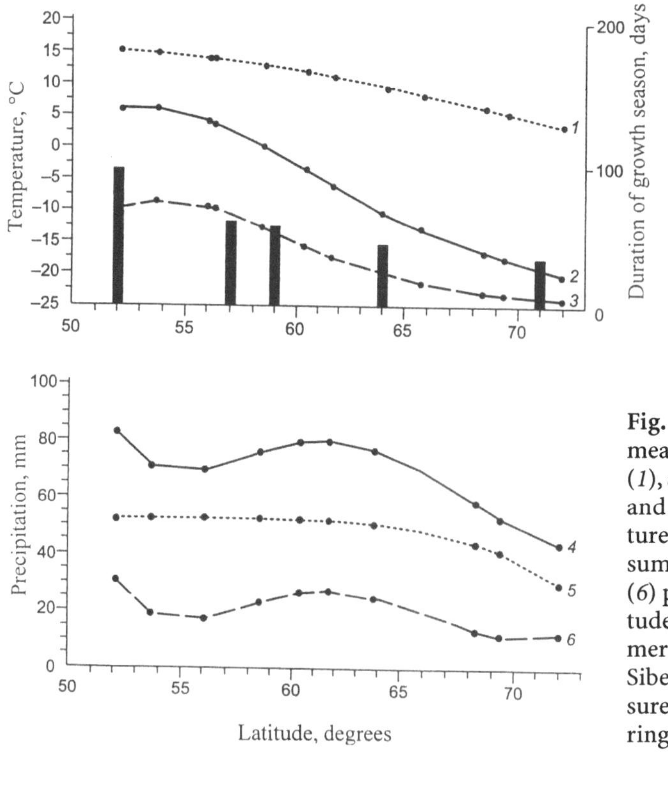

Fig. 1.8. Dependence of mean long-term summer (*1*), average annual (*2*), and winter (*3*) temperatures as well as annual (*4*), summer (*5*), and winter (*6*) precipitation on latitude along the Enisey meridian in central Siberia. *Columns* Measured duration of tree-ring growth season

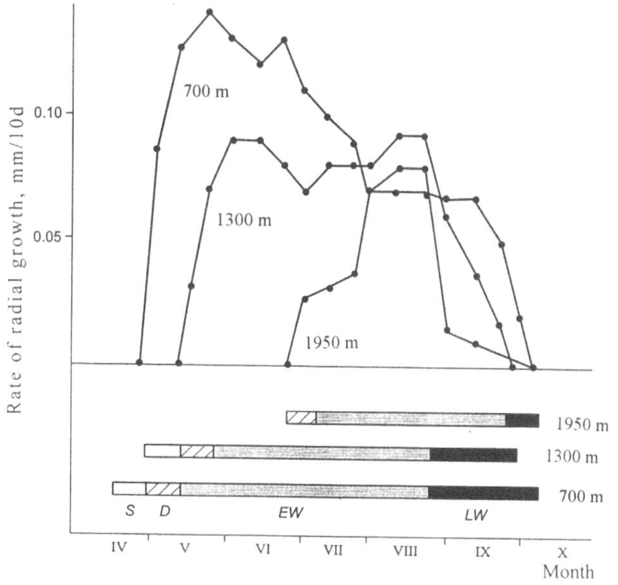

Fig. 1.9. Rate of radial growth in young larch trees and extent of phases of tree-ring formation at different elevations (700 m, 1,300 m, 1,950 m) in the Tirol mountains (Austria). *S* Cell swelling, *D* cell division in cambial zone, *EW* earlywood formation, *LW* latewood formation (Tranquillini 1979)

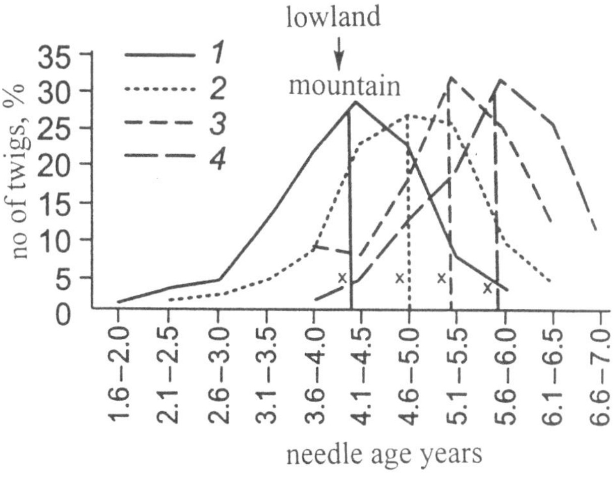

Fig. 1.10. Frequency distribution of average lifespan of spruce needles in northern Germany. The age of the needles increases as the elevation increases. *1–4* Regions from lower to higher elevations (Wachter 1985)

in terms of the beginning of cambial activation and cell divisions (Fig. 1.9). At the bottom of the mountains the growth begins at the end of April, whereas at 1,950 m above sea level meristem activity begins in the second half of June, i.e. more than 1.5 months later. These data also show that the termination dates of radial growth at all three elevations are similar. Therefore the growing season is shorter and the absolute tree-ring width smaller at the higher elevations. These results demonstrate the importance of the beginning of the growth season for growth and its integral outcomes, namely shoot lengthening and wood increment. In most conifers, growth rate and intensity of divisions in meristems are greatest in the first half of the season (Larson 1994). This is connected with the active production of growth hormones and a high concentration of assimilates (Sudachkova 1977). The unavoidably shorter seasonal growth duration in conifers at high elevations can be partially compensated by the retention of previous years' needles and the associated increase in photosynthetic efficiency. The number of years for which needles are retained in a spruce, for example, differs according to the elevation at which the trees grow (Fig. 1.10).

Local relief has a strong influence on the thermal regime (Gates 1980), so that eastern and southern slopes receive markedly more solar energy than western and northern slopes. This may result in very different seasonal courses of growth in shoots and stems of nearby woody plants of the same species (Fig. 1.11). As in the case of elevation differences, the main differences related to aspect result from differing initial dates for the growth season, for example, the later beginning of growth in trees on north-facing slopes.

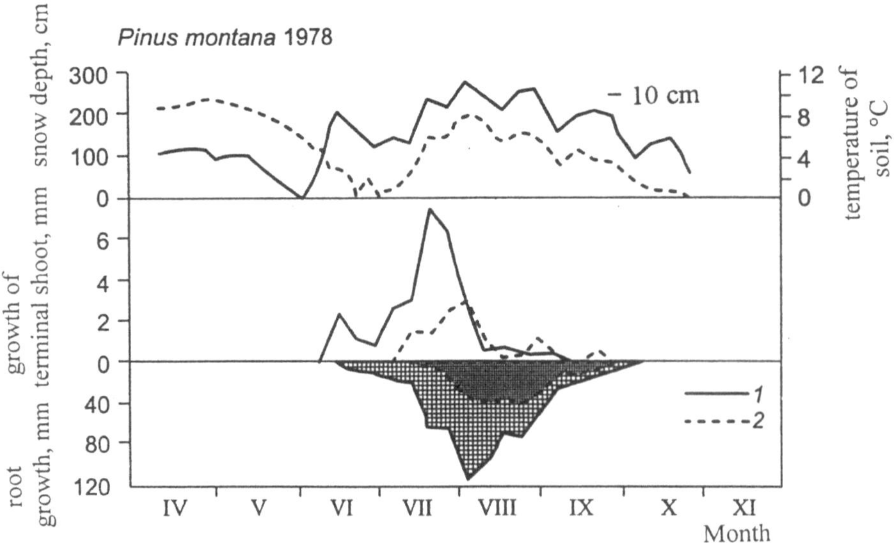

Fig. 1.11. Comparison of growth in the roots and shoots of two mountain pines with snow coverage and ground temperature on two sites during the 1978 season. The growing period lasted around 4.0 months on the more favorable eastern slope and only around 2.5 months on the cool northern slope. *1* East-facing slope, *2* north-facing slope (Turner and Streule 1983)

1.5 Soil Factors

We will consider only the major factors influencing tree growth, namely the temperature and water regime of the soil, its composition (mechanical, chemical, texture, etc.), and its content of mineral elements.

The seasonal dynamics of temperature in the upper layers of the soil are closely related to air temperature (Russel 1955; Pozdnyakov 1986; Zhang et al. 2000, 2001). These relationships may be strongly modified by the development of the forest floor and vegetation cover, which modify heat exchange and thus the gradient between air temperature and soil temperature. The influence of the soil thermal regime on woody plant growth may be seen clearly when permafrost soils are present. Moss cover interferes with heat exchange between air and soil, and so the boundary of the permafrost rises and suppresses growth both of the root system of trees and of other plant parts (Pozdnyakov 1986). When the forest floor or vegetation cover is disturbed (usually by fire), the seasonal warming of the upper soil layers of soil increases, the permafrost boundary recedes, and tree growth processes become more active. For example, after fire, larch tree-ring growth increases as the result of the degradation of the moss layer and improved soil thermal regime (Fig. 1.12). Growth is considerably suppressed in the pre-fire period, when thick moss cover allows the permafrost boundary to come close to the surface, and it sharply accelerates in the first post-fire years (Arbatskaya 1998).

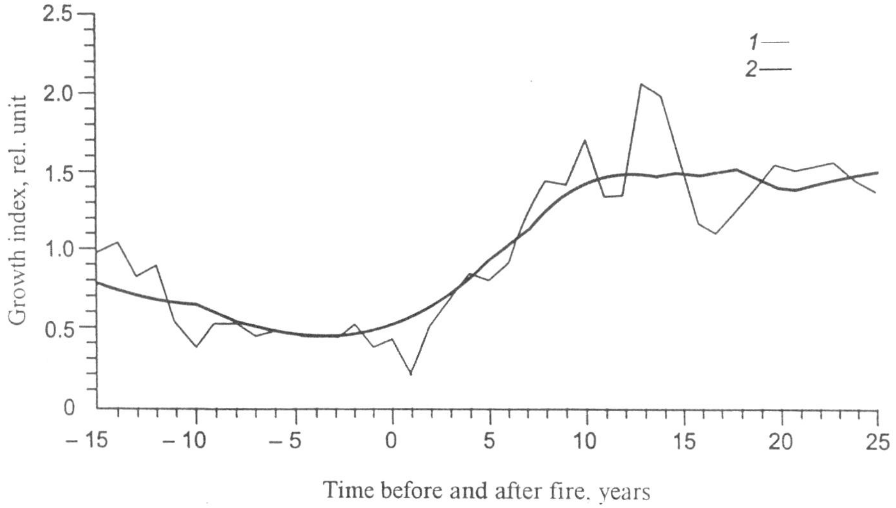

Fig. 1.12. Average pre- and post-fire response of larch radial growth for multiple trees and fires on permafrost (north taiga, Siberia). *1* Direct measurements, *2* smoothed curve (Arbatskaya 1998)

The seasonal warm-up of soil depends on its mechanical composition. Obviously, light sandy soils can get warm faster and to a greater depth than heavy clay. The mechanical composition of soil largely influences the mobility of moisture (Russel 1955; Spurr and Barnes 1980). The increase in field moisture capacity of soils (in a series from sandy soil to loam/sandy loam to clay) is accompanied by an increase of wilting point humidity (Table 1.2). Therefore in the seasonal cycle of growth, the reserves of water accessible to woody plants can vary over a wide range, depending on the timing and intensity of precipitation and transpiration The seasonal and intra-seasonal changes of transpiration play an important role in growth processes because they drive water movement, bringing mineral nutrients to assimilating tissues. Transpiration is largely limited by soil moisture content (Fig. 1.13).

Several authors have emphasized the acceleration of radial growth observed when the water table falls in marshy conditions (Melekhov and Melekhova 1958; Efremov 1987; Konstantinov 1981; Schultness 1990). This response to drying can be quite fast (Fig. 1.14), but can be extended for some decades (Efremov 1987). The detailed study of variability in radial increment in wet conditions due to water table fluctuations and the weather conditions of single seasons testifies that the direct influence of water regime changes is to a greater extent supplemented by an indirect one, namely by acceleration of the decomposition of organic substances and mineralization of the drier layers of soil cover and peat soil (Vaganov and Kachayev 1992). The response of growth acceleration at the expense of additional mineral assimilation follows with some delay in time.

Not surprisingly, much attention has been given to the effect of fertilization on tree growth and tree ring formation (Spiecker 1987, 1991; Buzykin 1977; Beets et al. 2001; Makinen et al. 2002). The outcomes varied over a wide range: from absence of response, up to practically direct connection with the growth rate change, depending upon the dosage of the introduced fertilizers.

Table 1.2 The field capacity, wilting point and available water in certain American soils (Russell 1955; 1 inch/foot = approx. 8.3 cm/m)

Soil	Field capacity (oven dry weight)	Permanent wilting point (oven dry weight)	Available water (inch/ foot)
Yuma sand	4.8	3.2	0.3
Delano sandy loam	9.1	4.2	0.8
Fresno sandy loam	11.1	3.1	1.3
Salinas fine sandy loam	28.2	20.0	1.3
Wooster silt loam	23.4	6.1	2.9
Aiken clay loam	31.1	25.7	0.7
Gila clay	30.4	16.0	2.4

Fig. 1.13. Impact of soil moisture on transpiration intensity of young oak trees. *1 Quercus alba* L., *2 Q. minor*, *3 Q. marilandica*, *4 Q. rubra* L., *5 Q. falcata*. *Arrow* shows wilting point (Kramer and Kozlowski 1983)

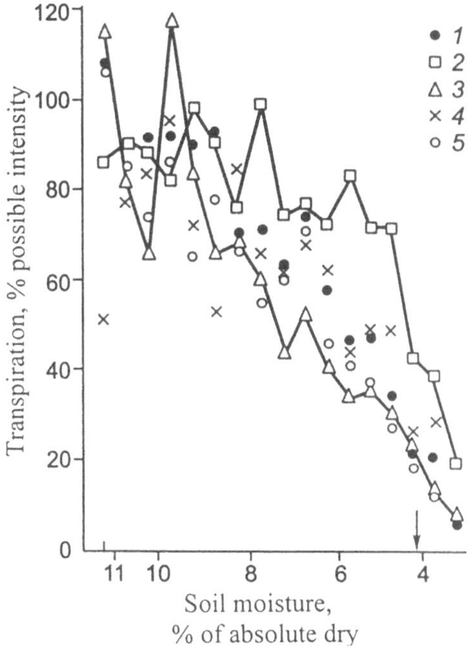

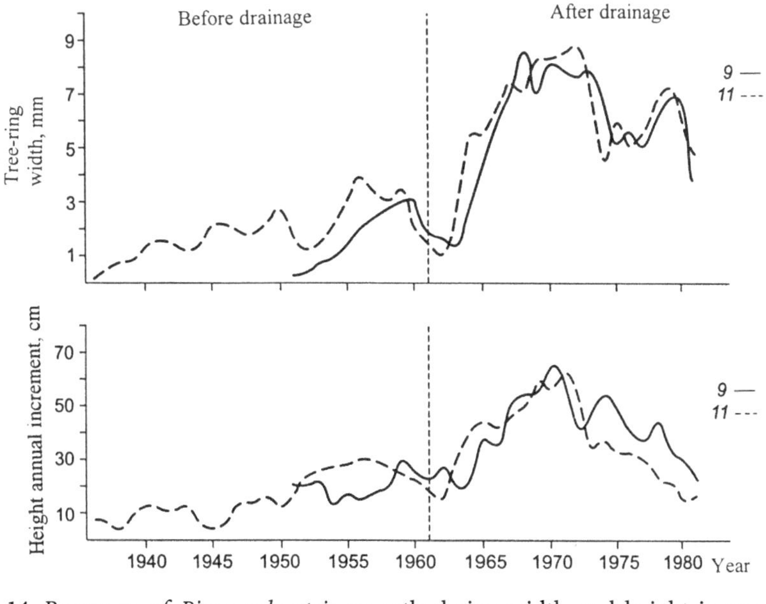

Fig. 1.14. Response of *Pinus sylvestris* smoothed ring-width and height increment to drainage of bog substrate (*9*, *11* site identifiers; Efremov 1987)

For example, the outcomes of a four-year experiment on the effect of a single application of extra nitrogen phosphorus and potassium (in 1970) on the course of radial growth of a pine stand on sandy soil in the Middle Angara valley (southern taiga) are shown in Table 1.3. Only additional nitrogen accelerates radial growth within the first period of years. Obviously, the additional mineral supply has a stimulating influence on woody plant growth only in those conditions where the particular element is at a minimum and essentially limits the processes of growth (Pechman 1960; Schulze et al. 1995). So, in larch stands in Middle Yakutia, nitrogen is the main limiting soil factor for

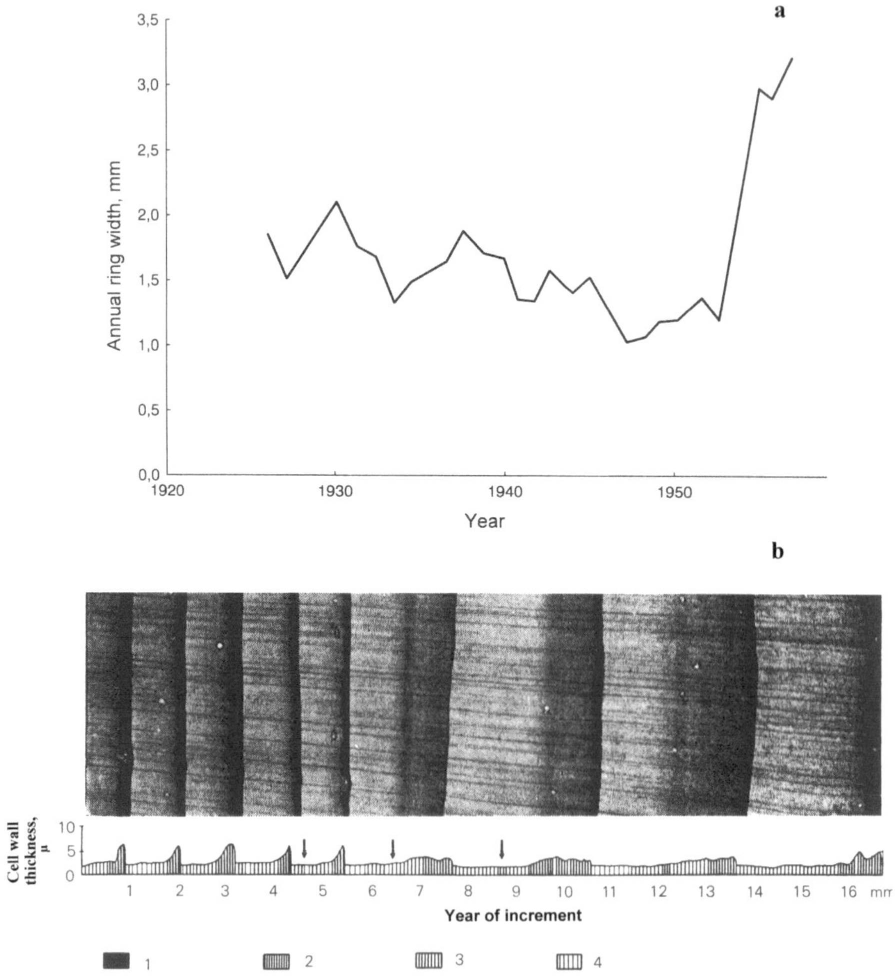

Fig. 1.15. Effect of fertilizers on tree-ring width (**a**) and on radial size and cell wall thickness (**b**) of spruce trees growing on the soils poor in nitrogen, Germany. Radial tracheid size: *1* >15 µm, *2* 16–30 µm, *3* 31–45 µm, *4* 46–60 µm. The *arrows* show the time of fertilizing (redrawn with changes from Von Pechman 1960)

Table 1.3 Change in stand diameter increment (%) in response to fertilization on a southern taiga *P. sylvestris* stand (Buzykin 1977)

Treatment	1969	1970	1971	1972	1973	Total incremental change
Control	0	0	0	0	0	0
N, P, K	0	3	4	47	19	73
N, P	0	1	4	42	17	64
N, K	0	7	15	45	32	99
P, K	0	−6	−2	1	−3	−10
N	0	10	14	49	28	101
P	0	−1	−7	−7	−9	−24
K	0	5	0	9	2	16

growth and accumulation of organic mass in stands (Schulze et al. 1995). The data in Fig. 1.15 show that the application of nitrogen to compositionally poor mineral soil produces a practically instantaneous response in growth acceleration and changes the characteristics of the tree rings that are being formed in a spruce. Ring width, earlywood width, and latewood width increase and cell wall thickness in latewood decreases.

1.6 Weather Factors

The weather conditions of single seasons may be summarized as the seasonal course of temperature and the intra-seasonal distribution of precipitation. These features determine the beginning dates of cell divisions in meristems, the growth rate in single intervals of a season, growth termination dates, and the overall seasonal course of the growth curve (Lobzhanidze 1961; Smirnov 1964; Kramer and Kozlowski 1983; Fritts 1976; Fritts et al. 1991). Temperature may be considered as the most important single factor in the initiation of meristem growth activity (Larson 1994). At the same time, low soil humidity can cause an earlier end of growth in a season (Smirnov 1964; Fritts 1956, 1976), or at least the onset of latewood formation (Zahner 1968b). A combination of temperature and humidity changes in particular intervals of a season produces acceleration or deceleration of growth processes (Vaganov et al. 1985; Schweingruber 1996; Horacek et al. 1999; Brauening 1999) and largely determines the overall result – the size and internal structure of the annual ring formed in that year.

More detailed consideration of the influence of weather conditions on the seasonal growth of woody plants may be found in later chapters.

1.7 Conclusions and Discussion

We described the aim of this book as "to introduce the way of thinking about the environmental control of tree-ring variability in conifers that is expressed clearly in its title: *Growth dynamics of conifer tree rings: images of past and future environments*". In particular, we noted that this is not a textbook of tree developmental biology, but rather an exposition of the thinking behind the simulation models of conifer tree-ring growth described in later chapters. In that thinking, a central role is accorded to the vascular cambium.

The brief analysis in this chapter has shown that the seasonal growth of woody plants depends on both internal and external factors. During seasonal growth, the genetic features of different species of woody plants and their age come into play. The processes of production and differentiation of cells in apical and secondary meristems are coordinated. Their intensity is influenced by a complex of more or less stable operating factors, such as geographical position, climate, and soil, as well as variable factors, such as weather conditions, moisture in soil, and mineral supply. The spectrum of factors influencing growth in woody plants is no doubt wider than considered by us here. Further discussion may be found in a number of more general books (Fritts 1976; Hughes et al 1982; Schweingruber 1988, 1996; Cook and Kairiukstis 1990; Bitvinskas 1974; Antanaitis and Zagreev 1981; Shiyatov 1986; Vaganov et al. 1985, 1996d). We have limited our discussion to the illustration of the influence of major factors so as to help the reader understand approaches to modeling seasonal growth and tree-ring structure, and realizing these as algorithms and computer programs.

2 Tree-Ring Structure in Conifers as an Image of Growth Conditions

2.1 Introduction

Direct observations of long-term environmental changes in natural ecosystems are extremely rare, and so it is necessary to use indirect indicators, or natural archives, of information about past environments. These include marine, lacustrine and terrestrial sediments, geomorphological features, speleothems, annual layers and other features associated with ice sheets and glaciers, and annual layers produced by living organisms such as corals and tree rings (Bradley 1999). Of the natural archives relevant to timescales from seasonal to several centuries, tree rings are probably the best understood, in the sense that their strengths and weaknesses have been thoroughly explored during the past several decades (Fritts 1976; Hughes et al. 1982; Cook and Kairiukstis 1990; Schweingruber 1996; Vaganov et al. 1996a, b; Jones et al. 1998; Hughes 2002).

As a result of their annual resolution and broad geographic distribution, it has been possible to subject the fidelity of tree rings as climate recorders to many thousands of objective tests by comparison with instrumental meteorological data. The accumulation of this mass of empirical experience has contributed to the development of the theoretical and methodological foundations of dendrochronology. A widely accepted set of procedures has been established, such as the selection of sites, species, trees and tree-ring variables to be sampled, cross-dating for chronological control, replication and standardization for noise reduction (Fritts 1976; Shiyatov 1986; Schweingruber 1988, 1996). These procedures differ radically from those used to measure forest growth, because the aim is different. Here the objective is to derive a reliable quantitative history of past environment, usually climate, not to derive a measure that is representative of the ecosystem itself. As with other natural archives, the use of tree rings in this way depends on the principal of uniformitarianism, which is the basis, for example, of all earth science. It is essentially the assumption that all the factors that controlled the formation of the archive in the distant past are known to the scientist interpreting it. Some of the most common practices specific to the use of tree rings as natural environmental archives are practicable because many trees may be sampled at any one location and multiple sites in a region, permitting the strengthening of a common external signal at the expense of tree-specific or site-specific noise by massive replication. This is based on the assumption that all trees in a stand, or even in a wider region, are subject to the same climate variability from year to year, and that most non-climatic factors influencing tree-ring formation are most likely to be specific to the individual stem, tree or stand.

While tree rings are integrated records of the influence of environmental conditions, their anatomical characteristics record growth rate changes produced by these changing conditions (Yatsenko-Khmelevsky 1954; Vaganov et al. 1985, 1996b; Schweingruber 1988, 1996). Tree rings not only integrate the outcomes of the growth process, but also register the process itself. As a result, the internal structure of a tree ring contains information on environmental conditions at seasonal, or even finer, timescales.

So, in this chapter, we will consider the features of conifer tree-ring structure, quantitative methods for the investigation of that structure (Sect. 2.2), and the major factors influencing variability in tree-ring growth and structure (Sects. 2.3–2.4). We go on to discuss how to derive chronologies of various tree-ring parameters (chronologies are means of many detrended time series) in Sect. 2.5 and to examine the relationships between the different parameters (Sect. 2.6). Readers familiar with the measurement of tree-ring widths, densities, and micro-anatomical features may wish to pass over Sects. 2.2.1, 2.2.2, and 2.2.3, respectively.

2.2. The Structure of Conifer Tree Rings and its Measurement

The microscopic structure of conifer wood consists of two types of cells: parenchyma, which have an oval or polyhedral shape with approximately identical dimensions in three directions, and strongly elongated tracheids (Borovikov and Ugolev 1989; Fahn 1990). Tracheids make up more than 90% of timber volume (Table 2.1). The parenchyma cells, rays, and resin ducts vary from 5% to 10% in various species. Tracheids are organized in rather regular files or rows extending through a part of one or several tree rings (Bannan 1955, 1957). The tracheids of earlywood formed at the beginning of a growing season have large radial sizes and smaller, thinner cell walls. Then, the first tracheids of the transition zone are formed, where the radial size of cells and thickness of their cell walls changes considerably. Finally, the latewood tracheids are formed, with small radial sizes and greater cell wall thickness. This is the basic pattern of the internal cell structure of conifer tree rings.

Table 2.1. Content of different elements in conifer wood (% of total wood volume; Borovikov and Ugolev 1989)

Genus	Tracheids	Rays	Resin ducts	Wood parenchyma
Pinus	91.0–94.0	5.3–8.4	0.5–11.0	–
Picea	92.5–95.0	5.0–7.2	0.2–0.3	Scarce
Larix	89.0	10.0	0.1	0.9
Pseudotsuga	92.5	7.3	0.2	Scarce
Juniperus	91.7	6.3	–	0.2
Sequoia	91.2	7.8	–	1.0

Given this basic structure as seen in transverse section, we will consider what quantitative structural measurements may be made of conifer tree rings (Fig. 2.1):

1. Tree ring width (L_T)
2. Number of cells in a radial file within a tree ring (N)
3. Radial size of cells (radial diameter; D_R)
4. Cell wall thickness (L_W)
5. Density of wood estimated using the relationship of the area of cell wall to the cross-sectional area of a tracheids (ρ_X).

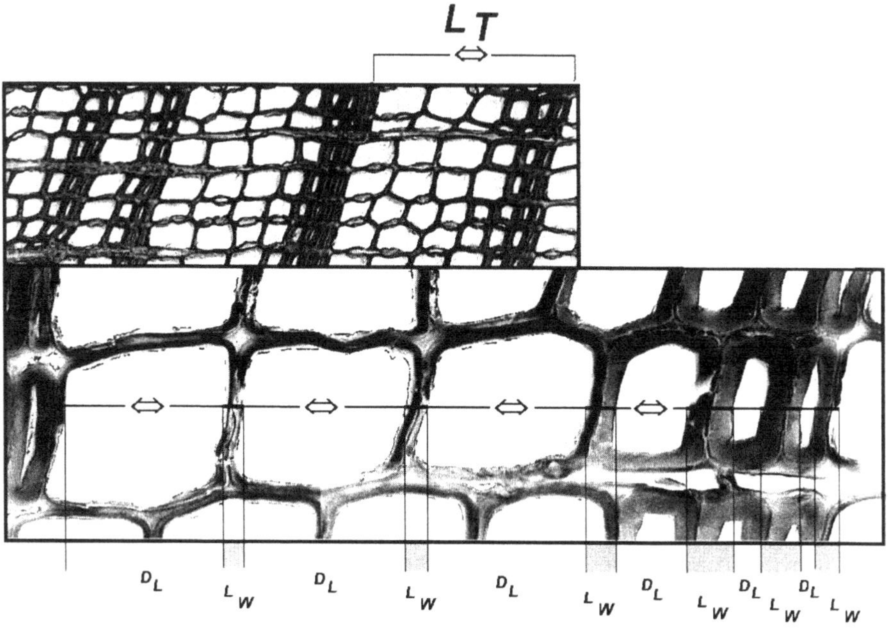

Fig. 2.1. Cross-section of wood: L_T tree-ring width, D_L radial cell lumen, L_W double cell wall thickness

The connection between these different characteristics may be presented as (Vaganov 1996a):

$$L_T = \sum_{i=1}^{N} D_R^i \tag{2.1}$$

where i is the serial number of a cell in a file and N is the number of cells in the given tracheid file of a tree ring.

$$\rho_X = \rho_W \sum_{i=1}^{N} 2L_W^i (D_R^i + D_T - 2L_W^i)/(D_R^i D_T) \tag{2.2}$$

where ρ_W is the specific gravity of the cell wall, D_T is the tangential size of a cell, the average value of which varies slightly within a ring (Vysotskaya et al. 1985; Borovikov and Ugolev 1989). Thus, tree ring width is a derivative of two main characteristics, and density of three:

$$L_T = f(N, D_R^i)$$

$$\rho_X = f(N, D_R^i, L_W^i)$$

2.2.1 Measurement of Tree-Ring Width

It is a rather simple task to measure the width of conifer tree rings when they have clear boundaries, each representing an arc whose radius is large relative to the ring width. It is more difficult to do this when the ring boundaries have strong curvature, as for example, close to branch traces, and is extremely problematic when the curvatures of the early and late boundaries are not parallel. Poorly defined ring boundaries with weak contrast between latewood and the succeeding earlywood may be characteristic of certain taxa, for example *Agathis australis* (kauri pine) in New Zealand (LaMarche 1982), or of many taxa in regions with weak seasonality. They may also result from the growth conditions of a particular year in species where ring boundaries are generally clear. For example, frost damage near the early or late boundary may obscure it, or extremely harsh conditions may result in a ring with only two or three cells in each radial file (a "micro-ring"). Such harsh conditions may also result in a ring being absent on the radius to be measured, or around a large fraction or all the circumference at that height above ground, or even on all wood available from that tree. To further complicate matters, intra-seasonal dynamics may result in "false rings". For example, a severe drought in the growing season may cause the untimely onset of latewood formation, which may be followed by the production of earlywood cells when rain returns (see below). In most cases, such false rings may be identified anatomically, as the return to earlywood cells is gradual, not sharp as in the annual ring boundary.

In order to best deal with these problems, it is essential that the wood be surfaced very well, so that the individual tracheids may be seen. It is also extremely important that cross-dating be conducted with the wood under the microscope. Cross-dating is the process of massively replicated pattern-matching of ring features (including but not limited to ring width), to unambiguously assign each ring to a specific year, first with reference to its neighbors in the stand and then with reference to calendar years. Only by careful cross-dating can missing rings be identified. Also, cross-dating often helps in the diagnosis of false rings and the discovery of micro-rings, when the dendrochronologist returns to the microscope to find the cause of a phase shift in the ring pattern of a sample relative to the rest of a collection.

There are two main methods of cross-dating in widespread use. In most laboratories in North and South America, cross-dating is done without first measuring ring widths. Only some *aide-memoire* of the strong features of ring pattern is needed – usually the skeleton plot as developed by Andrew Ellicott Douglass in the early decades of the twentieth century (Stokes and Smiley 1968). This permits the simultaneous comparison of several samples' ring patterns. After resolving anomalies due to missing and false rings, the calendar year assignment of each ring is marked permanently on the wood by a special code of pinholes for decades, half-centuries and centuries, as well as missing and micro-rings. The ring widths are then measured, with each width assigned to a known calendar year, and the measurements are subjected to rigorous statistical quality control designed to identify errors in both cross-dating and measurement. It is also common practice to remove markings from some samples so that they may be dated independently by another dendrochronologist. When practiced by an experienced worker, this method can be extremely rapid and has the very strong advantage that the scientist is in intimate contact with the wood at every stage of dating. In other laboratories, for example many in Europe, ring widths are measured first and patterns of ring width compared graphically and/or statistically to identify anomalies such as unusual phase shifts, which may then be resolved, as in the Douglass skeleton plot method, by reference to the wood (Baillie and Pilcher 1973). The same statistical quality control as in the Americas is then often applied to the dated ring-width series.

There are two main types of measurement system in use for ring widths, regardless of whether the rings have already been dated, or the measurement is part of the dating procedure. In one type, a stage carrying the wood is moved under a microscope or a camera with a macro-lens (Fig. 2.2). When the operator sees a ring boundary coinciding with cross-hairs in the microscope field of view, or a mark on a video screen if a camera is used, a button is pressed to send the measured ring width to a computer. The precision of the measurement device is commonly 0.01 mm or 0.001 mm, but this cannot always be applied to the measurements themselves. This is because the specific track along which measurements are made is determined by the opera-

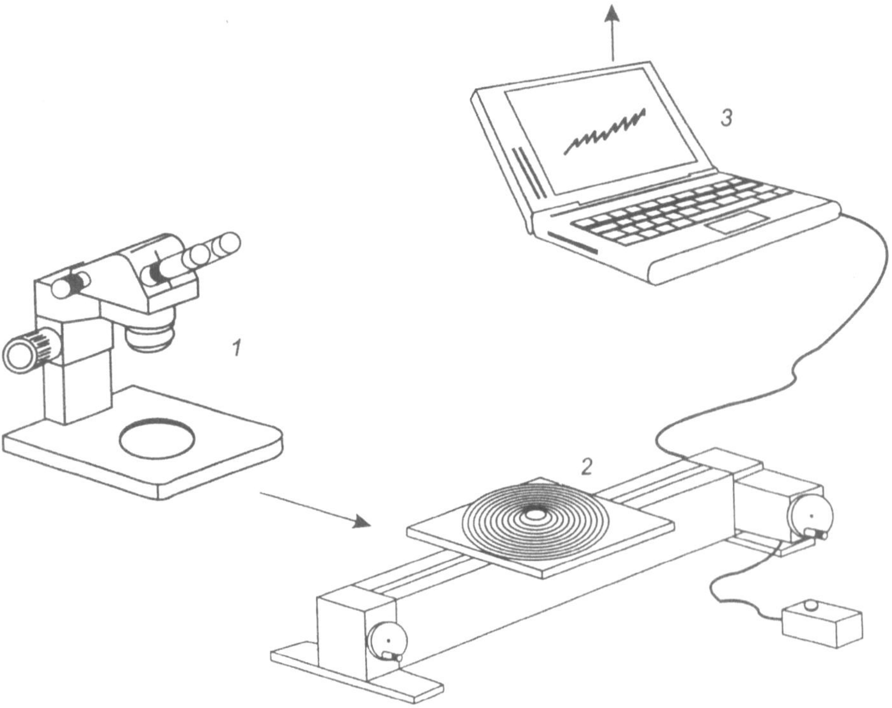

Fig. 2.2. Device for measuring and processing tree-ring width data: *1* stereomicroscope, *2* a specimen table with precision feed providing a linear sample displacement to 0.001 mm, *3* computer for compiling and processing data

tor, bearing in mind irregularities such as branch traces and resin ducts. The first widely used version of this approach based on a personal computer was developed by Robinson and Evans (1980) and there have been many refinements since (Vaganov et al. 1983; Schweingruber 1988; Rinn 1996), especially those based on linear optical scales so that the precision and accuracy of the measurement is independent of the translation mechanism used to move the wood sample.

In the other type of ring-width measuring system, software is used to derive ring widths, and perhaps other variables, from a digital image of the wood surface. In the simplest cases, a track is drawn across the image and peaks in the intensity profile along that track are identified as ring boundaries. Ring width is the distance between successive boundaries. Of course, the reliability of the measurement series depends on the relationship between the pixel size of the image and the width of the smallest rings, which can be as small as 20–30 μm. A further approach is to take the differential of the intensity profile, but care must be taken to know which peaks are produced by the earlywood–latewood transition and which by the ring boundary. Image-processing algorithms adapted from, for example, remote-sensing applications, have been used to identify ring boundaries reliably in many conifer species and to measure ring widths and other properties not only along a single track, but using the whole area of a defined region of interest – for example the whole width of an increment core (Conner et al. 2000).

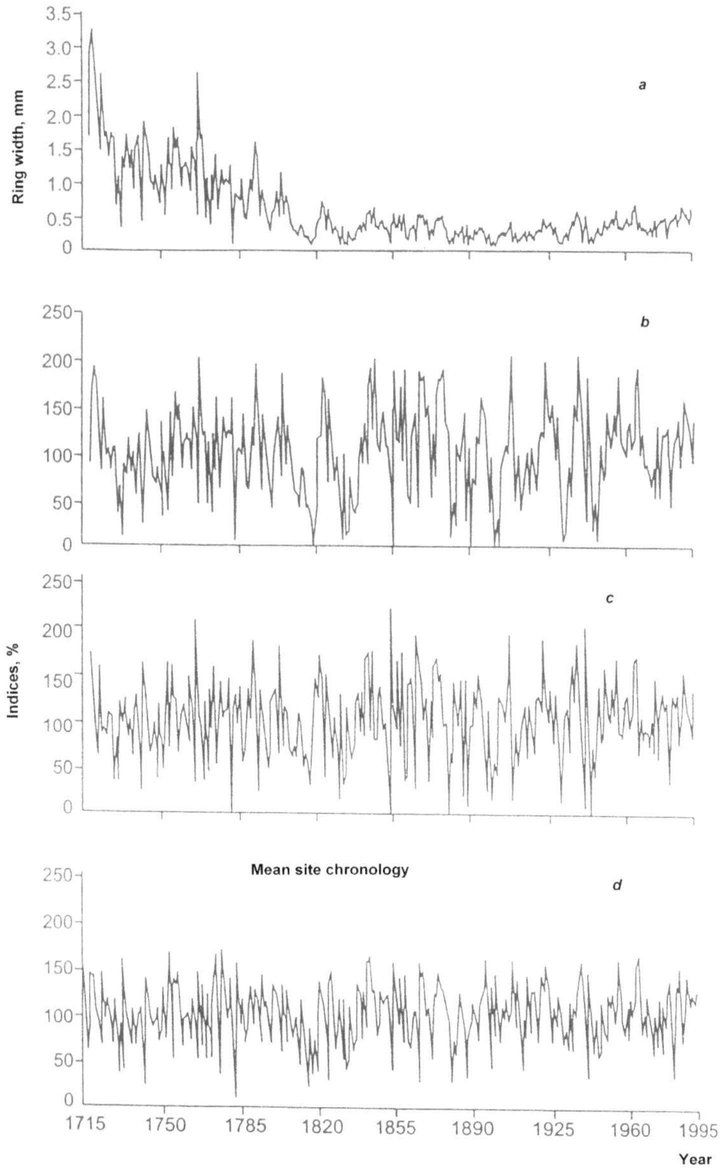

Fig. 2.3. Processing of initial tree-ring width series: **a** tree-ring width (TRW) series (tree JAH081), **b** TRW indices for JAH081 after age trend removal, **c** individual series after "prewhitening", **d** mean site chronology (Vaganov et al. 1996d)

Raw ring widths are rarely used in the dendroclimatological literature. Rather, the large effects of tree age and size are removed from the ring-width time-series for each sample, usually in an entirely empirical manner, so as to produce dimensionless indices. Then, a mean of these is taken, perhaps using a method that reduces the influence of outliers, and some corrections may be made to account for the changing number of samples available for different periods. The resulting time-series of ring-width indices is a representation of the common pattern of variability shared by the sampled trees. It may also be considered as a representation of ring width that is independent of tree age, size, stand position, and disturbance, and so maximizes external environmental influences. Although these methods are subject to constant revision and improvement, a good general introduction is given in Cook and Kairiukstis (1990).

A typical example of standardization of ring-width measurements is shown in Fig. 2.3.

2.2.2 Measurement of Wood Density Within Tree Rings

The methodological foundations of wood microdensitometry of tree rings were established by Polge (1966, 1969). Lenz et al. (1976) made many refinements of the procedures first proposed by Polge (1966, 1969). Parker and Henoch (1971) reported the first quantitative application of X-ray microdensitometry to conifer tree rings for dendroclimatology and Schweingruber et al. (1979) explored the application to density data of the multivariate statistical tools developed by Fritts (1976). There are two main stages, namely obtaining radiographs of the wood samples and measuring the optical density of the radiograph. This is then used to derive the specific gravity of the wood by using the simple curvilinear relationship that exists between the optical density of the radiograph and the specific gravity of the wood for a given thickness of the wood specimen and settings of the X-ray generator and the geometrical relationship of the source, the sample, and the X-ray film (Polge 1966). Although this procedure is, in principle, simple, great care must be taken at each stage of the preparation of the sample, the production of the radiograph, and its analysis (Figs. 2.4, 2.5, 2.6, 2.7). Thin laths are cut, usually to a standard thickness between 0.8 mm and 1.2 mm, using a special two-bladed saw. Not only is it important that the thickness be consistent along the length of the lath, but also that the cut surface is precisely perpendicular to the long axes of the tracheids. It is often necessary to make several laths from a single increment core, each overlapping the next, but cut at a slightly differing angle so as to maintain the perpendicularity to the tracheid long axes. It is necessary to remove mobile substances that may be X-ray opaque by boiling in organic solvents or water. Finally, because water is strongly opaque to X-rays, the moisture content of the wood must be

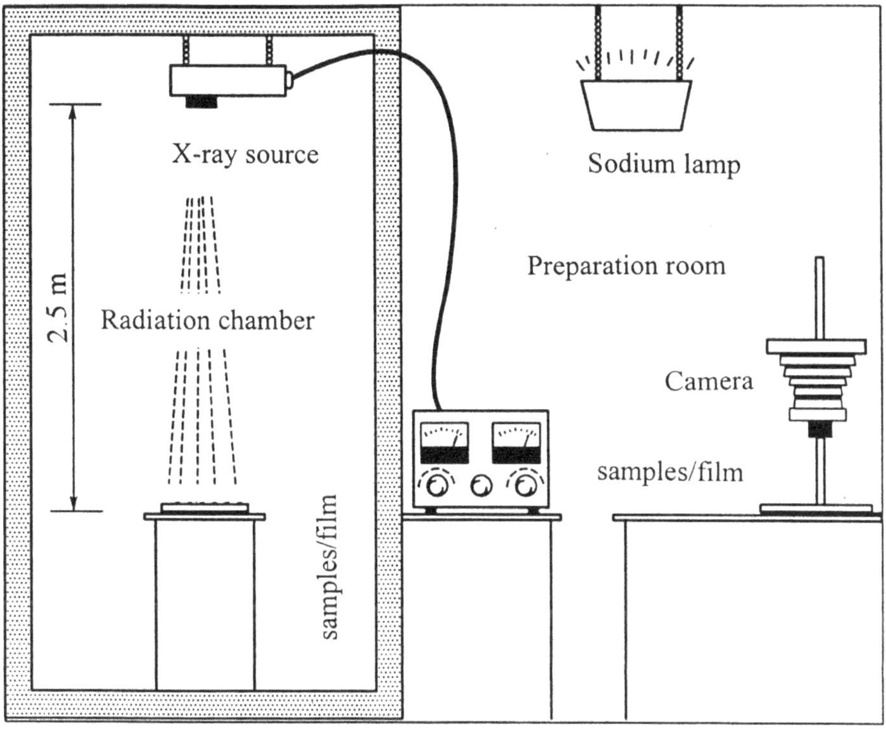

Fig. 2.4. Plan of radiation chamber. The radiation chamber is separate from the preparation and control rooms (Schweingruber 1988)

reduced to approximately 8–9% by weight before radiography (Lenz et al. 1976; Schweingruber 1990).

The radiographs are produced by exposure of the wood laths, along with a secondary density standard, placed directly on single-sided X-ray film with a fine-grain emulsion. The secondary standard is a step-wedge of a plastic with the same ratios of carbon, hydrogen and oxygen as wood. It may be calibrated against a primary standard in the form of a step-wedge of paper produced from the timber under investigation, whose density has been determined by the standard gravimetric–volumetric method. The precise spectrum and intensity of X-rays used is determined by the voltage (typically 8–10 kV) and current (typically 16–20 mA) used and by the window through which the X-rays pass on leaving the source. These settings produce soft X-rays similar to those used in medical applications. These provide the greatest contrast on the radiograph between the least and most dense zones (Polge 1966). Two main approaches are in use to optimize the geometry of the X-ray system. In one, X-raying of wood samples is done in a specially equipped room (Fig. 2.4) with a

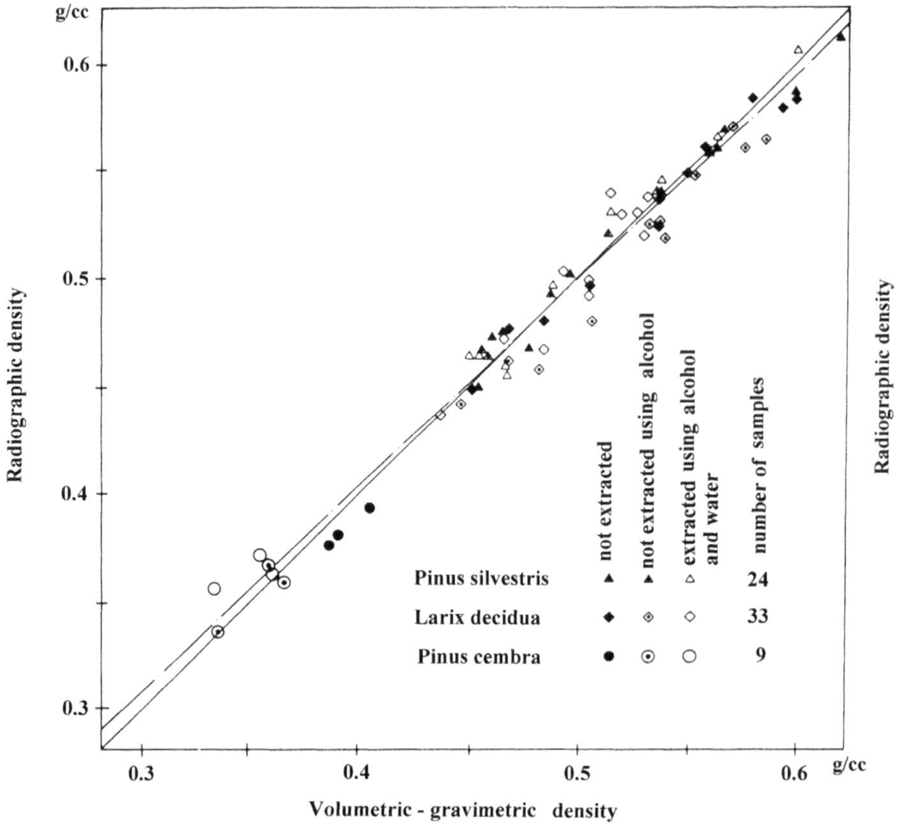

Fig. 2.5. The relationship between the radiographic-densitometric and volumetric-gravimetric densities. Differences according to species and whether or not mobile materials such as resin and heartwood substances have been extracted (Schweingruber 1988)

significant distance between the X-ray source and the samples and film (2.5-3.0 m). As a result, the X-rays arriving at the top surface of the wood are almost parallel to one another, reducing parallax problems that would cause the radiograph to be fuzzy rather than clear. The same objective may be achieved by passing the X-ray beam through a long slit and passing a stage carrying the sample, standard step-wedge, and film under the beam at constant speed. In this case, a special room is not needed, as the source and samples need only be a few centimeters apart, although the system must be encased in lead to protect the operator. Such an arrangement can produce geometry equivalent to having a distance of 27 m between the source and the subject (Milsom and Hughes 1978). It is, however, necessary to ensure that the speed of translation is constant and that the output of the X-ray source does not vary. A correctly prepared radiograph will provide a very sharp picture of

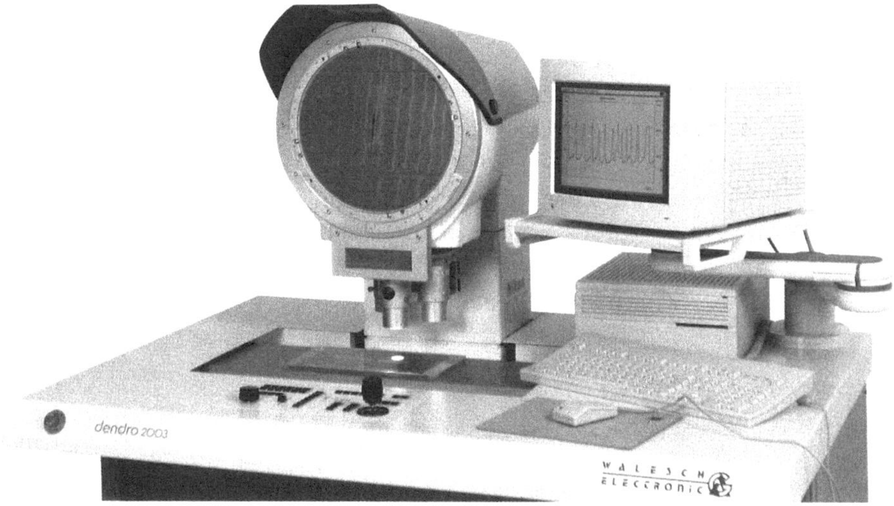

Fig. 2.6. Densitometer DENDRO 2003 (Walesch Electronics). The image of the radiograph is displayed on the large circular screen and its optical density is measured along a segment of the diameter of this screen, parallel to the ring boundary. These results are then converted to specific gravity and recorded as various ring width and density variables by the attached computer

the transverse view of the sample, with the same level of detail as may be seen on a properly surfaced wood sample; that is, the individual tracheids should be visible. It should not be possible to differentiate the sapwood and heartwood, because water has been removed; and there should be no patchiness associated with mobile substances such as resin. The optical density of the radiograph is measured using a microdensitometer specifically adapted for use with radiographs of wood (Fig. 2.6). First, the known density values of the step-wedge standard are used to calibrate each radiograph in terms of specific gravity. Then the track to be analyzed is chosen and the density profile of a belt along that track is obtained as a line of detectors traverses the image. It is essential that this line of detectors be held parallel to the ring boundaries if the density profile is not to be smeared (Fig. 2.7). As the peak value of density in the latewood ("maximum latewood density") is of particular interest and as it may be reached in only one or two latewood tracheids, it is important that the spatial resolution of the densitometer along the long axis of the tree-ring sample be as small as possible, ideally 10 µm or less. This is the approximate diameter in the radial direction of a latewood tracheid. The density profiles are displayed on the computer (Fig. 2.7b) and recorded in a format suitable for further processing and analysis. A tree-ring density profile has both regular elements and features specific to each year. For a quantitative evaluation of these constituents, the following quantitative parameters are used: minimum

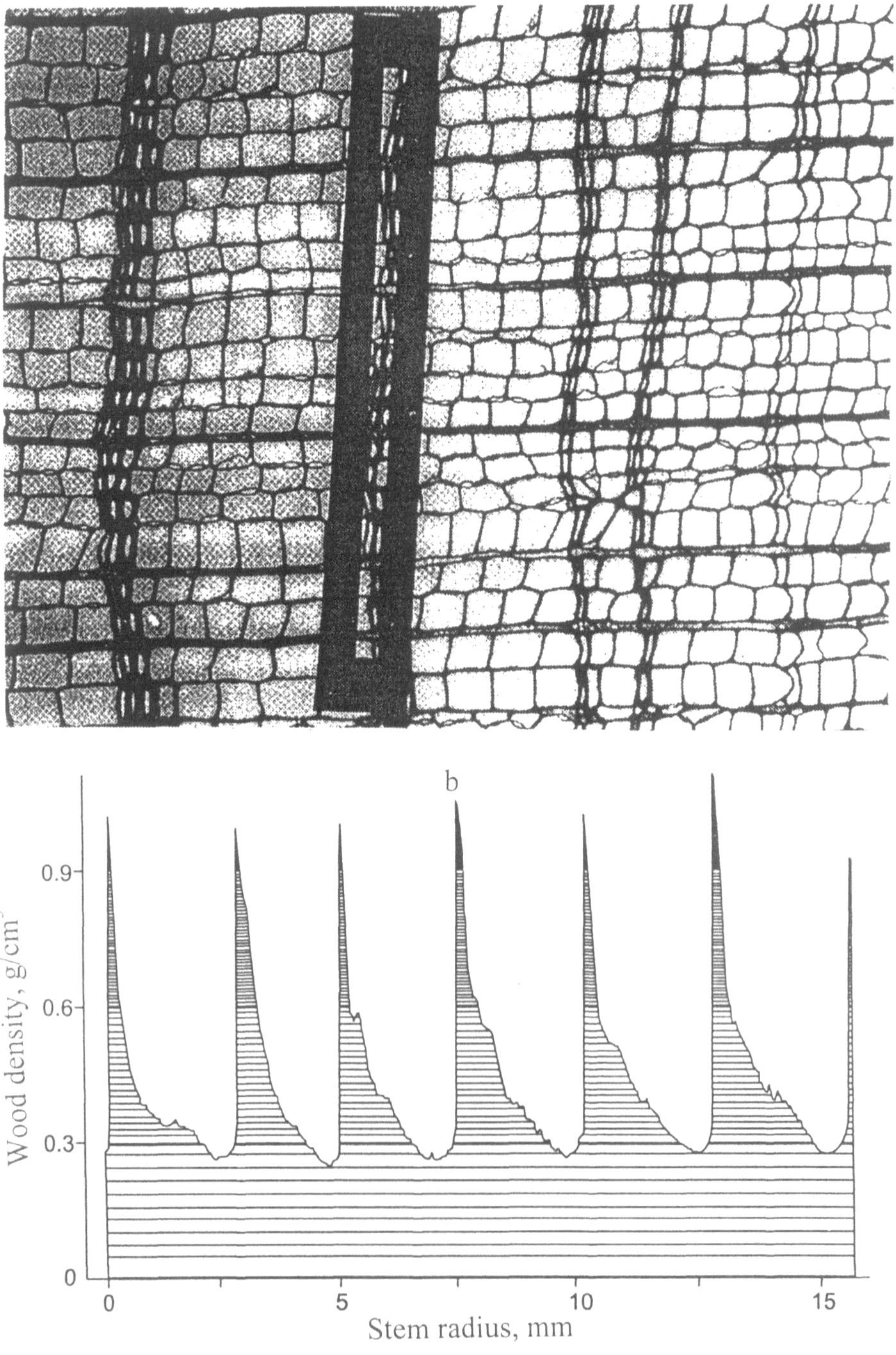

Fig. 2.7. Orientation and optical slit sizes (width 20 μm) for X-ray film scanning (**a**) and resulting curve of density (**b**) with marks of different density threshold values (Schweingruber 1988)

earlywood density, maximum latewood density, and mean density. The density of earlywood is calculated on a segment of the curve located below some empirically selected density level (for example 0.5 g/cm^3 or 0.6 g/cm^3; Fig. 2.7b). In a similar way, the density of latewood (a segment of the curve above the density level of 0.5 g/cm^3 or 0.6 g/cm^3) is also calculated. The widths of these two zones are also recorded. It can be demonstrated that, using the appropriate precautions, X-ray microdensitometric measurements reproduce precisely those made gravimetrically–volumetrically (Fig. 2.5).

Progress has been made recently with reflected-light microscopy to produce data closely analogous to those derived from X-ray microdensitometry (Sheppard and Graumlich 1996). This has the great potential advantage of removing the need for, and errors associated with, cutting laths normal to the tracheid direction, thus greatly simplifying the process.

2.2.3 Measurement of Radial Tracheid Diameter Within Tree-Rings (Tracheidograms)

Files of tracheids in tree rings represent regular variations in cell dimension and wall thickness. Their variability reflects the influence of internal and external factors on the seasonal growth of trees (Larson 1964; Richardson 1964; Fritts 1976; Ford et al. 1978; Melekhov 1979; Vaganov et al. 1985; Filion and Cournoyer 1995; Jardon et al. 1995a, b; Vaganov 1996a, b). Thus, it is valuable to have measured profiles of cell dimensions across tree rings. Most simply, such measurements may be made on thin cross-sections of wood (stained by safranin and set in glycerin jelly or balm) using transmitted light microscopy at a magnification of 250 to 400. A measuring microscope with an accuracy of 0.5 μm in the translation direction of the sample will accelerate measurement. Good optics permit the use of ambient light and more simply prepared samples, for example those prepared by careful polishing of the transverse surface, or cutting with a microtome or very sharp razor.

In order to combine the advantage of a measuring microscope for capturing primary data with the advantage of a computer for their accumulation and analysis, a device was constructed at the Institute of Forest in Krasnoyarsk (Vaganov et al. 1979, 1983, 1985). Rather as in a ring width-measuring machine, the operator identifies the boundary between two adjacent cells (the middle lamella) and the linear movement of the sample between two consequent boundaries thus indicated corresponds to the radial size of a cell (Fig. 2.8).

The development of image analysis software has permitted the development of systems for the semi-automatic measurement of the anatomic elements of wood (Jagel and Telewski 1990; Park 1990, 1993; Munro et al. 1996; Fig. 2.9). In addition to the sizes of cells and thickness of cell walls, such characteristics as the lumen area, cell wall area, and its perimeter (Jagel and

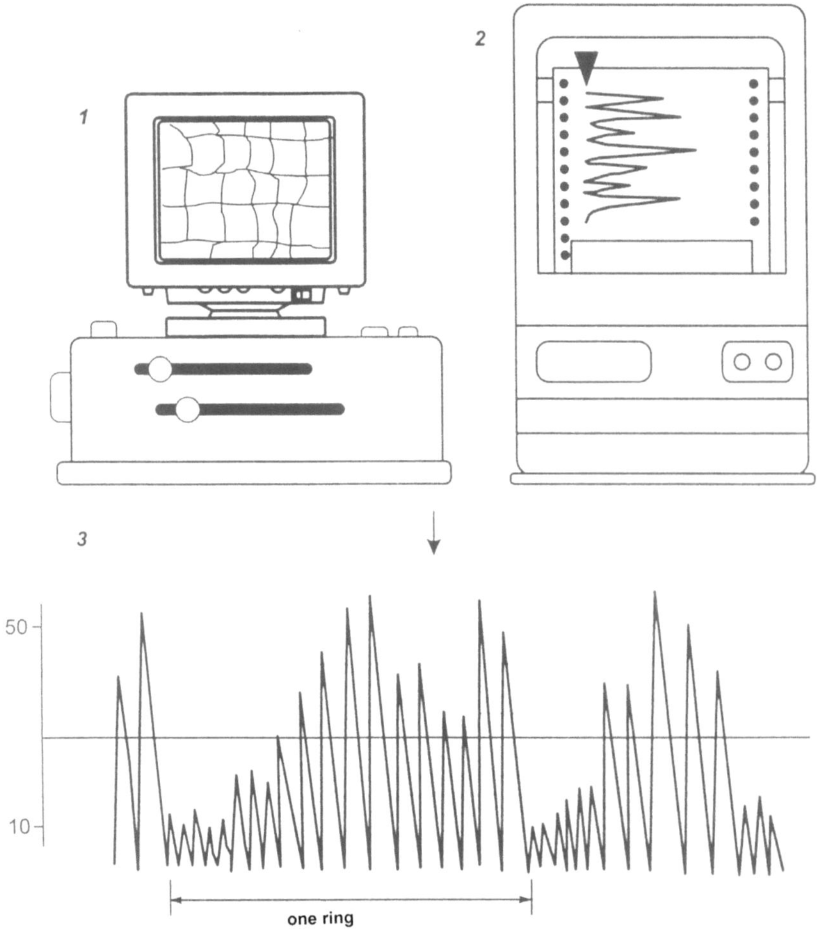

Fig. 2.8. Scheme of the automated setup for measuring and processing data on radial cell sizes within tree-rings (tracheidogram): *1* monitor showing microscope view, specimen table with precision sample translation, *2* computer and printer, *3* typical recording tracheidogram (Vaganov et al. 1985). Each peak represents a tracheid and its diameter is proportional to the radial dimension of the tracheid

Telewski 1990) are measured. The main problem is in getting surfaces or thin sections of sufficiently high quality. When this is done successfully, it is possible to obtain precise measurements of the cell characteristics from an image of the surface (Fig. 2.10; Munro et al. 1996). One problem is that, in such images, the change in image intensity from wall to lumen is not as sudden as might be wished. It turns out that it is not possible to apply the same threshold value to the determination of wall versus lumen in both earlywood and latewood, for example. This creates serious problems for approaches based on

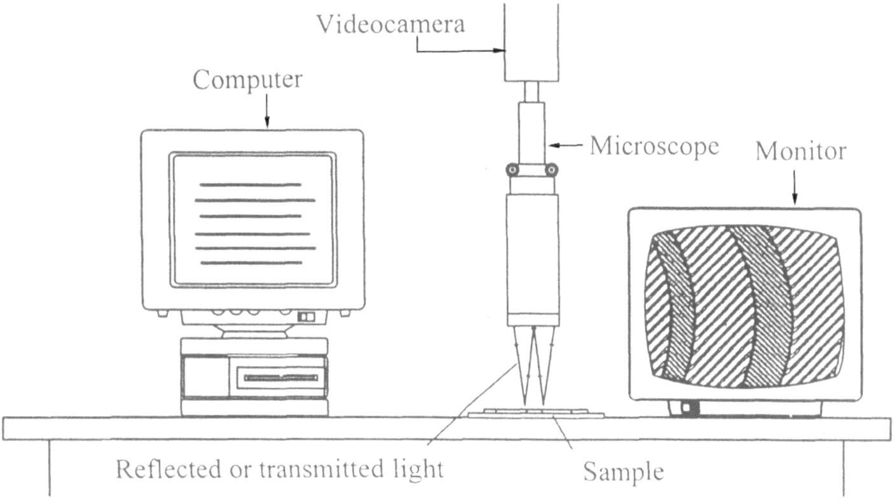

Fig. 2.9. Image analysis system for tree rings (Sheppard and Graumlich 1996)

transforming the image into a binary (black and white) one. Munro et al. (1996) approached this problem by permitting the operator to change the threshold value interactively (Fig. 2.10, upper panel). It was then possible to make repeatable measurements of cell size and double cell-wall thickness in rings of giant sequoia (*Sequoiadendron giganteum*), as indicated in the lower panels of Fig. 2.10.

Files of tracheids in a single tree ring, no less than in different tree rings, contain varying numbers of cells. Therefore there is a problem in the comparative analysis of micro-anatomical features. This problem is solved by means of normalization (standardization) of the number of cells to some standard number (in practice for narrower rings 15 cells, for broader rings 30). The procedure of standardization is done in such a way as not to deform the curve of variability of the cell sizes (Vaganov et al. 1985; Vaganov 1990). The normalization "compresses" or "stretches" the initial "tracheidogram" on an abscissa (number of cells), leaving its ordinate unchanged – the radial cell size. Thus, in the measured sequence (D_R^i), where i is the series 1, 2, ... N, each cell is repeated k times (where k is the number of cells in the standard tracheidogram), and then in the new sequence the sizes of cells are averaged in blocks of N cells. The resulting normalized tracheidograms for the several files of cells measured in an individual tree ring may be averaged to produce a mean tracheidogram. This normalization also allows a structural comparison of rings with different numbers of cells.

New possibilities arise from the combination of cell dimensions and wood density measurements. This follows from the simple relationship between measured density and calculated cell area (compare with Eq. 2.2):

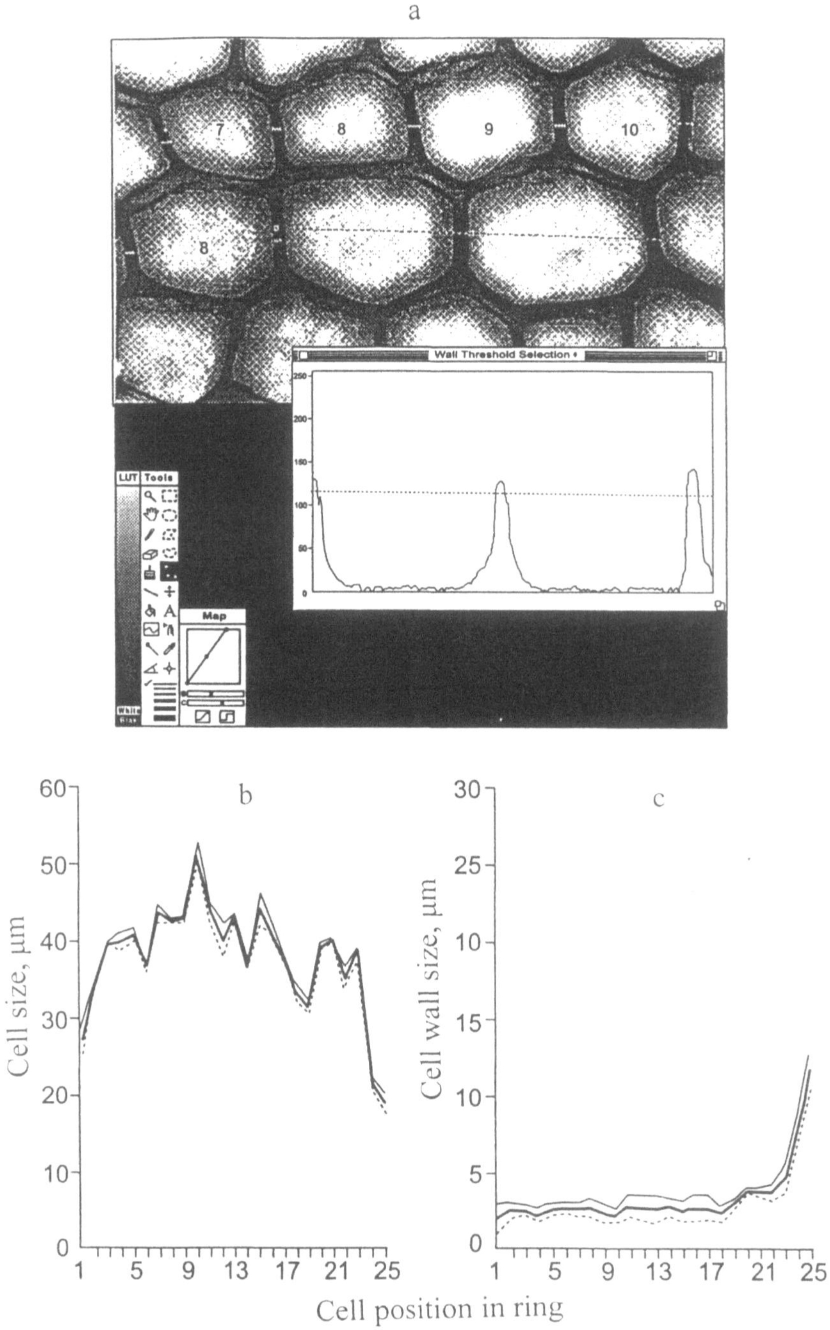

Fig. 2.10. Measurements of anatomical tracheid characteristics using an image analysis system: **a** enlarged wood micro-section and line of scanning, **b, c** measurement results of one tracheid file (**b** measurement of radial cell size, **c** measurement of cell wall thickness). The *insert* in **a** shows wall threshold selection as described in the text (Munro et al. 1996)

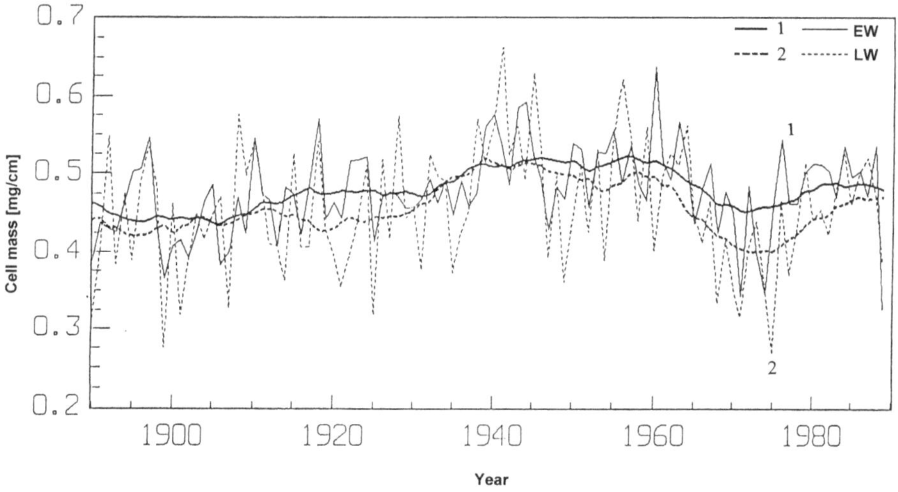

Fig. 2.11. Inter-annual variability of biomass of single tracheid in the early- (*1*, *EW*) and latewood (*2*, *LW*) of tree-rings in larch trees growing near the polar timber line in north-central Siberia. Multiyear variations are emphasized in the heavier lines showing 11-point running means (Silkin and Kirdyanov 1999)

$$M_W = \rho_X D_R D_T \tag{2.3}$$

where M_W in this case is mass of cell wall per linear unit (in micrograms). Silkin and Kirdyanov (1999) measured the density profiles and cell dimension for wood samples of larch (*Larix sibirica*) and then calculated wall mass per 1 mm of cell length using Eq. 2.3. The chronology of cell mass shows high variability and very similar average data of cell mass for early- and latewood tracheids (Fig. 2.11). If it is assumed that the length of early and late wood tracheids does not differ significantly (Benthel 1964; Skene 1969; Sastry and Wellwood 1971), then the mass of each cell (mass of cell walls) in earlywood and latewood seems to be the same, on average. This gives new opportunities for dendrochronology to reveal the influence of external factors on the accumulation of biomass in wood during a single season and on longer timescales.

2.3 Influence of Internal Factors on Tree-Ring Structure in Conifers

There is not an unambiguous answer to the question of the degree to which the growth rate of trees (tree-ring width) is under the control of internal (genetic, age, etc.) factors. For example, while there are many reports indicating that, for the majority of situations, the age–growth curves of individual trees correspond to the so-called biological growth curve (Terskov et al. 1981; Shiyatov 1986), there is at least an equal volume of reports that there are significant differences of individual age curves from the biological growth curve (Fritts 1976; Schweingruber 1988; Cook and Kairiukstis 1990). This may be produced by local environmental conditions or by interactions with neighbors during stand development. Mikola (1950) showed that the concavity of the growth curve may be directly related to the density of the forest stand. Naurzbaev et al. (2004) showed that the parameters of growth curves decreased with increasing latitude and elevation in two larch transects, suggesting a temperature-dependence of these parameters.

The structure (density of wood, size of cells, other characteristics) of tree rings is also under the control of internal factors (Chavchavadze 1979; Zobel and van Buijtenen 1989; Zobel and Jett 1995). This noticeable genetic control is even exhibited in phenological development. So, ecotypes of European spruce (*Picea abies*) that are distinguished by earlier or later opening of buds and main shoot growth show differences in tree-ring wood density (Worral 1970). That is, the timing of the opening of buds is a part of the genetic control of wood structure. For 150-year-old spruce trees, it has been shown that late-opening bud ecotypes have lower wood density (Mergen et al. 1964; Birot and Nepveu 1979). For Douglas-fir (*Pseudotsuga menziesii*), those ecotypes which commence latewood formation earlier and have a longer growing season also produce denser wood (Vargas-Hernandez 1990). Other research on Douglas-fir shows that early bud opening is associated with higher wood density in tree rings (McKimmy 1959; Kennedy 1970). The data for two kinds of pines (*Pinus taeda, P. elliottii*) show that wood density is negatively correlated with the start date of cell divisions in the lateral meristem and the date of transition to the formation of latewood and is positively correlated with the date of growth cessation (Vargas-Hernandez and Adams 1994).

The existence of species-specific relationships between tracheid formation and the growth of certain other plant parts supports the case for a strong genetic control of tracheid formation in both earlywood and latewood. Mitchell (1961) found a relationship between growth of the terminal shoot and the transition to latewood formation in Douglas-fir. Integrating the data on seasonal growth of conifers in the Moscow area showed that earlywood tracheids in Scots pine (*P. sylvestris*) are formed during growth of the main

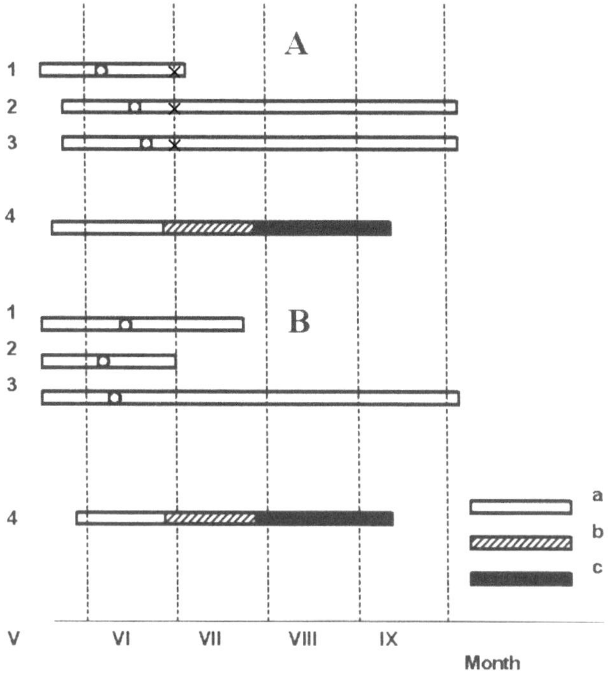

Fig. 2.12. Bars showing growth period of needles, shoots and tree-ring formation of pine (**a**) and spruce (**b**): *1* growth of terminal shoot, *2* increment of fresh needle mass, *3* increment of dry needle mass, *4* formation of xylem elements in the tree-ring. *Circles* mark the time of maximum relative growth rate, *crosses* mark the excess of relative growth rate of dry needle mass increment over fresh. Periods of tracheid formation in the early (*a*), transitional (*b*), and late (*c*) wood zones (Vaganov and Terskov 1977)

shoot and in European spruce (*Picea abies*) they form during needle growth (Smirnov 1964; Vaganov and Terskov 1977; Fig. 2.12). In jack pine (*Pinus banksiana*), the first latewood tracheids are formed within three weeks of completion of height growth (Kennedy 1969). Therefore ecotypes from a milder climate adapted to a longer season finish terminal growth later and form tree rings with lower density.

For a quantitative evaluation of the influence of genetic factors, the value of heritability (h^2) is used, which is the ratio of genotypic variability to phenotypic variability. For example, for Douglas-fir, low values are obtained for wood density (h^2 values of 0.2, 0.3; Vargas-Hernandez 1990; Vargas-Hernandez and Adams 1994). In contrast, the density and ratio of early- and latewood have a close connection, so that the percentage of latewood closely correlates with density for the majority of coniferous species (Pereligin 1969; Poluboyarinov 1976; Zobel and Jett 1995). The range of values of heritability (h^2) obtained for different kinds of conifers is great: from 0.15 to 0.92 (Goggans 1962; Nicholls et al. 1964, 1980; Kennedy 1966).

There have also been attempts to evaluate heritability values for other anatomical features of tree rings. Van Buijtenen (1965) showed significant genetic control of tracheid diameter and cell wall thickness for several conifer species in North America. In loblolly pine (*P. taeda*), $h^2 = 0.84$ for latewood cell wall thickness and $h^2 = 0.13$ for earlywood cell wall thickness. For Caribbean pine (*P. caribaea*), similar values of heritability were found for early- and latewood cell wall thickness ($h^2 = 0.50$; Barnes et al. 1983). There is also evidence for strict genetic control of radial tracheid dimensions in European spruce (*Picea abies*; Boyle et al. 1987). It has been suggested that the significant differences found between heredity values for clones of European spruce (*P. alba*) are connected with differences in the architecture of the cell walls (Kennedy 1966).

Unfortunately, work on the evaluation of the relative influence of genetic and external factors has largely been done on young trees or for the tree rings of juvenile wood (Zobel and Jett 1995). This limits the relevance of these findings to tree-ring chronologies constructed using ring-width measurements, characteristics of wood density, sizes of cells, and cell wall thickness (Vaganov et al. 1985, 1996b; Schweingruber 1988).

2.4 Influence of External Factors on Tree-Ring Structure in Conifers

It is necessary to consider the influence of external factors on each of the parameters of tree-ring structure (L_T, N, D_R, L_W, ρ_x; Sect. 2.1) separately and in combination. The width of a tree ring (and the number of cells, which is strongly correlated with it; Gregory 1971; Vaganov et al. 1985, 1992) has been a subject of dendrochronological research since the early twentieth century. Many papers and a number of monographs have dealt explicitly with the influence of external factors on variability in tree rings (Fritts 1976, 1991; Hughes et al. 1982; Schweingruber 1988, 1996; Cook and Kairiukstis 1990; Shiyatov 1986; Vaganov et al. 1996b, c). A broad spectrum of external factors, such as temperature and moisture availability, wind, fires, insect outbreaks, industrial pollution, snow avalanches, forest management, and many others, have been considered. The broad use of tree-ring chronologies in ecology at various spatial scales, climatology and hydrology, as well as human history, has promoted the development of the theoretical and methodological fundamentals of dendrochronology and dendroecology (Fritts and Swetnam 1989; Schweingruber 1996). For example, Fritts et al. (1965), examined ring-width variability along a vegetation gradient in the semiarid southwestern United States. They identified patterns of variability in tree-ring width connected with a gradient of dryness, increasing inter-annual variability in precipitation, and an increase in the proportion of the year for which moisture was the limiting factor for ring-width growth (Figs. 2.13, 2.14):

1. Tree-ring width decreases proportionately as total precipitation decreases.
2. The correlation between tree-ring width series from different parts of a tree, and between trees, increases as total precipitation decreases over most

Fig. 2.13. Schematic diagram of the relationship between precipitation and temperature, plant hydration, physiological processes, and formation of a narrow ring in a dry climate (Fritts 1976)

of the vegetation gradient, but then decreases rather dramatically in the immediate neighborhood of the lower forest border.

3. The decrease in correlations near the lower forest border results from an increased frequency of missing rings at the most extreme levels of moisture deficit.
4. Inter-annual variability in precipitation increases as the total precipitation decreases (with decreasing elevation); and so the inter-annual variance of the ring-width series is greatest in the lower part of the gradient where there is least precipitation.

Fritts (1976) used a conceptual scheme (Fig. 2.13) of the influence of moisture deficit on the formation of narrow rings in trees to explain these observed patterns. The physiological mechanisms known to cause increased water stress in the tree and to decrease hormone production and the produc-

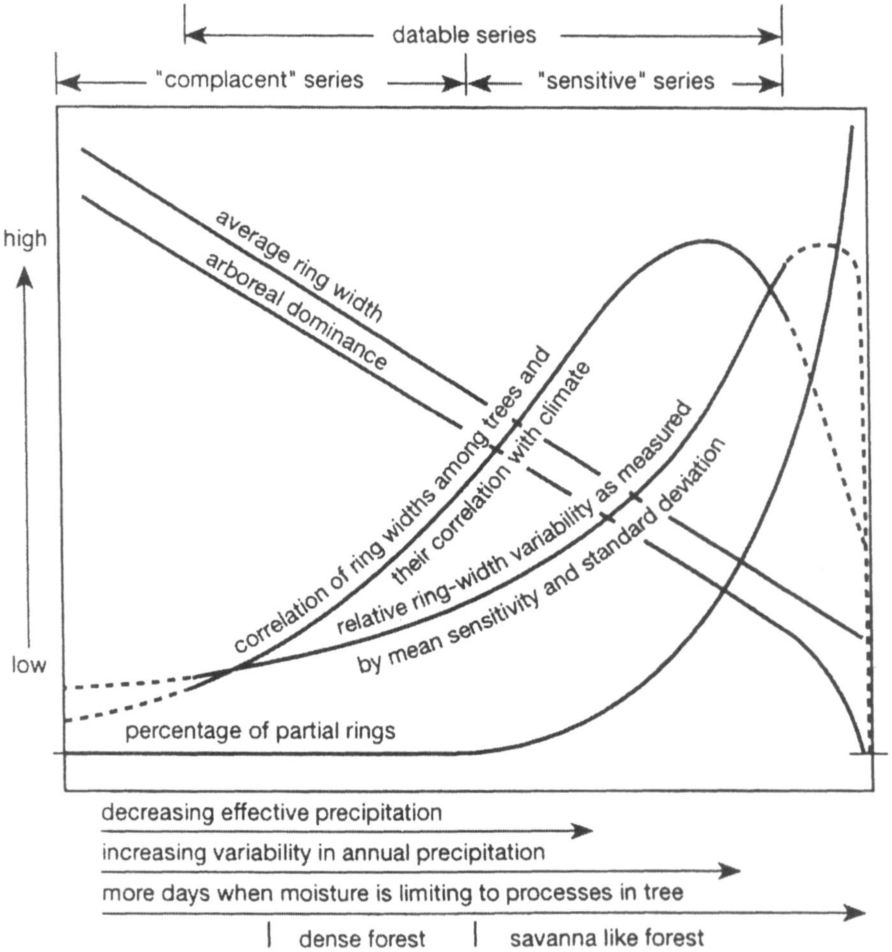

Fig. 2.14. The relationship between the statistics of a time-series of tree rings along changing altitude profiles, beginning at the lower, arid timber line of the south-western United States. For details see text (Fritts et al. 1965)

tion of assimilates are aggregated. Finally these mechanisms result in reduced formation of new xylem cells and hence a narrow ring is produced.

The patterns shown in Fig. 2.14 have rather general applicability to the variability of tree-ring width along an environmental gradient of a factor limiting growth. A similar scheme may be devised for temperature (Vaganov 1996a; Shashkin and Vaganov 2000). Minimum tree-ring width, maximum dispersion (inter-annual variance), sensitivity, and percentage of missing rings are observed in trees growing at polar or upper timber lines (Shiyatov 1986; Vaganov et al. 1996b).

The external factors essentially influence the size of cells, thickness of cell walls, and finally the density of tree rings. We will now consider some results illustrating the effects of physical and biological external factors on the formation of wood in conifer trees.

2.4.1 Light (Intensity and Photoperiod)

Even if we may reasonably assume that temperature and water availability have direct effects on cambial activity, in the case of light the effect is undoubtedly indirect, being mediated through the photosynthetic tissues. The control of growth by light intensity and photoperiod (day-length) has been examined in several monographs and textbooks (Alexeev 1975; Howe et al. 1995; Kozlowski and Pallardy 1997). It is not possible here to review much of the work in which the effect of light on the height or diameter growth (and tree-ring width) of trees was studied. We will briefly consider some results concerning the influence of light intensity and photoperiod on the anatomy of conifer tree-rings.

Richardson (1964) showed a clear positive effect of light intensity and day-length on tracheid cell wall thickness and mean lumen diameter in xylem formed in stems of Sitka spruce (*Picea sitchensis*) seedlings (Fig. 2.15). Larson

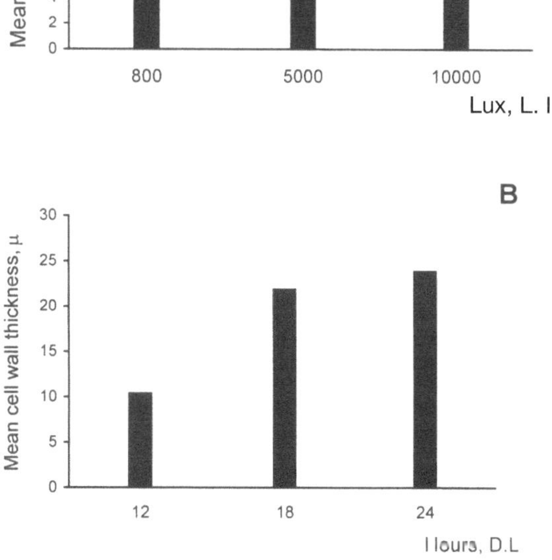

Fig. 2.15. The influence of light on anatomical characteristics of seedlings of Sitka spruce (*Picea sitchensis*) and pine (*Pinus radiata*): **a** effect of light intensity (*L.I.*) on cell wall thickness (means of four spruce seedlings), **b** effect of daylength (*D.L.*) on cell wall thickness (means of 27 spruce seedlings), **c** (see page 44)

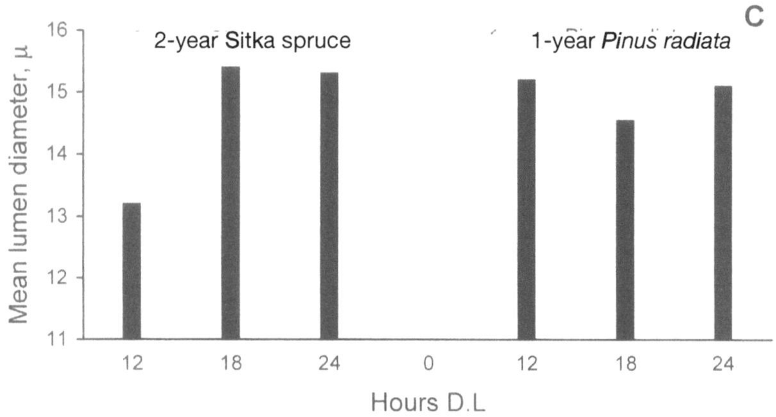

Fig. 2.15. c effect of D.L. on radial lumen diameter (means of 12 spruce and 18 pine seedlings; Richardson 1964)

Table 2.2. Mean radial diameter of tracheid produced in apical stem segments (from Larson 1962). *LD* Long day, *SD* short day

	Diameter (µm)	Needle length (cm)
Experiment 1	(during active needle growth)	
LD-M	15.2	
LD-L	14.5	
SD	11.3	
Experiment 2	(at the termination of needle growth)	
LD-M	14.0	14.1
LD-L	13.5	11.7
SD	7.0	8.0

(1962, 1964) showed an obvious effect of photoperiod on anatomical characteristics of tracheids in xylem formed in loblolly pine (*Pinus resinosa*) seedlings. He found that, on a long day, the formation of larger tracheids was associated with greater needle length (Table 2.2.). He showed that this effect was clearly photoperiodic rather than simply photosynthetic. Day-length influences wood density by controlling the formation of earlywood and latewood cell types through the types of cambial derivatives, rather than the width of the xylem increment (Waisel and Fahn 1956). The growth and formation of earlywood was resumed if the trees were returned to the long day-length condition (Wareing and Roberts 1956).

2.4.2 Temperature

Dendrochronologists have made many investigations of the effect of temperature on radial tree growth and wood density (Fritts 1976; Hughes et al. 1982; Schweingruber 1996). Significantly less data are available for the temperature effect on the anatomical features of tree rings. Much of the relevant research was conducted on trees growing at high latitudes where temperature is presumably a leading limiting factor. Some research concerning anatomical structure was made on seedlings in controlled conditions. There are some common findings:

1. Temperature is the most important factor for growth initiation in boreal and temperate climates (Creber and Chaloner 1984; Iqbal 1990). At high latitudes, growth (new cell production) ceases in about the middle of August and the duration of the season of wood production is mainly determined by the starting date of cambial activity (Mikola 1962). Leikola (1969) showed that a 0.5 °C deviation in the mean April–May temperature caused significant shifts in the starting date of cambial activity. For the Siberian Subarctic, we found early summer (mid-June to mid-July) temperature and snow melt timing to be very important for the variation in radial growth of larch trees (Hughes et al. 1999; Vaganov et al. 1999; Kirdyanov et al. 2003). Wood production ceases at a much higher temperature than is necessary for its initiation (Denne 1971a).
2. At high latitudes, tree-ring width variations correlate well with average summer (June–August) temperature, but maximum density shows a significant correlation with temperature for a larger part of the growing season – for example, May–September (Fig. 2.16; Briffa et al. 1990, 1992a, 1998; D'Arrigo et al. 1992). A longer growing season due to high temperature will definitely increase the percentage of late wood in tree rings (Larson 1964).
3. Denne (1971a), in experiments with Scots pine (*P. sylvestris*) seedlings, showed that a temperature increase from 17.5 °C to 27.5 °C produced only a 10 % increase in tracheid diameter. Contrary to this, we found a significant increase of tracheid diameter in the earlywood of larch tree rings near the northern timber line associated with a long-term summer temperature increase (Vaganov 1996b). Note that these effects of early summer temperature on earlywood tracheid diameter occur in the temperature range 5–14 °C. In more southerly sites where the early summer temperature was higher (between 12 °C and 19 °C), the effect of temperature on tracheid diameter was diminished because the temperature was no longer in the range where it was clearly limiting (Vaganov 1996b). Under conditions close to optimal temperature (as in Denne's experiments), the limiting effect of temperature on tracheid diameter is probably small.
4. Temperature and tracheid wall thickness were inversely correlated in several conifer species (Wodzicki 1971). Similar data were obtained by Antonova and Stasova (1993, 1997). This contradicts results from maxi-

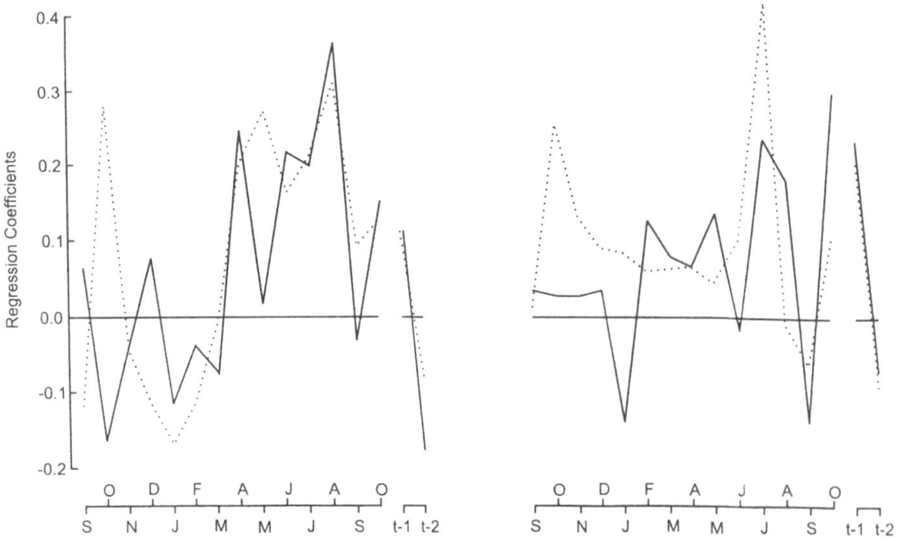

Fig. 2.16. Results of a regression analysis to determine the pattern of growth response in Scots pine (from northern Fennoscandia) to temperature in the months September (in year t-1) to October (in year t-0). Values plotted are coefficients from multiple regression analysis of tree-ring maximum latewood density (*left*) and ring widths (*right*) in relation to instrumentally recorded temperatures in the region. Two periods were used for the analysis to examine the stability of the relationships: 1876–1925 (*dotted lines*) and 1926–1975 (*solid lines*). In addition to the monthly climate variables, the ring-width values from the two preceding years (t-1, t-2) are also included to assess the importance of biological preconditioning of growth in one summer by conditions in preceding years. The analysis shows that maximum latewood density is increased by warm conditions from April to August of year t-0; and a similar, but less strong signal is found in tree-ring widths. Growth in the previous year is also important (Briffa et al. 1990; reprinted by permission from Nature, copyright 1990, Macmillan]

mum latewood density, which is mainly determined by cell wall thickness (see Chap. 5). At the upper elevation or northern timber lines, rings with thin-walled cells in the latewood (so-called "light rings"), are produced by a cold autumn or sharp cooling at the end of summer (Fig. 2.17; Filion et al. 1986; Schweingruber 1993).

Many of the contradictions in publications on the effect of temperature on tracheid dimensions (diameter, wall thickness) are caused by other uncontrolled but important external factors, such as water supply or light intensity, and by uncertainties in the ranges of strong limitation of one factor and alteration of the limit by another factor. For example, Denne (1971a) chose a temperature range which is close to optimal for growth. Thus there is no pronounced effect of temperature on anatomical structure. In the case of

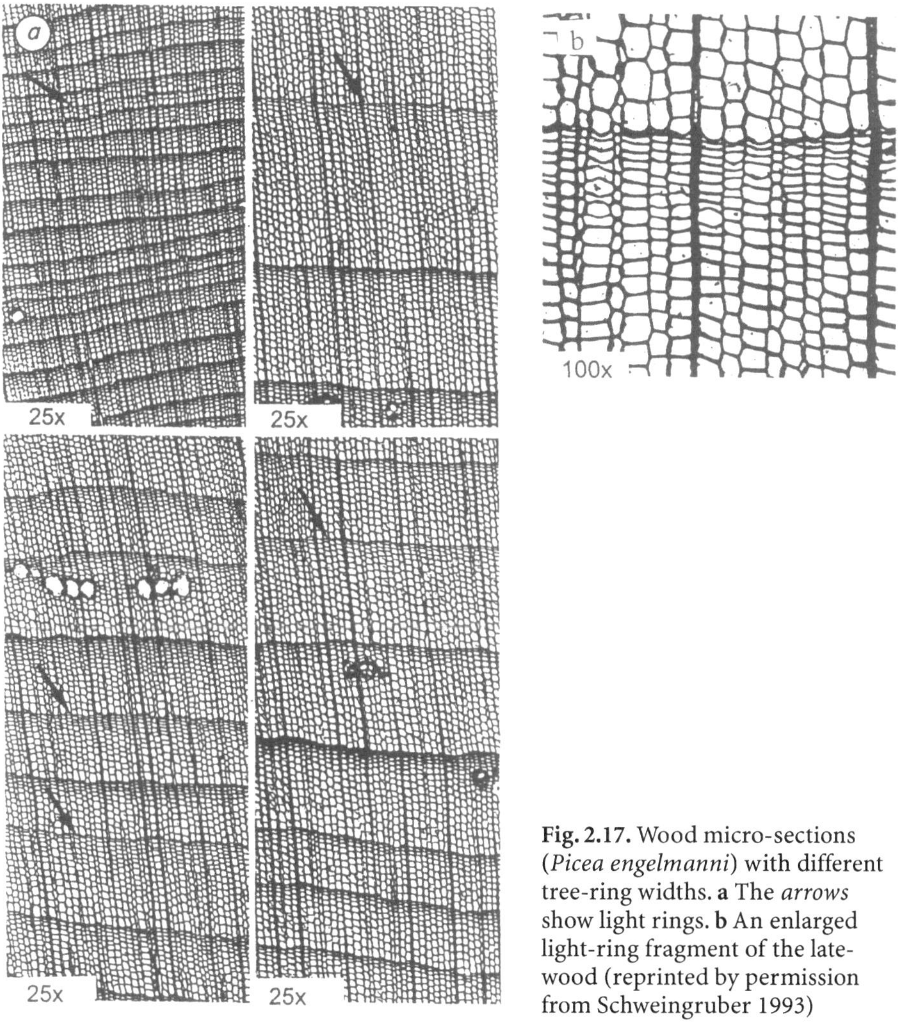

Fig. 2.17. Wood micro-sections (*Picea engelmanni*) with different tree-ring widths. **a** The *arrows* show light rings. **b** An enlarged light-ring fragment of the latewood (reprinted by permission from Schweingruber 1993)

Antonova and Stasova (1993, 1997), there was no control of the soil water content during the production and formation of latewood tracheids. Hence the apparent negative effect of temperature could come from its indirect effect on water loss from soil due to increased evapotranspiration.

2.4.3 Water

The availability of soil water may affect the growth rate and formation of wood, both at long timescales and within a season (Zahner and Oliver 1962; Kozlowski 1968; Creber and Chaloner 1984; Brauning 1999). For example, pine in marshy conditions forms not only narrower tree rings, but rings with smaller absolute size and proportion of latewood (Fig. 2.18). In periods of suppressed growth, a tree ring may have only one or two cells of earlywood and one small-sized thin-walled latewood cell. Considerable variability of cell

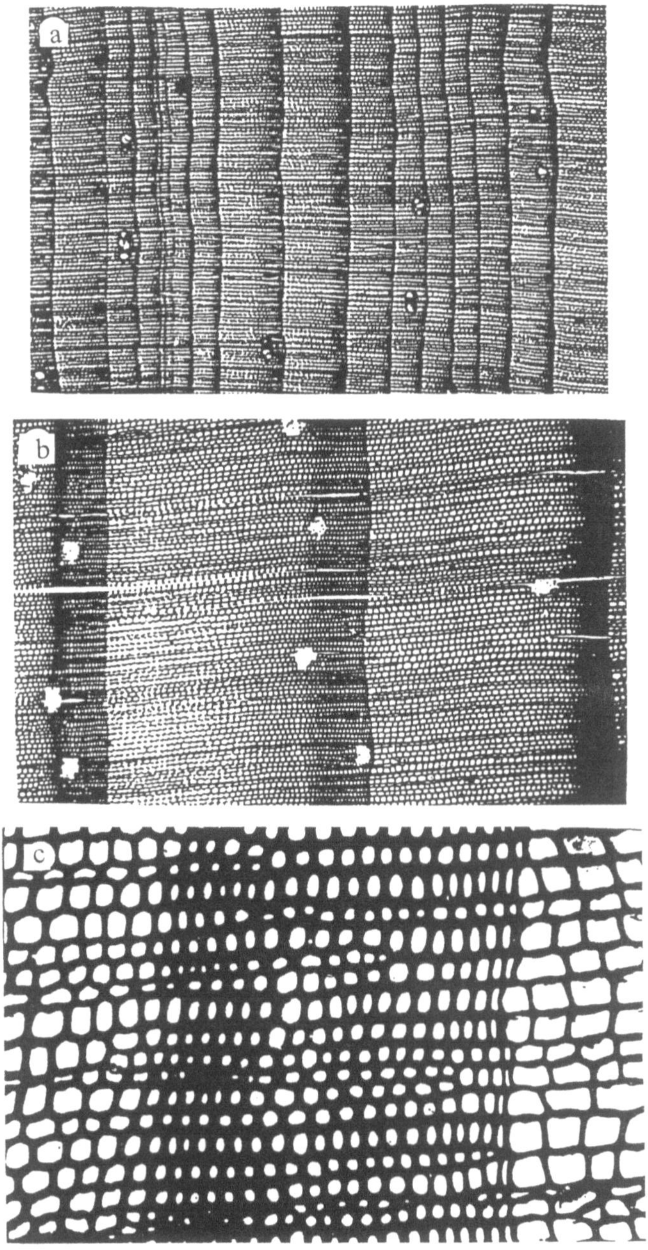

Fig. 2.18. Cross-sections of pine wood (*Pinus sylvestris*) from different site conditions: **a** peatland, **b** forest–steppe transition zone, **c** enlarged micro-section fragment showing formation of false ring in the latewood (reprinted by permission from Schweingruber 1993)

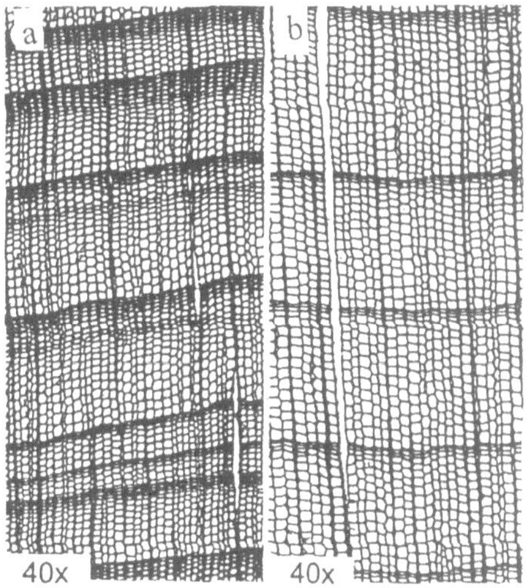

Fig. 2.19. Micro-sections of *Pinus ponderosa* wood (**a**) from the semi-arid zone and (**b**) from a relatively moist site (reprinted by permission from Schweingruber 1993)

size and cell wall thickness may occur within wide tree rings grown in favorable conditions. "False" rings may be produced when small, thick-walled cells are seen at the beginning of the latewood zone and are followed by larger, thin-walled cells (Fig. 2.18). The formation of "false" rings is a common phenomenon when intra-seasonal droughts occur in a precisely designated rainy season, as seen, for example, in the mountains of the American southwest (Fig. 2.19). The layer of larger cells is produced in response to the arrival of the rains at the northern fringes of the Mexican monsoon in early July. However, during certain years, the trees can completely stop growth in the "pre-summer" drought period; and in such a case a "false" ring identified on anatomical features does not differ from the annual ring.

An integrated view of tree-ring structure may be gained by plotting frequency distributions of the radial sizes of tracheids (Vaganov et al. 1985; Vysotskaya et al. 1985; Vysotskaya and Vaganov 1989). Pines growing in moist, moderately moist, and dry conditions (Fig. 2.20) show characteristic changes in frequency distributions. Pines from moist conditions have a weakly negatively skewed distribution (a preponderance of large cells), whereas those from moderately moist conditions show a bimodal distribution, which is even more pronounced in the case of trees from dry conditions. As the water deficit increases, the relative number of small tracheids increases and the average value and mode of their sizes decreases. Asymmetric and bi-modal curve distributions can be described as the sum of two normal distributions (Dunin-Barkovski and Smirnov 1955):

$$f(x) = v_1 \frac{1}{\sigma_1 \sqrt{2\pi}} e^{-\frac{(x-d_1)^2}{2\sigma_1^2}} + v_2 \frac{1}{\sigma_2 \sqrt{2\pi}} e^{-\frac{(x-d_2)^2}{2\sigma_2^2}}, \tag{2.4}$$

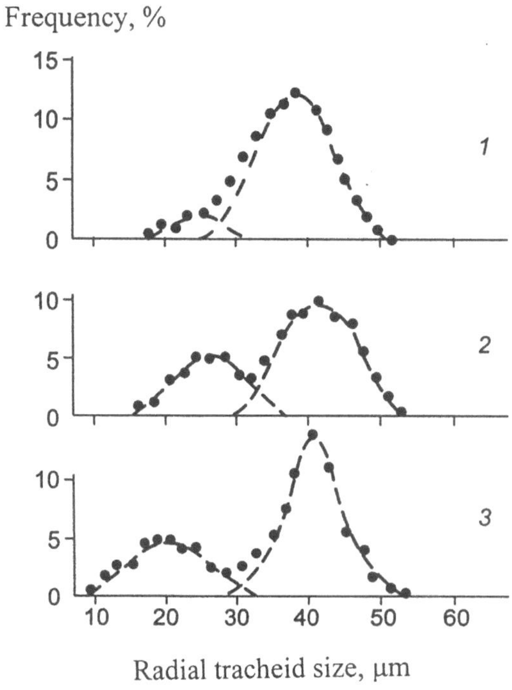

Fig. 2.20. Frequency distribution of the tracheid radial dimensions in *Pinus sylvestris* tree rings from the hydric (*1*), mesic (*2*), and xeric (*3*) sites. See text for futher explanation (Vaganov et al. 1985)

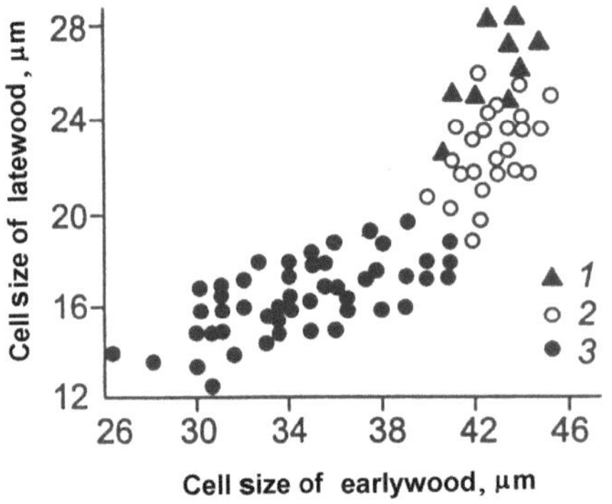

Fig. 2.21. Cell size of latewood (tracheid radial dimension) vs cell size of earlywood in pine tree-rings from the hydric (*1*), mesic (*2*), and xeric (*3*) sites (Vaganov et al. 1985)

where $f(x)$ is the density probability, d_1, d_2, ρ_1 and σ_2 are respectively the mean sizes and standard deviations of "contributing" distributions, and v_1 and v_2 are the relative shares in the common distribution of cells with large (earlywood) and small (latewood) sizes. If we designate A_1 and A_2 as appropriate amplitudes for the "contributing" distributions, then for the normalized curve distribution v_1 and v_2 are proportional to the product of early- and latewood. Therefore on the diagrams indicating the connection between the mean sizes of cells in "contributing" distributions, the measurements for tree rings in pine trees from conditions with different moisture levels are divided into pre-

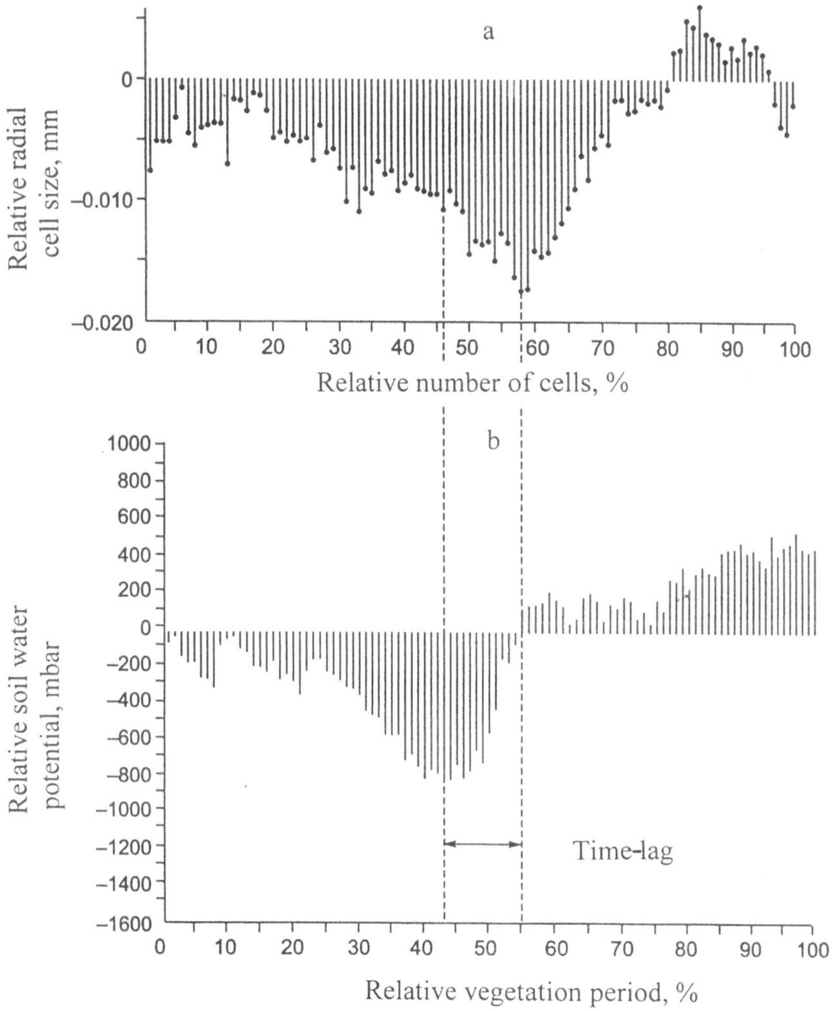

Fig. 2.22. Time-lag between minimum soil water potential (**b**) and minimum tracheid diameters (**a**) in spruce tree rings on a "Parabraunerde" site during the summer drought in 1976 (Wilpert 1991)

cise groups (Fig. 2.21). These groups show that, with the increase in moisture deficit, the mean size of earlywood cells decreases as it does in latewood. The dispersion of the cell sizes in latewood increases.

The reduction of tracheid sizes in connection with water stress within a growing season was explicitly investigated by Wilpert (1990, 1991). Measuring the water potential of soil and the dynamics of seasonal growth of tree rings in European spruce (*Picea abies*), he showed precisely that the decrease in the tracheid radial sizes was produced by an intra-seasonal drought (decrease in water potential; Fig. 2.22). In fact, there is a broad range of water potential values over which tracheid radial diameters do not vary significantly, but below some critical value of water potential there is an accelerating decline in cell sizes (Fig. 2.23). This is described by a curvilinear equation which seems to be very similar to those obtained in direct measurements (Kramer and Kozlowski 1983).

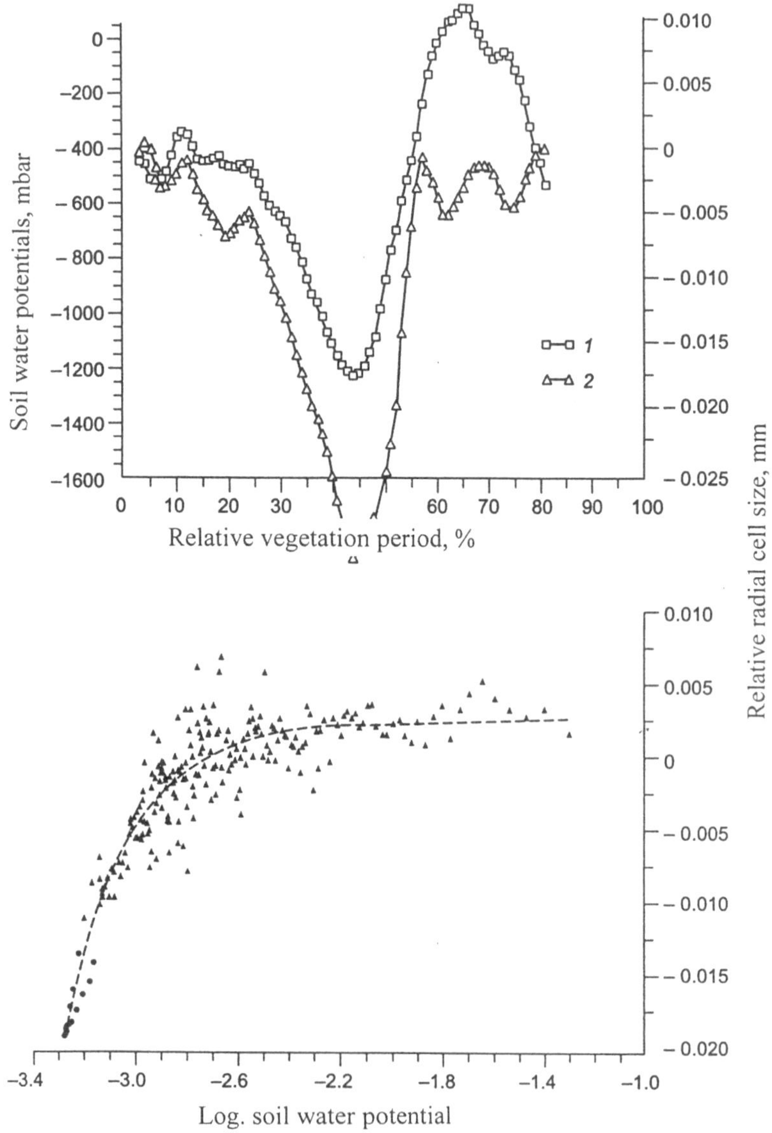

Fig. 2.23. Relation between soil water potential (*2*) and tracheid diameter (*1*). **a** Single year (1976) at a "Pseudogley" site. **b** Nonlinear regression of all data on the "Parabraunerde" site (Wilpert 1991)

2.4.4 Some Other Factors

Ionizing radiation is one of the factors rendering the strongest effect on the structure of tree rings in growing trees (Musaev 1993, 1996; Kozubov and Taskaev 1994). The 1986 emergency at the Chernobyl nuclear power station provided a unique opportunity to study the response of woody plants to different doses of irradiation (depending on distance from the station; Fig. 2.24). Lethal damage to cells is observed at high doses, first of all in the cambial zone and in other meristems; and there are signs of various kinds of anatomical anomalies in the structure of tree rings (Fig. 2.25). It is possible to classify and recognize the most frequent anatomical anomalies on wood cross-sections:

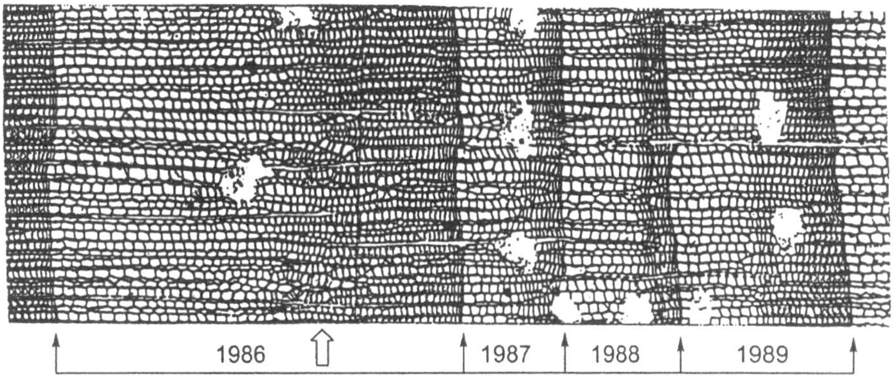

Fig. 2.24. The typical anatomical structure of pine tree rings formed after the Chernobyl atomic power station emergency. The *open arrow* shows a visible disturbance in tracheid structure. The *thin arrows* indicate tree-ring boundaries (Musaev 1996)

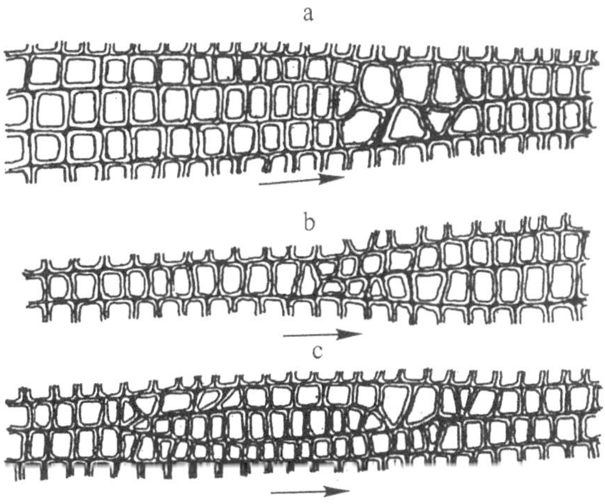

Fig. 2.25. The most typical radiation-related disturbances in tracheid file structure of tree rings: **a** interruption of file, **b** dichotomous files, **c** included file. Growth from pith to bark is indicated by *arrows* (Musaev 1996)

(1) interrupted files indicating loss of cambial initials, (2) dichotomous rows created at the expense of the increase in a number of initials by means of pseudo-transverse division, (3) built-in short rows derived by division of xylem mother cells of limited potential. Formation of large cells at the expense of suppression of the last divisions of xylem mother cells before exit to the zone of enlargement is also quite frequent. Musaev (1996) analyzed in detail the frequency of such anatomical disturbances in annual tree-rings for 1986 (the year of the April emergency) according to distance from the station. The normal, or "background" frequency of each type of disturbance was about 10 %, but their frequency in tree rings near the station rose to 40–45 %. At the distance from the station where the dose of ionizing irradiation was 2–3 times, less the frequency decreased to 20 %; and at the distance of 30 km from the station it did not differ much from the normal frequency (Fig. 2.26).

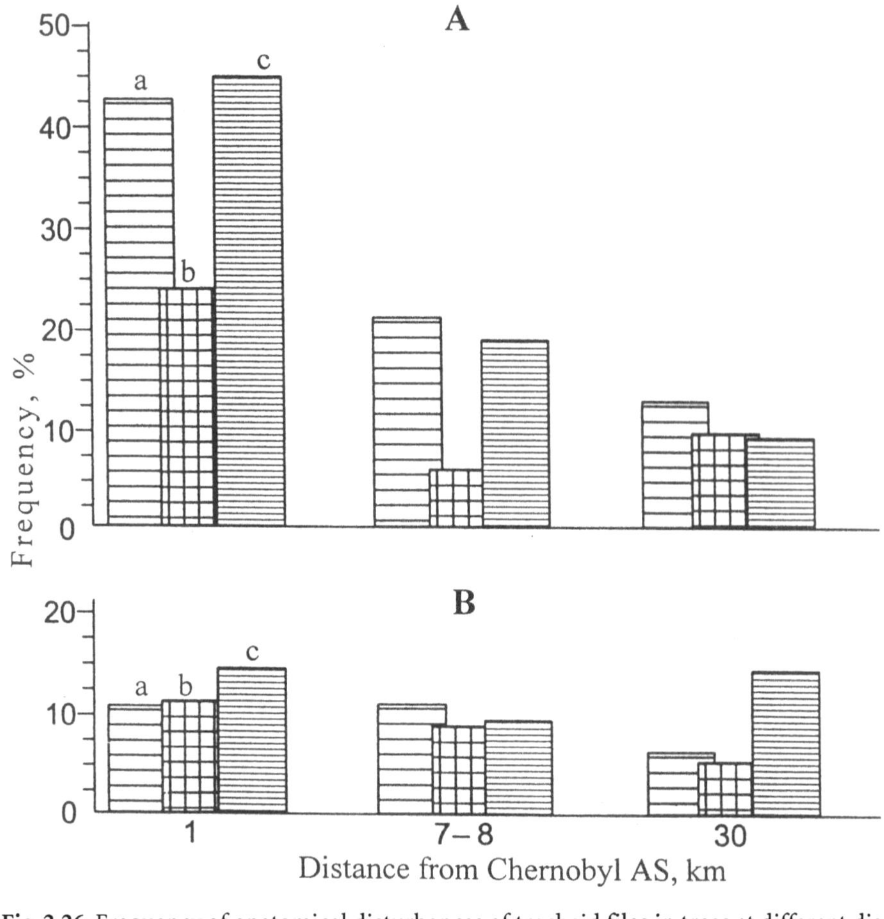

Fig. 2.26. Frequency of anatomical disturbances of tracheid files in trees at different distances from Chernobyl atomic power station (*AS*) in 1986 (**a**) compared with 1985 (**b**). For *a*, *b*, *c*, see Fig. 2.25 (Musaev 1996)

Musaev (1993, 1996) considered anomalies in tracheid anatomy in tree rings from different parts of the stem (Fig. 2.27) and showed that cell damage in tree rings formed in 1986 near the top of the stem is further from the beginning of the ring than in the lower parts of the stem. This results from more intensive production of cells at the beginning of the season in the upper part of the stem. Comparing tracheidograms with the climatic data, Musaev reconstructed the seasonal dynamics of annual ring growth in 1986 and found when the activity of the emitted radioactive isotopes produced maximal disturbance of the xylem cell-growth process.

Defoliation is an important biotic factor influencing the structure of annual rings (whether artificial or by insects; Vaganov and Terskov 1977; Vaganov et al. 1979; Schweingruber 1979; Jardon et al. 1994b; Filion and Cournoyer 1995). For the first year of damage, the width of the tree ring

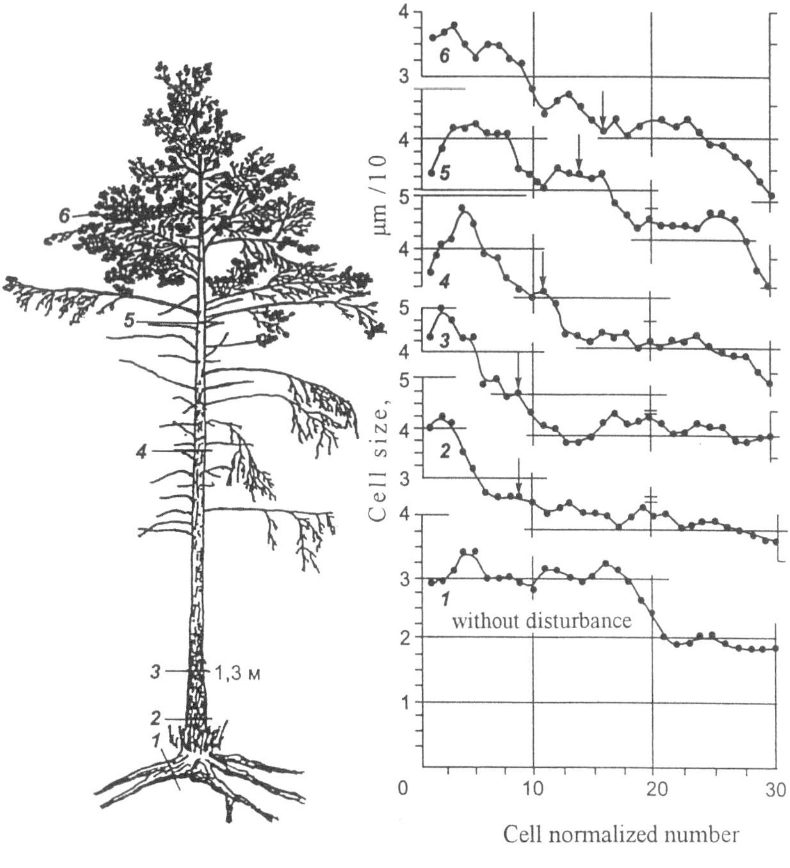

Fig. 2.27. Dynamics of cell size in 1986 tree ring near Chernobyl (1 km). The *arrows* show initial disturbances in tracheid files. *1–6* Sample positions in stem and roots (Musaev 1996)

decreases and cells with thin walls are formed in the latewood zone. The sizes of the cell lumina (i.e. internal empty spaces of cells) decrease, as does tree-ring density (Fig. 2.28). The next year, the ratio between the size of earlywood and latewood shifts strongly towards latewood (Vaganov and Terskov 1977). Then recovery to a normal level of the anatomical characteristics of tree rings (size of cells, cell wall thickness) accelerates, with recovery of tree-ring width (Fig. 2.29). It is possible to classify the influence of defoliation on tree-ring structure as indirect, operating through the level of growth hormones and assimilates.

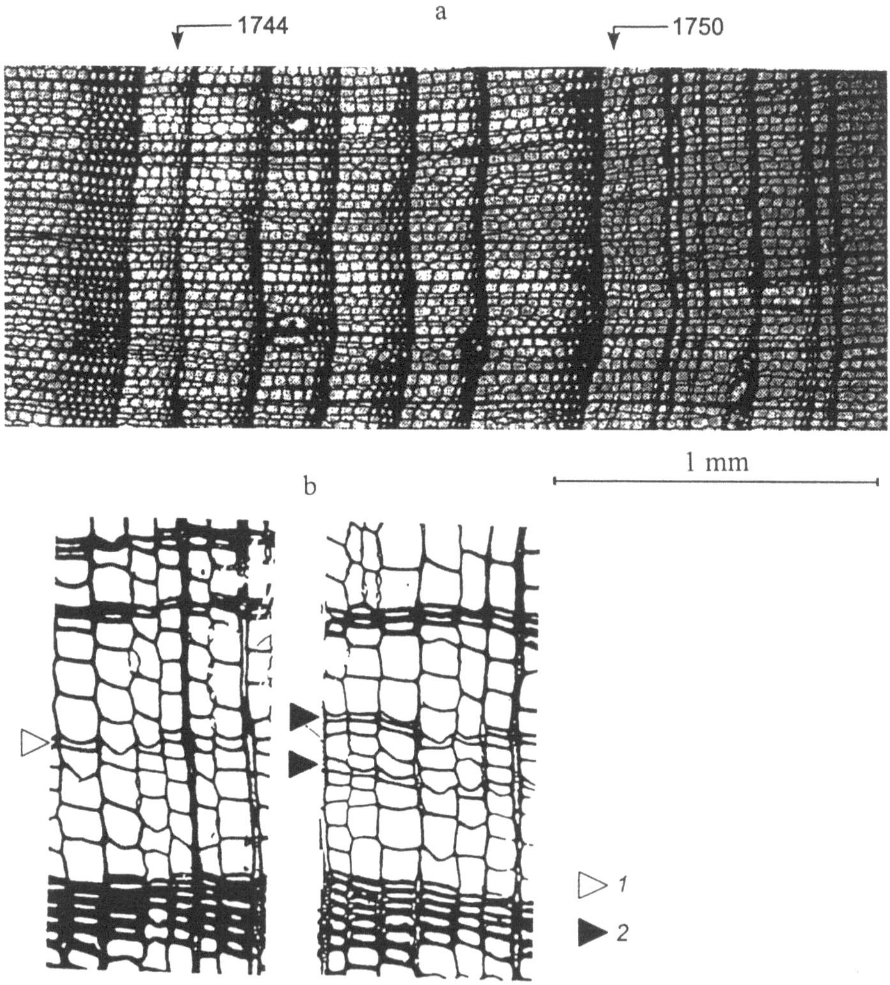

Fig. 2.28. Wood micro-section of *Larix laricina* with tree-rings formed in the indicated insect attack years (**a**; Jardon et al. 1994b) and microscopic illustration (**b**, enlarged 200×) of tree rings in a larch showing anomalous anatomical properties as the result of damage done by larch bud moth. *1* Tree-ring sequence with one probable attack phase, *2* tree-ring sequences with two probable attack phases (Schweingruber 1996)

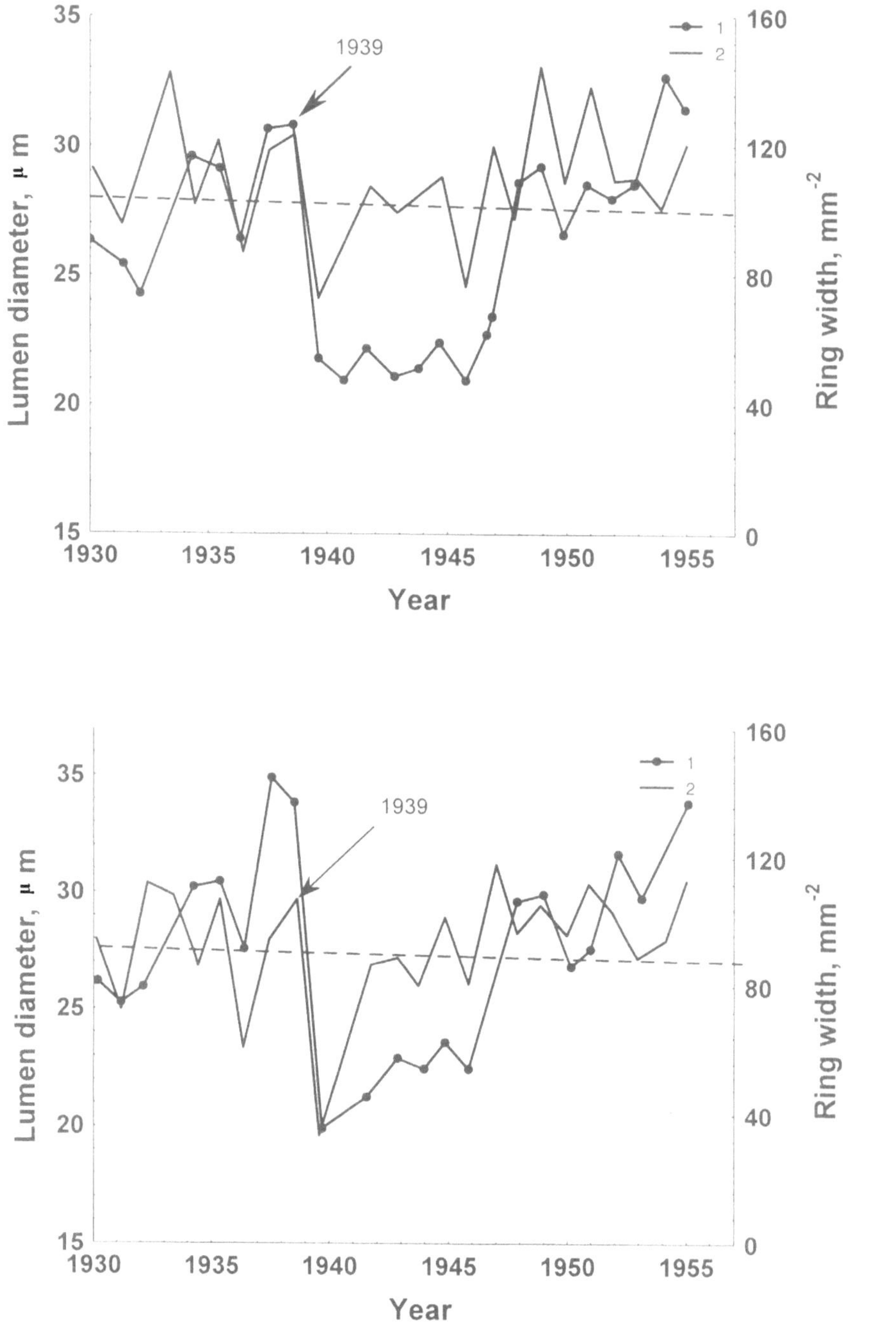

Fig. 2.29. Variation in cell lumen diameter and ring width at two study sites between 1939 and 1955 (North Quebec, Canada): *1* tree-ring width, *2* mean lumen diameter (Filion and Cournoyer 1995)

2.5 Deriving Chronologies for Parameters of Tree-Ring Structure

2.5.1 Variability of Radial Cell Sizes, Cell Wall Thickness, and Wood Density Within Tree Rings

The main patterns of variability of cell wall thickness and wood density within tree rings are best considered for separate files of tracheids (Fig. 2.30). In the earlywood zone, where the radial sizes of tracheids exceed 30 μm, the cell wall thickness is small (≤2 μm) and is practically independent of variations in the radial size of cells. This may be seen especially well in ponderosa pine (*Pinus ponderosa*; Fig. 2.30b). In the first half of the tree ring in the given

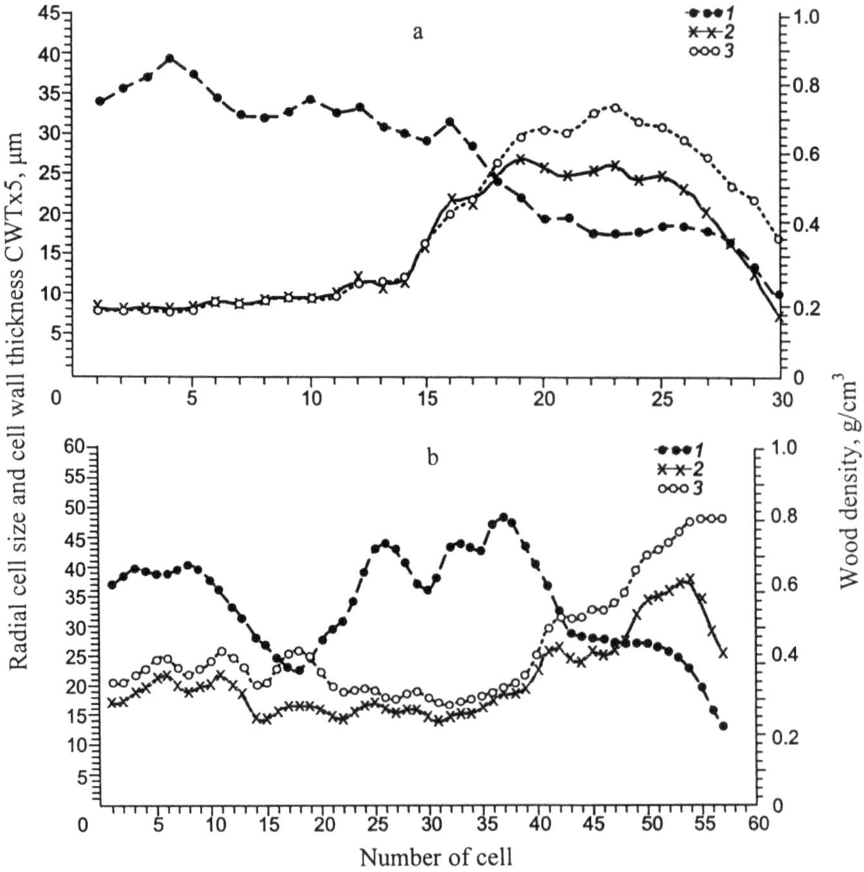

Fig. 2.30. Variability of radial tracheid size (*1*), cell wall thickness (*2*) and wood density (*3*) within a tree ring of Siberian larch (**a**) and ponderosa pine (**b**)

year, a downturn in the sizes of cells (i.e. formation of "false" rings) is observed, while the cell wall thickness hardly changes at all. In the second half of the season with the beginning of a transition zone of tracheid formation (in larch from 14–15 cells, in ponderosa pine from 37–40 cells), the radial size of cells decreases in a stable way and the cell wall thickness increases up to a certain limit, and then for the last cells to the boundary of the tree ring again decreases. A typical set of relationships may be seen in the diagram of the radial size of cells versus cell wall thickness (Fig. 2.31). The relation of cell wall thickness to the radial size of a cell shown here is probably generally applicable. The zone of the tree ring, for which the cell wall thickness only weakly depends on the size of the tracheids, is formed right at the beginning of the growing season. It can be rather broad, but can involve only two or three cells (as may be seen in the diagram for a larch tree ring). In the latter case, these cells hardly differ from those in the transition zone.

As the density of wood is a derivative of change in the radial tracheid size and cell wall thickness, its changes within a tree-ring can differ little from the character of the changes of these parameters. For example, in larch tree-ring density, the cell wall thickness reaches its maximum in the 18th cell, while

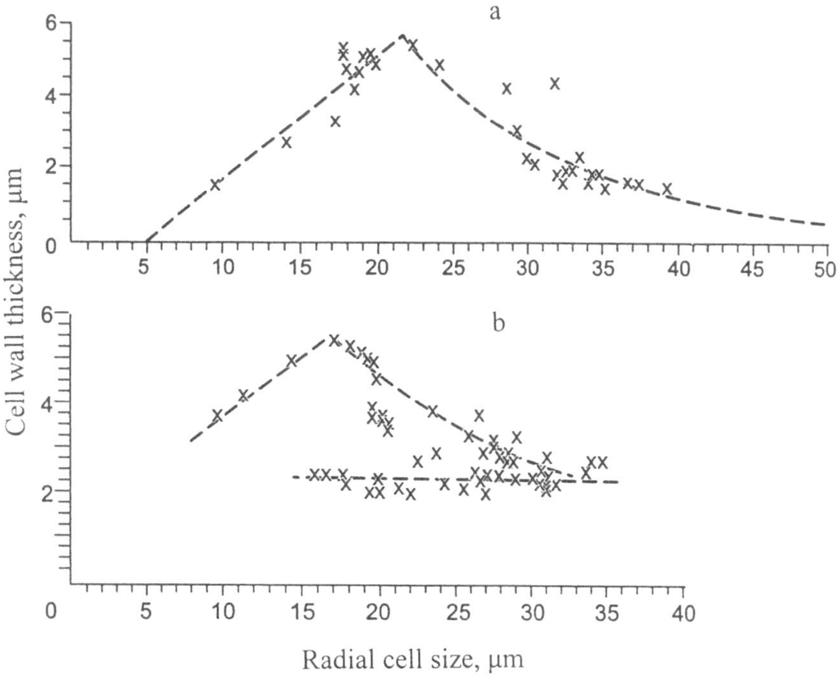

Fig. 2.31. Relationship between radial cell size and cell wall thickness in tree rings of Siberian larch (**a**) and ponderosa pine (**b**). *Broken lines* indicate general form of relationships. See text for further explanation

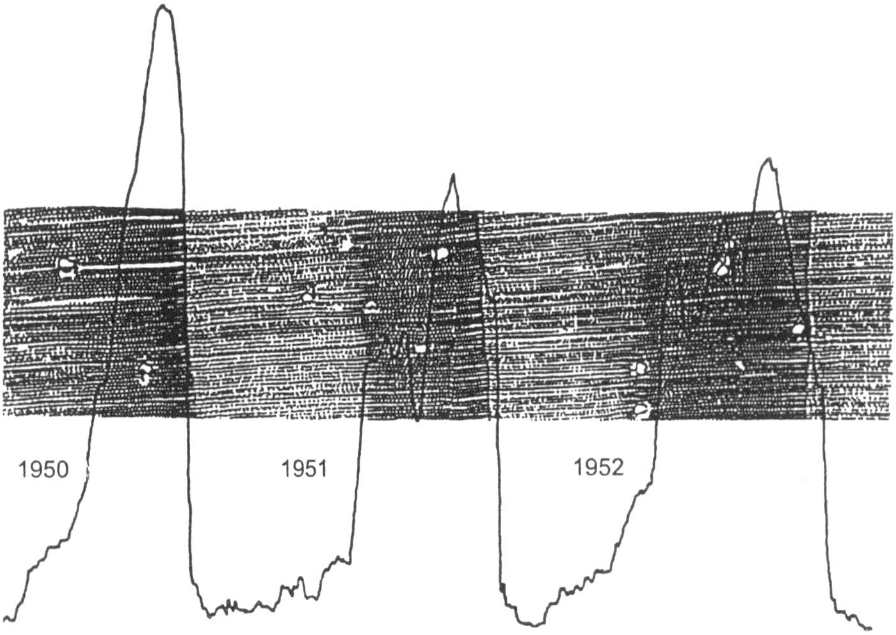

Fig. 2.32. Micro-section and the corresponding densitogram of pine (*Pinus sylvestris*) from the lowlands of Northern Germany (Schweingruber 1996)

maximum density is seen in the 23rd. In most conifers, density and cell wall thickness do not reach a maximum in the last cells of the ring, but more often in the middle of the latewood zone. The diagrams of cell size versus cell wall thickness show that cells with radial sizes between 20 μm and 25 μm have the highest density. Numerous examples are given as densitometric curves (Fig. 2.32).

2.5.2 Acquisition and Statistical Characteristics of "Cell Chronologies"

The use of the variability of tree-ring width in dendrochronology is based on two main principles (Fritts 1976): (1) the synchronicity of inter-annual variability of tree-ring series from different trees in a stand, (2) the stability of the response of trees to external effects (climate) during the greater part of their life. Tree-ring width is the characteristic of tree growth that integrates external effects for the whole season. The synchronicity of inter-annual changes in increment between single trees suggests that the common component of response considerably exceeds the individual (Cook and Kairiukstis 1990). This common component is an integral of changes taking place in different intervals of the season, affecting differing components of ring size and structure. Therefore it may reasonably be assumed that the parameters of tree-ring structure dependent on seasonal growth kinetics will also show synchronicity of inter-annual changes. Densitometric methods provide convincing evidence that the parameters of tree-ring density (especially maximum density) are dendrochronological characteristics, i.e. their inter-annual variability is rather synchronous in different trees in a stand and even within a region (Parker and Henoch 1971; Polge and Keller 1969; Schweingruber et al. 1979; Schweingruber 1987, 1988, 1996). Figure 2.33 shows an example of good synchronism among individual curves of maximum wood density of tree-rings in larch (*Larix sibirica*) trees at one site in subarctic Siberia (Kirdyanov and Zakharjewski 1996). If the common variations in growth rate during a season predominate over the individual ones, it is reasonable to expect that inter-annual variations in tracheid dimensions will also be synchronous in tree rings of different trees at one site. We will consider this issue further.

The procedure for compiling "cell chronologies" differs a little from that for tree-ring width or maximum density. This is because tracheidograms of separate cell files even inside one annual ring (and tree rings of one year of formation in different trees), contain differing numbers of cells. Therefore, at the first stage, all measured tracheidograms are standardized to a standard number of cells (see Sect. 2.2.1). Statistical evaluations have shown that, to get reliable data on the variability of cell sizes within tree rings, it is necessary to measure five or six files (Vaganov et al. 1985). Then, after standardization of numbers of cells, for each tree-ring the measurements of different files are averaged, giving a mean tracheidogram for that ring in that tree. Then, as the number of cells in normalized tracheidograms is identical in different trees

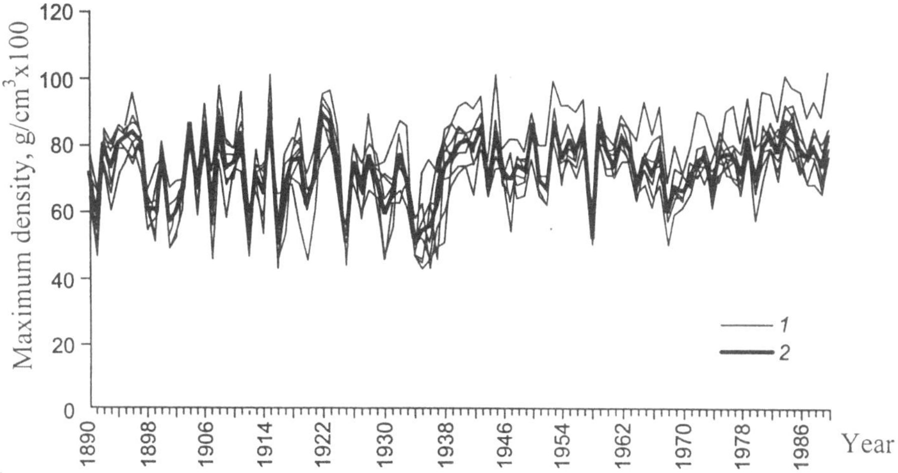

Fig. 2.33. Annual variability of maximum latewood density in tree rings of Siberian larch for individual samples (*1*) and the average for the site (*2*)

and in different years of growth, it is possible to construct a "cell chronology" for inter-annual variability of the sizes of cells for this or that item (for example, from the first to the 15th cells in a standardization to 15 cells). So, the variability of conditions right at the beginning of a growing season characterizes the size of cells 1–2, in earlywood it characterizes the size of cells 4–5, in the transition zone cells 9–10, and in the latewood zone cells 12–13. A time-series of one of these (for example the mean cell size for cells 1–2) for all samples for each year represents a "cell chronology" (Vaganov et al. 1994, 1996 c). For example, individual "cell chronologies" are shown in Fig. 2.34 for latewood of Siberian spruce (*Picea obovata*) from the northern limit of distribution close to the Norilsk region. The individual chronology shows good synchronicity (72 %) until 1978. After that the synchronicity decreases sharply, owing to the heavy influence of pollution from the Norilsk metallurgical plant.

The main statistical characteristics of "cell chronologies" and tree-ring width chronologies are compared in Table 2.3. The sites where the trees for measurement were selected are located at the northern tree limit and at various distances to the south of it. The standard deviations of inter-annual variations in cell size in the different zones of the tree ring are proportionately less than in chronologies of tree-ring width. However coefficients of correlation of individual chronologies with mean chronologies are high and differ little from those for typical tree-ring chronologies. There is no pattern in the variability of mean cell size, for example, in the earlywood or latewood zones according to latitude. In fact, the statistical parameters of "cell chronologies"

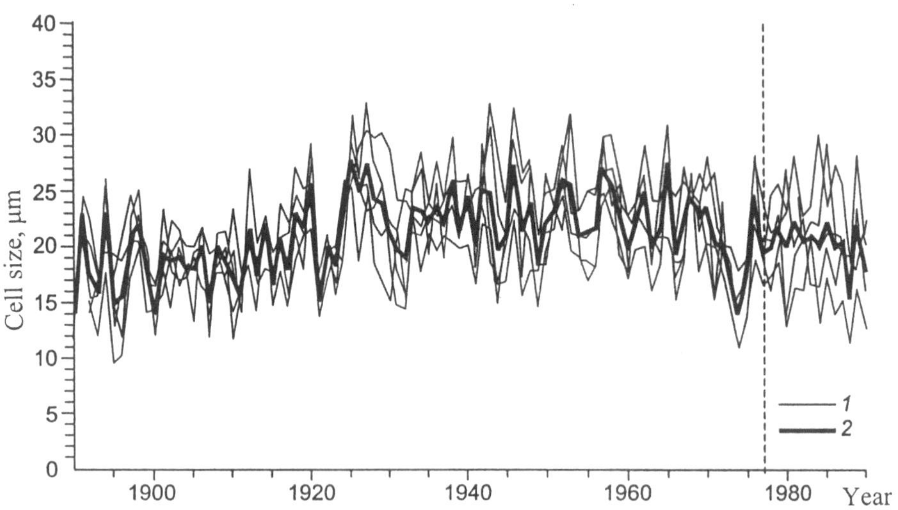

Fig. 2.34. Individual cell chronologies for cell sizes in the latewood of individual samples of Siberian larch (*1*) and the average for the site (*2*). The *vertical broken line* indicates 1978. See text for further discussion

Table 2.3. Statistical characteristics of larch cell chronologies. *EW* Earlywood (positions 4–5 in tree ring), *TZ* transitional zone (positions 9–10), *LW* latewood (positions 12–13), *TRW* chronology of tree-ring width

Chronology identifier and variables	Mean cell size (μm)	Standard deviation	Coefficients Variation	Sensitivity	Correlation of individual sample with mean chronology
Polar timber line					
IKO					
EW	45.3	5.29	0.117	0.107	0.77
TZ	30.2	6.01	0.199	0.224	0.77
LW	16.3	2.54	0.156	0.176	0.74
TRW index		0.35	0.350	0.450	0.79
GKA					
EW	42.0	4.65	0.110	0.090	0.69
TZ	27.4	4.70	0.171	0.176	0.66
LW	16.7	2.65	0.159	0.135	0.64
TRW index		0.39	0.390	0.490	0.76
ALL					
EW	45.5	3.06	0.069	0.064	0.67
TZ	30.8	5.63	0.182	0.183	0.74
LW	16.4	2.58	0.157	0.161	0.81
TRW index		0.31	0.310	0.370	0.83
250 km to the south of polar timber line (KUR)					
EW	50.1	6.20	0.124	0.130	0.77
TZ	26.5	5.26	0.198	0.234	0.76
LW	19.2	3.17	0.166	0.156	0.75
TRW index		0.32	0.320	0.400	0.74
500 km to the south of polar timber line (KHL)					
EW	46.8	5.80	0.124	0.134	0.68
TZ	25.6	5.26	0.205	0.252	0.65
LW	15.0	2.59	0.173	0.157	0.76
TRW index		0.28	0.280	0.300	0.72

show that the cell sizes in annual rings of different trees of one site change rather synchronously. This suggests an important role for environmental variations (usually climatic) in controlling such structural parameters. The data in the table testify either that cell sizes are under more rigid control by internal factors than the production of cells during a season and, therefore, the width of an annual ring, or they depend on a kinetic parameter of seasonal formation of a tree ring that changes less than the production of cells during the growing season.

2.6 Long-Term Relations Between Different Anatomical Characteristics of Tree Rings

The patterns seen in the variability of the radial sizes of tracheids, their cell wall thickness and wood density inside separate annual rings are also seen in the long-term variability of these parameters. The relationship between tree-ring width and the number of cells in files (Fig. 2.35) is most clear and may be considered as a straight line. It shows that the increase in annual increment is mainly determined by increased cell production in the cambial zone during the growing season. The straight-line relationship between the width of annual rings and the number of cells is characteristic of all conifers without exception and of a broad spectrum of ecological conditions. Deviations from it, as a rule, testify to gross infringements of the structure of annual rings and are observed with extreme damage, such as produced by high doses of ionizing radiation or damage of cambial cells by frost ("formation of frost rings"; Glerum and Farrar 1966; Fritts 1976; Schweingruber 1996).

The total annual production of cells for a season (and the mean rate of production of cells during a season) influences other quantitative parameters of tree-ring structure. This may be seen in the case of a typical example of tree-ring width, cell size, and maximum wood density variations in larch and spruce at the northern limit of their distribution (Fig. 2.36). For rather wide

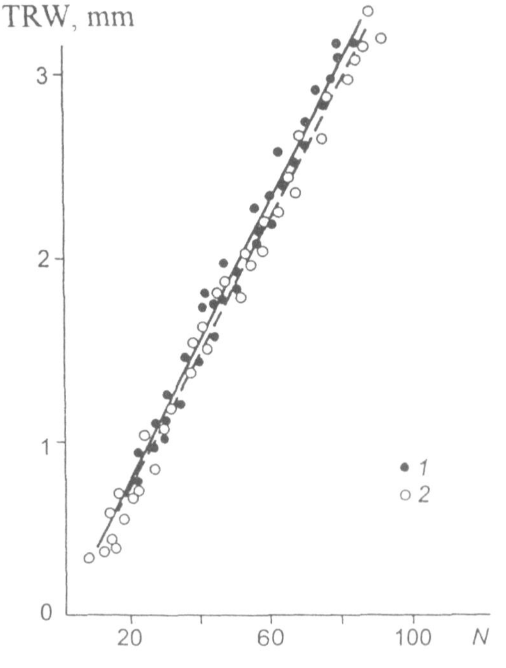

Fig. 2.35. Cell number vs tree-ring width of Siberian larch (*1*) and Scots pine (*2*) in the southern taiga

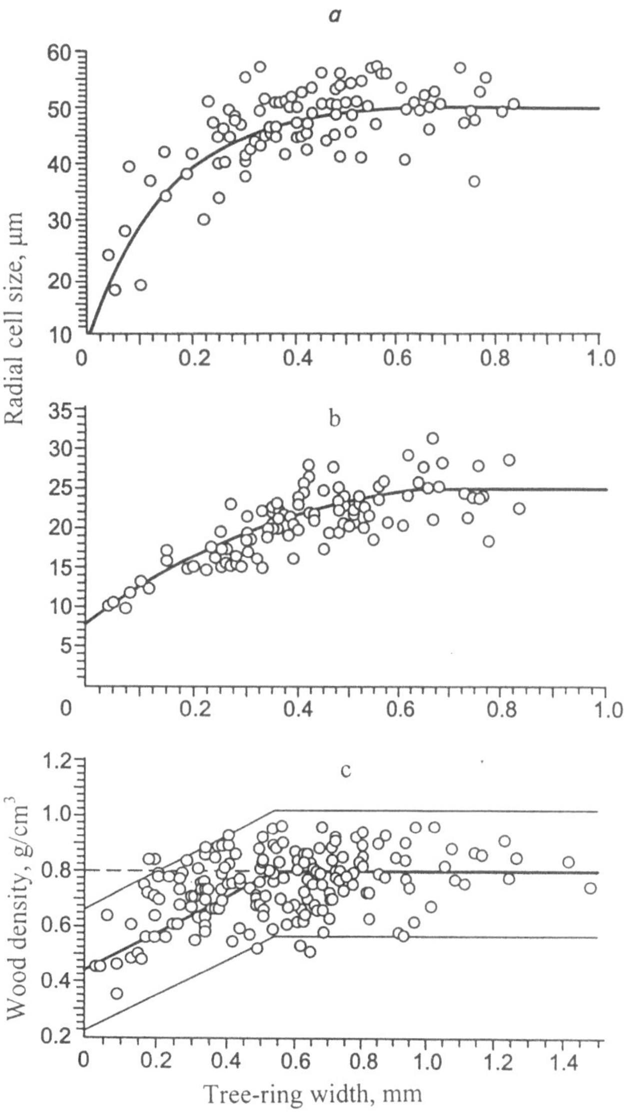

Fig. 2.36. Relationship between tree-ring width and cell size in the early- (**a**) and latewood (**b**) and between tree-ring width and maximum wood density (**c**) of a Siberian larch tree growing at the polar timber line

rings (with an increment more than 0.5–0.6 mm) the correlation of the radial cell sizes in earlywood and latewood with tree-ring width is practically absent. However, if the growth rate decreases (narrow rings formed), cell sizes in earlywood and latewood, and maximum wood density, decrease. The lower limit for the size of cells in earlywood and latewood of larch, for example, is approximately 8 µm, the mean size of a cell in the cambial zone. Maximum density decreases to a limit of 0.45 g/cm^3, corresponding to a minimum cell wall thickness of about 1.5 µm.

The comparison of the long-term variability in cell sizes with both minimum and maximum density also shows interesting outcomes that confirm the relationship between these parameters of tree-ring structure (Table 2.4, Fig. 2.37). For example, the standard deviation of wood-density chronologies appears not to be higher than that for cell chronologies. The dynamics of earlywood cell sizes show changes that are inverse to those in minimum density (Fig. 2.37). In contrast, the size of cells in latewood has a positive association with changes in maximum density. The data in Table 2.5 show that the relationships between cell size and wood density in annual rings of larch trees are stronger than in the annual rings of spruce trees. The observed relationship in the long-term variability of in cell size and wood density can be explained using the diagrams of cell size and cell wall thickness (Fig. 2.38). It can be seen that, for earlywood and transition zone cells, changes in cell wall thickness are inversely proportional to cell size. Such relationships may be approximated by a negative exponential curve. The second part of the diagram describes changes in cell wall thickness and latewood cell size. The relationship between parameters is positive and close to linear. When recalculated as density, the maximum wood density values correspond to maximum sizes of cell wall thickness, but with cell sizes between 15 μm and 25 μm. Therefore minimum wood density is negatively correlated with earlywood cell sizes; and maximum density shows significant correlation (especially when smoothed) with latewood cell sizes (Table 2.5). Seasonality modifies the diagram of relationships between cell size and cell wall thickness. For narrow tree rings (low rate of production of cells) the maximum cell wall thickness is lower than for wide

Table 2.4. Inter-annual variability of cell sizes (μm) and density (g/cm^3) in conifer wood at a site in the northern taiga of central Siberia

Chronology	Average value	Standard deviation	Coefficient	
			Variation	Sensitivity
Larch earlywood				
Cell size	43.5	5.38	0.124	0.130
Minimum density	0.245	0.026	0.107	0.099
Larch latewood				
Cell size	16.7	2.75	0.165	0.156
Maximum density	0.796	0.081	0.102	0.108
Spruce earlywood				
Cell size	39.9	2.95	0.074	0.057
Minimum density	0.315	0.029	0.083	0.077
Spruce latewood				
Cell size	21.0	3.22	0.153	0.145
Maximum density	0.715	0.049	0.069	0.069

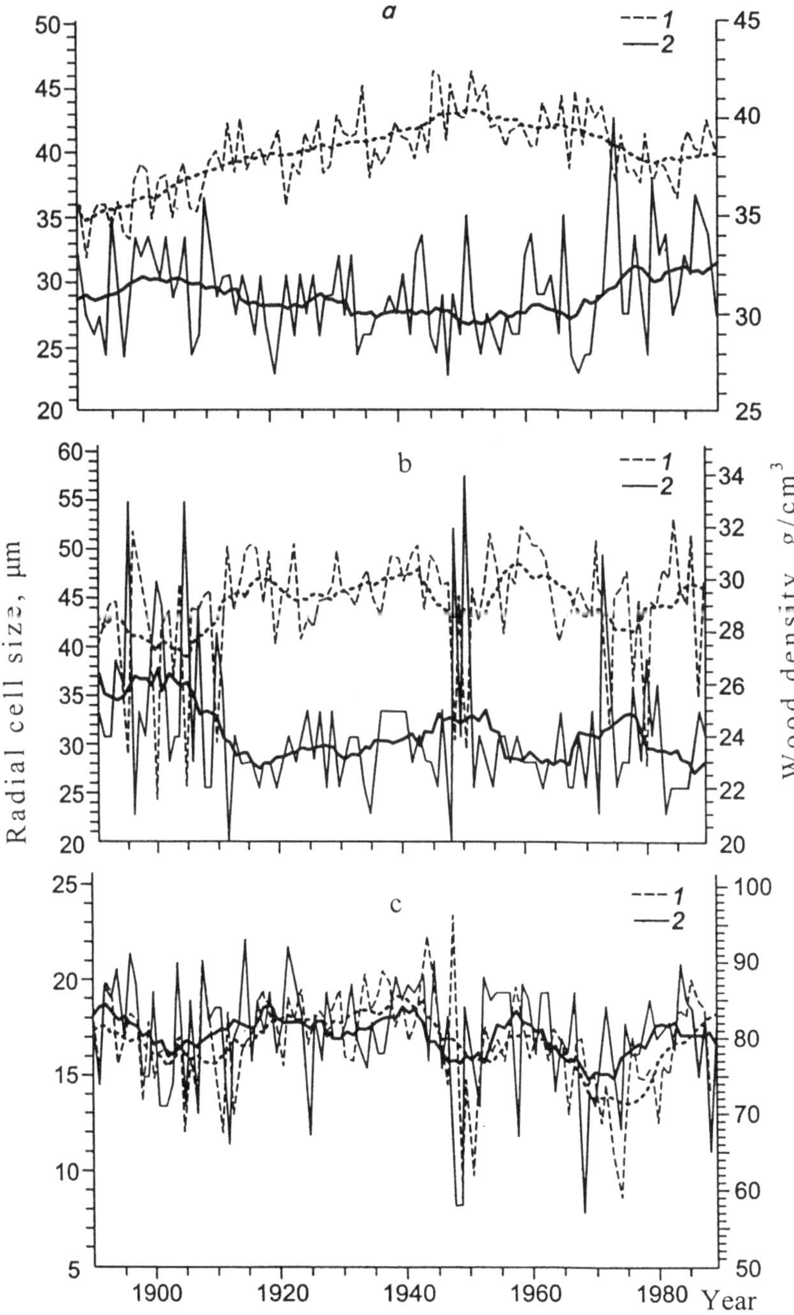

Fig. 2.37. Variability of representative site chronologies of radial cell sizes (*1*) and wood density (*2*) in tree rings. **a** Cell size of earlywood and minimum density of Siberian spruce from the northern taiga, **b** cell size of earlywood and minimum density of Siberian larch from forest-tundra, **c** cell size of latewood and maximum density of Siberian larch from the forest-tundra. Smoothed curves are also given

Table 2.5. Correlation between radial tracheid sizes and wood density in inter-annual variability of tree-ring structure indices in conifer wood at a site in the northern taiga of central Siberia

Index	Earlywood Larch	Spruce	Latewood Larch	Spruce
Raw data	−0.753**	−0.259*	0.309*	0.169
Smoothed data	−0.872**	−0.722**	0.698**	0.236*

* significant at $P<0.05$
** significant at $P<0.01$

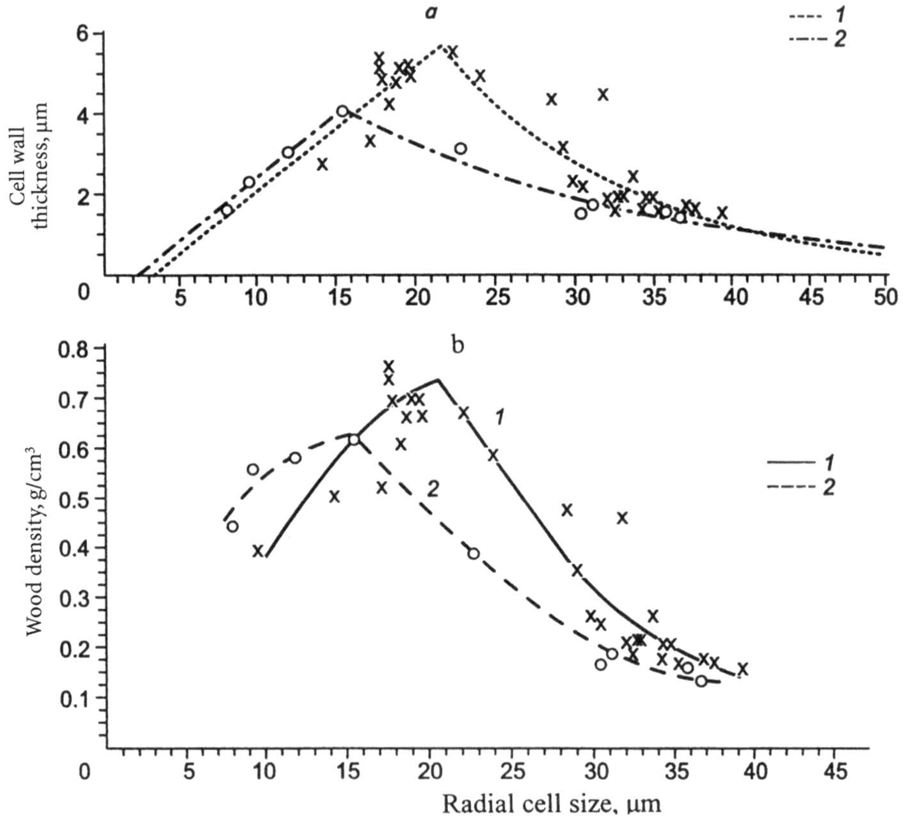

Fig. 2.38. Dependence of cell wall thickness (**a**) and wood density (**b**) on radial tracheid size in representative samples for wide (*1*) and narrow (*2*) tree rings of larch from a site in the northern taiga of central Siberia

tree rings. Thus, all quantitative parameters of tree-ring structure (tree-ring width, number of cells, cell size, cell wall thickness, density of wood) depend on one another to some degree. The relationships between them can be obvious, as for example, between ring width and number of cells, but it may also be characterized by more complicated relations. For example, cell size versus tree-ring width, or cell wall thickness versus radial cell size. Apparently, such relations reflect a correlation of processes of cell production, cell enlargement and thickening of the cell wall, which on the one hand are under the control of the internal factors of differentiation of xylem within the seasonal cycle (Savidge 1996; Vaganov 1996a), and on the other reflect the direct and indirect influences of environmental conditions on these processes (Zimmermann 1964; Denne and Dodd 1981; Vaganov 1996a).

2.7 Conclusions and Discussion

The characteristics of tree rings as natural archives of environmental variability have been discussed and the existence of a mass of empirical experience and data described. Some of the strengths and limitations of present approaches to using these data have been considered, with special reference to the use of tree rings as natural archives of information on climate variability. This archive is based not only on information on the size of the tree ring, but also on its chemical composition and its microanatomical structure. The internal structure of the tree ring contains information on environmental conditions at seasonal, or even finer, timescales.

A notation for describing the main measurements of tree rings in use was introduced and the general principles for making these measurements surveyed. The concept of the tracheidogram was introduced. Consideration was then given to evidence for the effects of the various internal and external factors discussed in Chap. 1 on the size and internal structure of tree rings. Particular attention was given to the handling of micro-anatomical data so that they may used as "chronologies" analogous to those based on ring width or density data. This then permits the examination of long-term relations between different anatomical characteristics of tree rings. There is a rich variety of possible relationships among these variables and between them and the external environment, giving considerable versatility in recording weather and climate phenomena.

3 Seasonal Cambium Activity and Production of New Xylem Cells

3.1 Introduction

During the seasonal formation of tree rings, many interrelated processes are realized at different hierarchical levels: (a) at the level of the whole tree, the growth and development of apical meristems and coordination between them and the lateral meristem, (b) at the tissue level, the production and differentiation of new cells, depending on hormone control and nutrient availability, (c) at the single cell level, the organized biochemical transformation of matter, based on enzyme activity under the control of differently activated genes, etc. (Wilcox 1962; Gamaley 1972; Roberts 1976; Barnett 1981; Catesson 1990; Savidge 1996). Most of these processes are affected by changing environment if we use this word in a broad sense, for instance, for the tissue the environment is the whole tree, for the cells the environment is a tissue or organ. The result of these processes is the unique anatomical structure of a tree ring formed in a certain year in a certain place in the tree. In spite of much research conducted to study the control of the seasonal activity of the meristem (in our specific case – the lateral meristem or cambium) and the cell differentiation of xylem, there is as yet no common quantitative theory based on known mechanisms of internal and external control of xylem growth and differentiation that takes into account the hierarchy of processes involved (Denne and Dodd 1981; Savidge 1996; Vaganov 1996a). Even so, the need for such a theory to forecast the results of the combined effect of internal and external factors on tree growth and wood formation is unquestionable.

This chapter is central to the line of argument we pursue in this book. We will introduce definitions of the basic terms needed for the study of cambial activity and the production of new xylem cells by examining the seasonal course of these phenomena (Sect. 3.2). The methods needed to study the seasonal kinetics of tree-ring growth and structure will be described (Sect. 3.3) and the cell organization of the cambial zone introduced (Sect. 3.4). After considering seasonal patterns of cambial activity (Sect. 3.5), we will introduce a phenomenological approach to patterns in the kinetics of cambial activity (Sect. 3.6). This provides the basis for some basic conclusions on the control of the kinetic parameters of the cambial zone relevant to cell production (Sect. 3.7).

3.2 Seasonal Activity of Cambium and Xylem Differentiation (Basic Definitions)

Cambium is generally considered as one layer of cells, which has polypotent abilities for an unlimited number of divisions and further differentiation. During a season, cambium will derive (for xylem on the one hand and phloem on the other) some layers of cells which are capable of limited division. These are called xylem and phloem mother cells, respectively. The zone where cells divide is defined as the cambial zone. It contains xylem mother cells, cambial initials and phloem mother cells. Wilson et al. (1966) offered the most successful scheme and nomenclature describing the various types of cells and tissues connected with the activity of secondary meristem – cambium (Fig. 3.1). Cells pass through these stages during the course of the season as they develop (Fig. 3.2). Each xylem cell is a derivative of a fusiform cambial cell and each xylem cell is an affiliated cell of the last divided xylem mother cell located at the boundary of the cambial zone. After losing the ability to divide, the affiliated cell moves into a zone of radial expansion and then into a zone of cell wall thickening, where the formation of the secondary cell wall mainly happens. According to the nomenclature in Fig. 3.1, cambium is a layer of initial(s), that is, polypotent cells. Differentiated xylem includes some layers: a layer of xylem mother cells capable of restricted division, a layer of radially enlarging cells, and a layer of cells that have finished radial expansion, but are forming a secondary cell wall. Mature xylem is made up of cells in which all

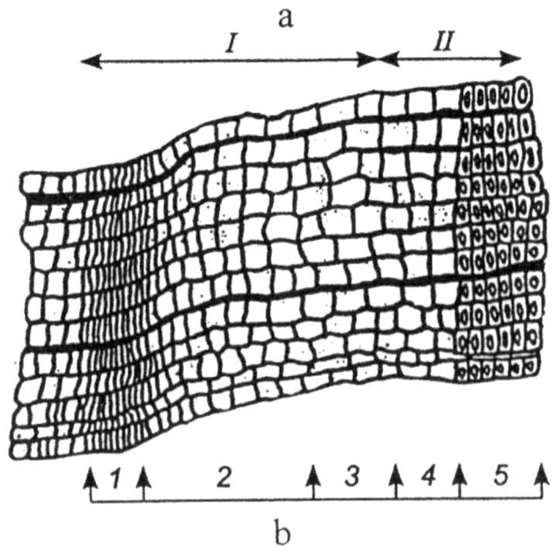

Fig. 3.1. A tree ring being formed (**a**): *I* cells in maturing phase, *II* mature tracheids); and (**b**) classification of xylem after Wilson et al. (1966): *1* cambial zone, *2* enlarging zone, *3* secondary wall thickening zone, *4*, *5* mature tracheids of the current and previous years accordingly

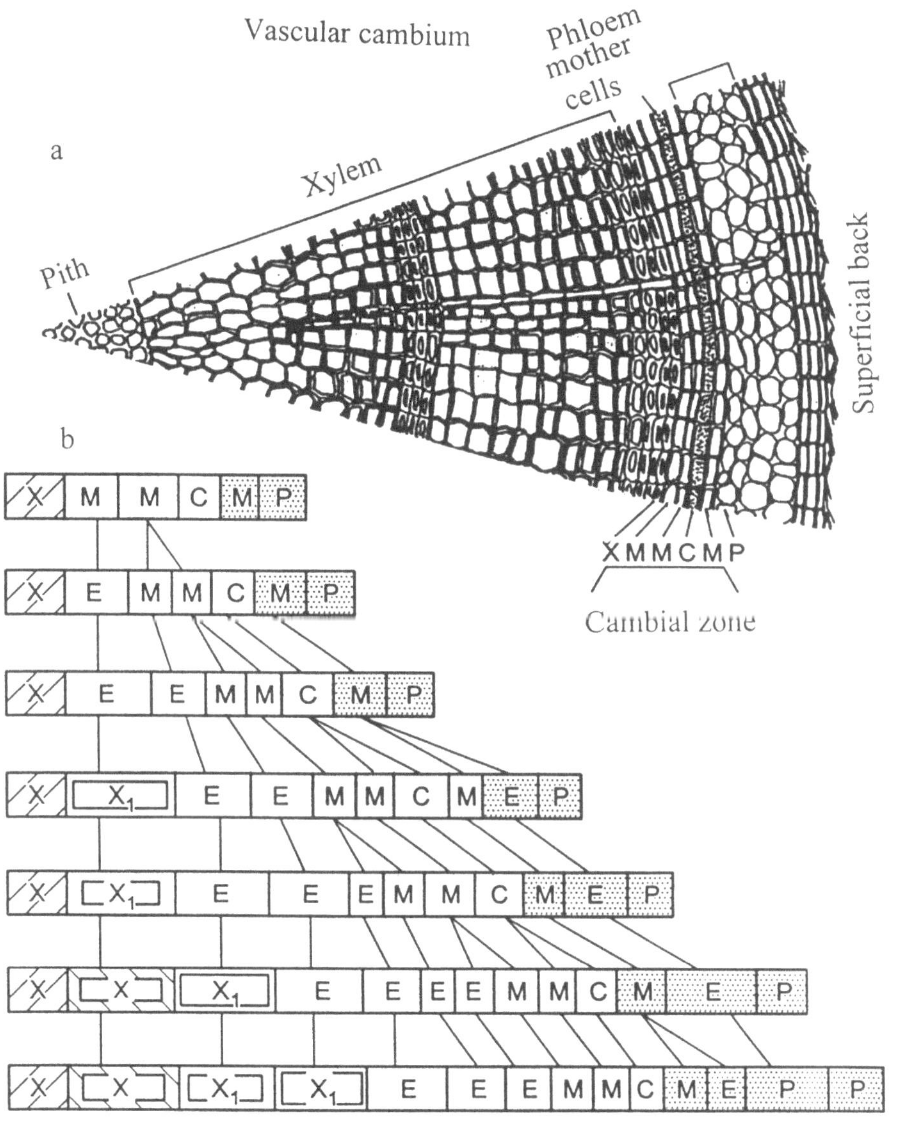

Fig. 3.2. Wood cross-section of pine (**a**) and scheme illustrating cell production in the cambial zone (**b**): C cambial initial, M mother xylem and phloem cells, X mature xylem cells, P mature phloem cells, E cells in the enlarging phase (Wilson and Howard 1968, as modified by Schweingruber 1996)

these processes are complete and the protoplast is practically absent (dead cells). The features of seasonal dynamics are reflected in the rates of each of these processes (division, enlargement, cell wall thickening) and in the number of cells located in each interval of a season during each stage of differentiation and maturation.

3.3 Methods for Studying Seasonal Kinetics of Tree-Ring Growth and the Formation of Their Structure

Techniques for investigating the seasonal dynamics of tree-ring growth may be considered as either direct or indirect. Direct methods include the periodic withdrawal of small wood samples from the trunk of a growing tree during the season (Lobzhanidze 1961; Wodzicki 1971). Thin cross-sections are prepared for each wood sample, stained to enhance the observation of the layers mentioned above, and mounted in glycerin gel or balsam. Such cross-sections represent instant temporary "images" of the xylem as it is being formed (Fig. 3.2) and it is possible to measure the number of cells in the cambial zone (N_c; by cytological determination of the divided cells, or by the minimum radial sizes of cells), the number of cells in the zone of enlargement (N_e), and the number of cells in the maturing zone (cell wall thickening; N_t; Wilson 1964; Gregory 1971; Vaganov et al. 1985).

In order to separate the different developmental zones of maturing secondary xylem it important to have characteristic and well defined anatomical features for the different developmental stages. Although these are described in detail elsewhere (Wilson 1964; Larson 1963, 1964; Whitmore and Zahner 1966; Wodzicki 1971), we give some general indication on how the borders of developmental zones on a transverse section of maturing secondary xylem may be defined. At the transition from the cambial zone to the expansion zone, the radial dimensions of the cells exceed those observed in the cambial zone, relative to their tangential size. As mentioned above, this distinction may also be observed by detailed cytological examination. The border between the radial expansion zone and zone of secondary wall formation is recognized by the thickening of cell walls and the presence of lignified walls. The transition from secondary wall-forming to mature tracheids can be localized by the presence or absence of cytoplasm. Quantitative information on the cambial, elongation, and maturation zones can be obtained using light microscopy of stained transverse thin sections of maturing secondary xylem. A wide range of stains and combinations of stains has been used. Toluidine blue, safranin, and methylene blue are often used to stain the cell wall. Sometimes differential staining is used to distinguish lignified and non-lignified walls. For example methylene blue/azur A may be used to distinguish between lignified (blue-green) and non-lignified (red-purple) walls or safranin/astra blue, when safranin is displaced by astra blue from non-lignified walls, which will become blue. Of course, all of these techniques require the development of skill and a sufficient degree of replication in order to provide robust, repeatable results.

Such repeated wood sampling transforms the picture of seasonal formation of a tree ring from a static to a dynamic one. It is then possible to calculate the rate of division and the rate at which cells pass through the separate

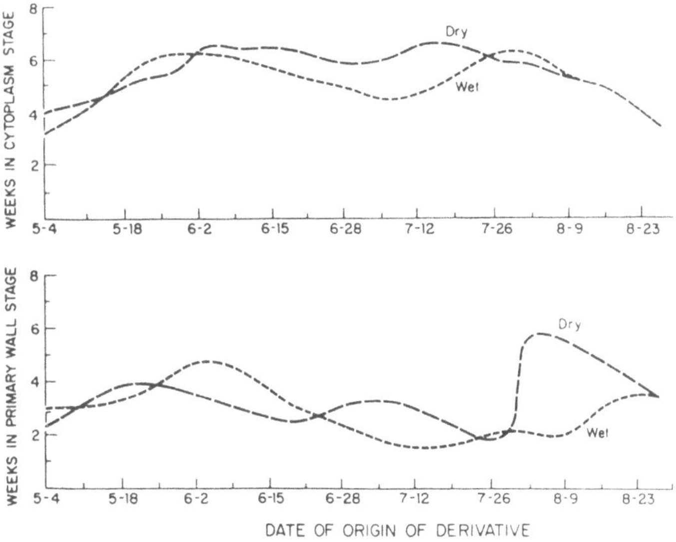

Fig. 3.3. Time in cytoplasmic zone (*top*) and time in primary wall zone (*bottom*) for xylem derivatives at the lower stem position, wet and dry treatments. A derivative originating at any date on the abscissa remains in each zone for the period of time indicated on the ordinate (Whitmore and Zahner 1966)

periods of differentiation and maturation (Whitmore and Zahner 1966; Skene 1969; Wodzicki 1971; Antonova and Stasova 1993, 1997). Evaluation of these dynamic characteristics of the processes of cell differentiation and maturation shows how long a cell takes to pass through each stage of development to become a mature xylem cell. For example, enlargement and cell wall-thickening can take 3–4 weeks each (Whitmore and Zahner 1966; Skene 1969; Vaganov et al. 1985; Fig. 3.3). The main defect of this method is the wide range of values obtained for each quantity, because each successive sample is taken from a different place on the tree's circumference. It has been shown that strong replication, perhaps including as many as 100 trees, is needed to make a reliable evaluation of the kinetic characteristics of seasonal growth (rates of division, enlargement, wall-thickening; Wodzicki 1962, 1971; Wodzicki and Peda 1963).

Another approach, which may be named the method of point-wounds, is based on periodic local damage of living (still possessing protoplasm) cells with the help of a thin needle or pin introduced into a trunk of a tree through the bark (Wolter 1968; Denne 1977; Kuroda 1986; Wilpert 1990, 1991). At the end of a season or the next year after completion of growth, it is possible to sample this point on the tree and to distinguish three layers (Fig. 3.4):

1. A layer of cells which was derived by the cambial zone and has partially or completely generated a cell wall (lignified) and has been disrupted by the pinning
2. A layer of xylem cells that had already left the cambial zone at the time of pinning, but, because they were still living, show some deformation

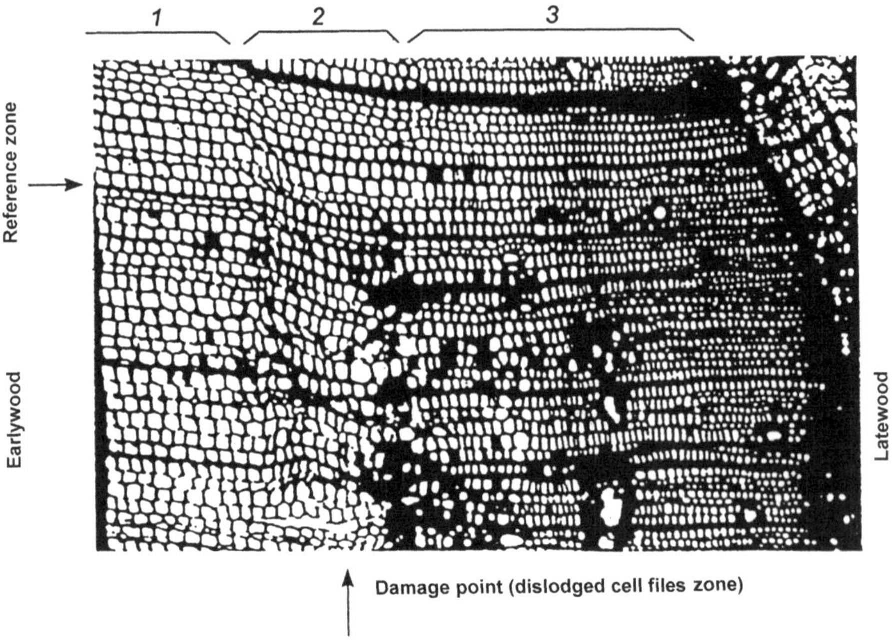

Fig. 3.4. Photomicrograph of spruce section (50 × magnification) cut near the insertion point of a pin: *1* already differentiated tracheids, *2* tracheids in differentiation, *3* tracheids separated from cambium after pinning. The reference zone marked on the photo was not damaged (Wilpert 1991)

3. A layer of cells, derived from the cambial zone after pinning and showing noticeable disturbance in the structure of tracheid files and the sizes of cells.

Thus, the method permits measurement of the number of cells and the width of cell layers located in the zone of enlargement and in the zone of cell wall-thickening (Nobuchi et al. 1995). For the precise interpretation of the data and measurement of appropriate layers of cells, it is necessary to make the wood cross-sections precisely in the zone of damage. It is not easy to produce such a needle- or pin-shaped injury and to recover it at the end of the season. A modified method has been developed in which a sharp knife is used instead of a needle or pin, (Kuroda and Kiyono 1997; Fig. 3.5).

Dendrometers are a widely used indirect method of measurement of the seasonal dynamics of tree-ring growth. There are several designs for measurement of the circumference of a stem with the help of flexible bands (Fritts and Fritts 1955; Liming 1957; Smirnov 1964; Palmer and Ogden 1983; Vaganov et al. 1985). These methods are indirect because, alongside the measurement of actual growth produced by the increasing number of cells and their increasing sizes during the season, they also measure seasonal changes in the

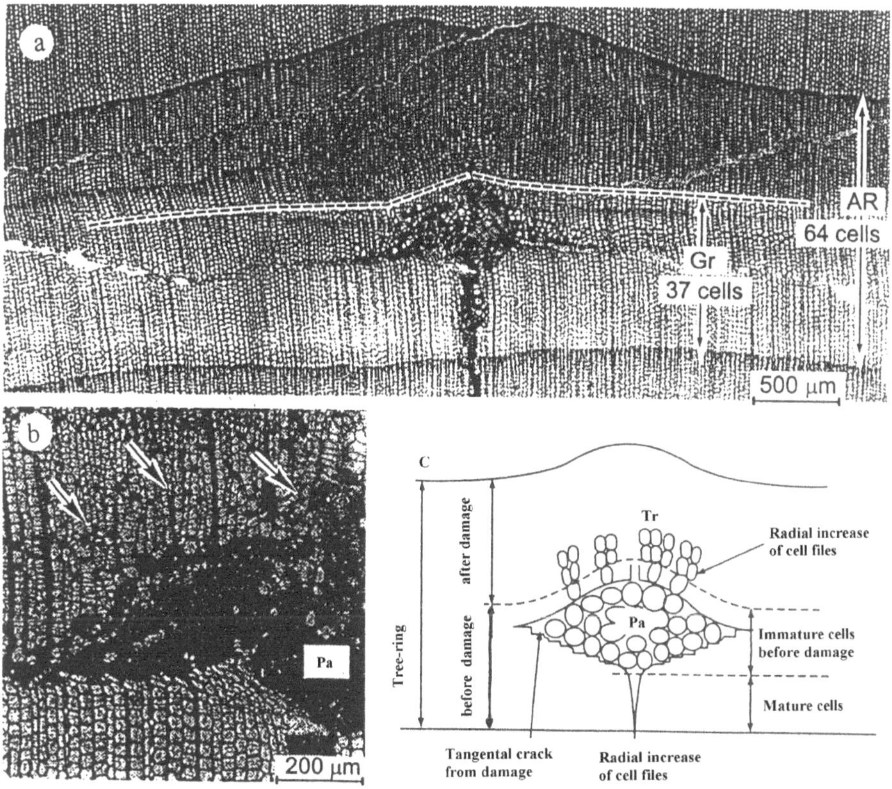

Fig. 3.5. Micro-section of annual ring formed after intra-seasonal damage by a wound: **a** general view of cross-section with wound, the *dashed line* shows the probable cambial zone location at the moment of damage, **b** disrupted tissue (*Pa* parenchyma cells), **c** the wound and definition of individual zone boundaries (Kuroda and Kiyono 1997)

water content of the tissues, which can considerably exceed the increment of new cells (Kozlowski and Winget 1964; Kozlowski 1968; Hinckley and Bruckerhoff 1975; Vaganov and Terskov 1977). Dendrometer data may considerably overestimate growth rate in the first part of the season because of the high water content of tissues and may show a "negative" increment at the end of the season, when the water losses exceed uptake (Fig. 3.6).

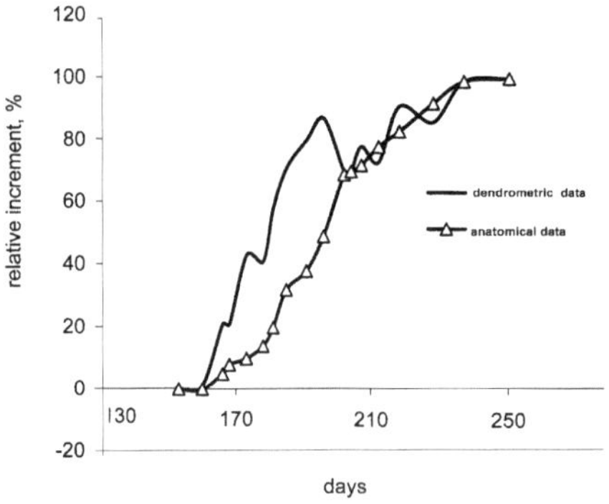

Fig. 3.6. Within-season mean dendrometric and anatomical data for *Pinus sylvestris* radial growth in the southern taiga of central Siberia

3.4 Cell Organization of the Cambial Zone

The formation of mature xylem cells is usually considered as a process that occurs in three stages: the division of cells, the growth of cells by radial expansion, and the maturation of tracheids, when cell walls thicken and the protoplast is autolysed. All stages are divided in space and time, although they may overlap partially (Fig. 3.7). In the majority of plants, xylem is a complex tissue consisting of differentiated cells of more than one type, generated by a secondary meristem, the cambium. Plant meristems, and cambium in particular by virtue of its accessibility, became a subject of research with the appearance of the first microscope. The history of cambium research is well described, for example, in the book by Larson (1994). However, until now our knowledge of the factors initiating and regulating the origin of tissue from meristems and the physiology of meristem has been rather fragmentary. This has led to the absence of a uniform nomenclature (Wilson 1966; Schmid 1976; Catesson 1984; Larson 1994).

Cambium has common features intrinsic to all meristematic tissues and has, as a highly specialized secondary meristem, specific features:

1. It is a self-sustaining cell-like system; that is, it retains its functions for a long time, frequently throughout the life of the plant; which may last centuries or millennia (see Table 1.1).
2. In woody plants, the cambium grows at the expense of the growth of the tree. An increase in the number of cambial cells happens at the expense of the division and differentiation of those cells and the primary (apical) meristem.
3. Cambial derivatives can be differentiated into various types of xylem and phloem cells.
4. Cambium has a strictly ordered spatial organization.

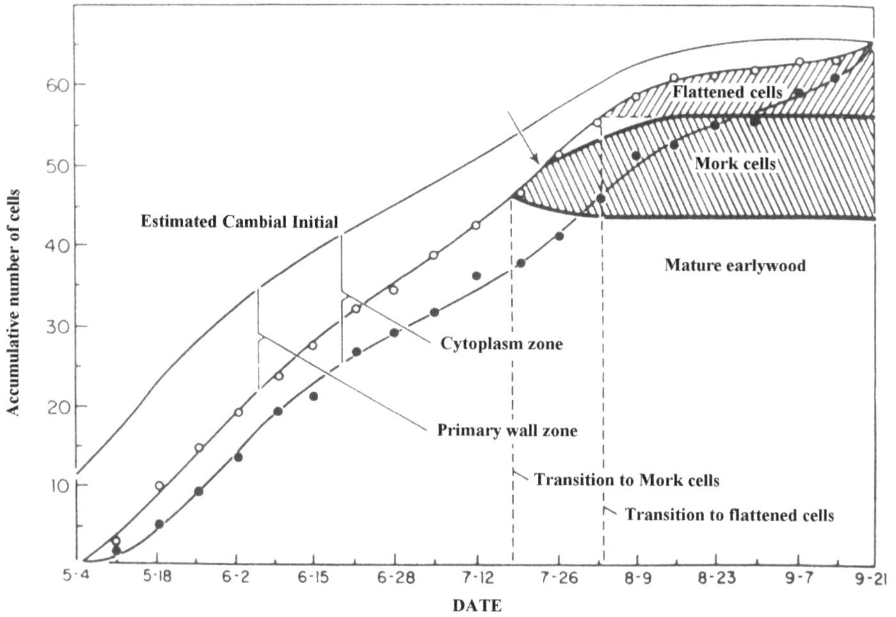

Fig. 3.7. Example of smoothed curves from plotting of cell counts in the various tissue zones. This set of curves is for xylem development at the lower stem position, clipped treatment. Points plotted along the primary wall zone and cytoplasmic zone boundaries represent the actual mean number of cell in these zones and were located with reference to the upper curve, representing the estimated cambial initial. Also shown is the graphical determination of the times of inception of flattened cells and Mork latewood cells (Whitmore and Zahner 1966)

The cambial cells form a continuous layer, covering the trunk, branches and roots. Therefore the cambium, on the one hand, is distributed in space, and on the other hand, is a downlink system where the adjacent cells are in direct contact. The spatial organization of the cambial zone is important from the point of view of regulation of its activity, as it imposes a number of specific requirements on regulation and control. Another, but no less significant, aspect of the spatial organization of the cambium is that it is the basis of the spatial cell-like organization of xylem and phloem. For example, it orders radial tracheid files, the formation of the vessel system, and so on.

As radial tracheid files are structural elements of a tissue, the question of the cell-like structure of the cambium is reduced as a rule to the cell-like structure of cambium in one radial file. For more than 100 years, two conflicting concepts have existed concerning the formation of radial cell files from cambium. The main difference between these points of view relates to the number of cambium cells initiating cell files and, accordingly, the different cell kinetics that result from this. According to the first (stated by Sanio in

1873; see Larson 1994), each radial file originates from one cambium cell – an initial. The key moment here is the division of the initial. It can be symmetric, or it may be multiplicate, when both cells produced become initials and, correspondingly, one of them gives a beginning to a new radial file of cells. In asymmetric division (differentiated), one of the cells produced retains the properties of an initial, while the other is transformed into a cell element of xylem or phloem (Newman 1961; Wilson 1966; Mahmood 1968, 1971; Schmid 1976). Because of the impossibility of precisely identifying an initial cell and the rigidity of the scheme of its differentiation, the second hypothesis appeared, according to which there can be more than one cell capable of being an initial in a radial file (Catesson 1964, 1980, 1984; 1990, 1994; Savidge 1996). In this case, the process of differentiation is gradual and may be modified in the early stages. One of the main reasons for the existence of only one initial is the multiplication of its division and the appearance of new files simultaneously in both xylem and phloem. Such anatomical observations are most completely described in Bannan's research (1955, 1957). It is improbable that all initials in a file divide simultaneously in such a way, although Savidge (1996) presents a microphoto in which some cells divide radially at the same time. The controversy about the number of initials continues. There are also different nomenclatures associated with the two points of view on the structure of the zone of divided cells in a radial file (Wilson 1966; Larson 1994).

We will adhere to the more generally used nomenclature offered by Wilson (1966), according to which the cambial zone consists of one initial and includes all its derivatives that still have the ability to divide, that is, the mother cells (Fig. 3.1). In this case the term "cambium" is only applied to the initial. The orientation of the plane of division and the consequent direction of growth of cells of the annual ring determines the anatomical structure of xylem. In coniferous species, cambial cells have a spindle-shaped form and are elongated along the axis of the stem (Lewis 1935; Meeuse 1942; Dodds 1948). The radial size of these cells and their length and tangential size vary in various species. Variability is determined genetically, though the frequency of divisions and growth rate of cells also influence the sizes of cells in the cambial zone. The cells of the cambium can differ in size in different parts of one tree. In many species, there is an increase in the length of cambial cells near the end of the growing season (Iqbal and Ghouse 1985a, b; Ajmal and Iqbal 1987a, b; Siddiqi 1991). In conifers, the mean length of cambial cells is about 5 mm and the volume is about 10^{-6} μm^3 (Clowes and Juniper 1968). Each cell has some contacts with similar cells and with ray cells. For example, in *Pinus sylvestris* each cell has up to 14 contacts with adjacent cells and more than four contacts with ray cells (Dodds 1948).

There are three types of division of cambial cells. Increase in the number of cells in a file is produced by periclinal divisions of cambial cells, when the plane of division is perpendicular to a radius of the trunk. In anticlinal division (multiple division) the plane of division passes along a radius. Bannan

(1955) estimated that the frequency of anticlinal divisions comprises 1–2 % of all divisions observed in the cambial zone. Also, there are cross-sectional (transversal) divisions, from which are derived rays, parenchima cells, and cells of the resin ducts. In this case, the plane of division is perpendicular to the principal axis of the trunk. Intermediate versions are also possible. So far as cells produced by division always have identical sizes, the size of xylem elements is stable and the cambial cells are capable of growing widthwise and lengthwise. The mean sizes of cells in the cambial zone are thereby conserved.

3.5 Seasonal Activity of the Cambial Zone (Basic Quantitative Results)

The growth of a tree ring is the result of periclinal divisions of cells in the cambial zone and their differentiation. The growth rate depends on the number of cells in the cambial zone and their rate of division. In coniferous species, the growth of a tree ring during a season is always acompanied by a change in the number of cambial cells, which has characteristic dynamics that are general for all species (Wilson 1966; Gregory 1971; Skene 1972; Kutscha et al. 1975; Vaganov et al. 1985). In dormancy, the size of the cambial zone reaches a minimum and usually includes four or five cells but can reach up to ten (Table 3.1). The radial diameter of cells in the cambial zone is equal to 5–6 μm on average but does not exceed 10 μm (Bannan 1955; Alfieri and Evert 1968; Vaganov et al. 1985). Activation of the cambial zone starts with a swelling of cells and then the first divisions appear. After activation, the size of the cambial zone is increased and the number of cells in it increases and reaches maximum values up to 20 (15–16 on average for different species; Table 3.1). There is evidence for a relationship between the number of cells in the cambial zone during the dormant period (and at the starting date) and the

Table. 3.1. Number of cells in dormant and active cambium for different conifer species (extracted from Tables 11.2, 11.4 in Larson 1994)

Species	Number of cells in cambium		Reference
	Dormant	Active	
Juniperus californica	1	8	Alfieri and Kemp (1983)
Abies alba	1–3	5–10	Fink (1986)
Thuja occidentalis	1–4	10–15	Bannan (1962b)
Picea glauca	2–8	12–16	Gregory (1971)
Cryptomeria japonica	3–6	6–11	Itoh et al. (1968)
A. balsamea	7–8	15–16	Kutscha et al. (1975)
Pinus strobus	6–10	15–20	Murmanis (1971)

total annual production of xylem. So, Gregory (1971) found that this relationship for Alaskan white spruce is described as $N_c = 3.82 + 0.05\,N$ ($r^2=0.75$, $n=37$, $P<0.001$). Sviderskaya (1999) obtained similar results from seasonal observations of tree-ring formation in Scots pine (*P. sylvestris*) and Siberian fir (*Abies sibirica*) in the Siberian taiga (Fig. 3.8).

It is very difficult to experimentally determine the boundary between the zone of dividing cells (cambial zone) and the zone of radial enlargement. Even the most direct method of counting the dividing cells (Wilson 1964) does not give reliable results, because it counts the dividing cells in different radial files, which are not identical in their kinetic characteristics. There can be a deviation of two or three cells between files. Indirect methods are mainly based on the small radial diameter of cambial cells (Imagawa and Ishida 1970, 1972; Vaganov et al. 1985). To determine the variation in the number of cells in the cambial zone in Japanese larch (*Larix leptolepis*), Imagawa and Ishida (1970, 1972) used the ratio between the radial and tangential dimensions of cells (D/T), which is less than 0.3 for cells of the cambial zone (Imagawa and Ishida 1970, 1972). The same approach was applied to count the variations in cambial cell number of different coniferous species during the course of several growing seasons in the southern taiga of Central Siberia (Vaganov et al. 1985; Sviderskaya 1996, 1999).

Depending on the conditions of growth and the species of a tree, the number of cells in the cambial zone can be sustained for the rather long period during which most of the tracheids are formed. Then the size of the cambial zone gradually decreases. During the growth of an annual ring, the main prin-

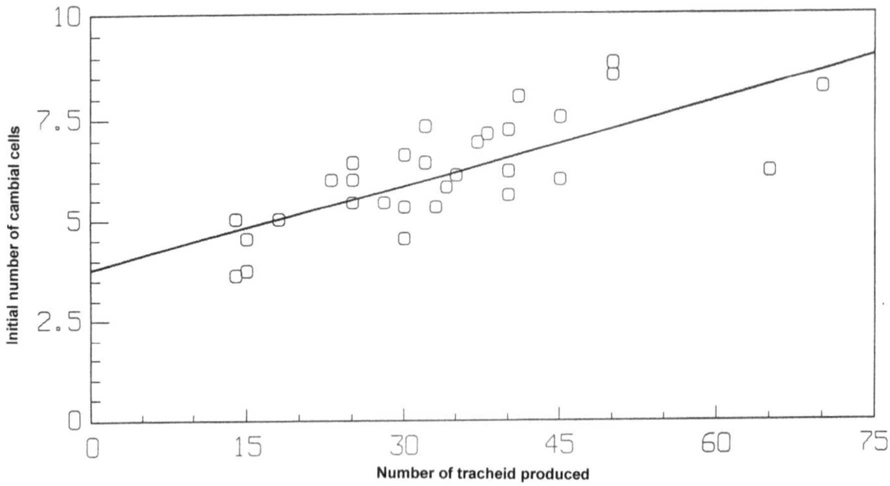

Fig. 3.8. Dependence between initial number of cells in the cambial zone at the beginning of the season and total number of tracheids produced in pine (*Pinus sylvestris*) tree-rings from the Siberian taiga

ciple is always maintained: an increase in the rate of formation of new tracheids is always accompanied by an increase in the number of cells in the cambial zone (Fig. 3.9). However, the relationship between the rate of tracheid production and the size of the cambial zone also depends on site conditions. In similar conditions, the maximum size of the cambial zone is always greater in dominant trees (or fast growing, or those with higher vigor) than in suppressed trees (Bannan 1955; Wilson 1966; Gregory and Wilson 1968; Lebedenko 1969; Gregory 1971; Vaganov et al. 1985). In white spruce (*Picea glauca*), the relationship between the number of cells in the cambial zone and the total annual xylem increment is curvilinear (Fig. 3.9; Gregory and Wilson 1968; Gregory 1971). The same curvilinear relationship occurs between the annual xylem increment and the maximal number of cells in the cambial zone, as seen in measurements of seasonal tree-ring formation of *Pinus*

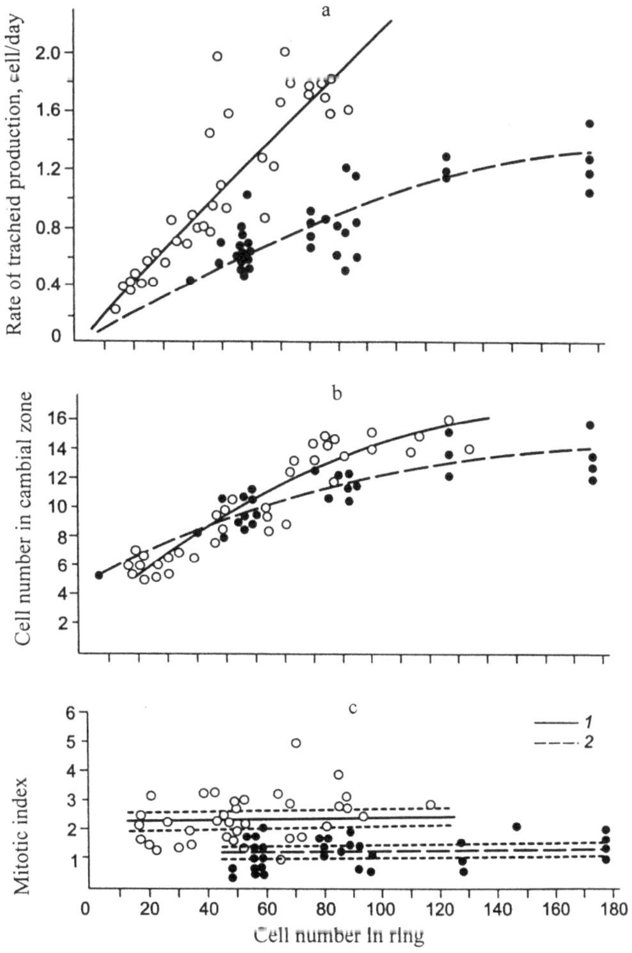

Fig. 3.9. Radial rate of tracheid production (cells per day) by tree vigor (**a**; annual radial tracheid increment, i.e. cell number in ring), **b** radial fusiform cell number in the cambial zone by tree vigor, and **c** mitotic index (percentage of cambial zone fusiform cells in mitosis) by tree vigor for white spruce of New England (*1*) and Alaska (*2*; Gregory and Wilson 1968)

sylvestris in Siberia (Fig. 3.10). This suggests that the formation of the widest rings is more strongly associated with an increasing growth rate of cambial cells (a shorter cell cycle on average) than with increasing numbers of cells in the cambial zone. Unfortunately, these are the only studies on this topic (Gregory and Wilson 1968; Sviderskaya 1999). They detected, for example, that in white spruce growing in completely different climatic conditions (in Alaska and in New England), the relationship between the maximum number of cells in the cambial zone and tree-ring and tracheid production is practically the same, though the rate of tracheid production differs by almost a factor of two (Fig. 3.9).

An important conclusion follows from the seasonal dynamics of the number of cells in the cambial zone and the relationship between maximum size of the cambial zone and annual xylem increment. It is that one of the mechanisms of growth rate control is connected with the changing number of dividing cells in the cambial zone.

Theoretically, a change in the number of cells in the cambial zone can happen in different ways. For example, with a constant rate of cell division, the number of divisions of mother cells can be changed before they pass to the next stage of tracheid development – radial expansion. Another possible version is that the number of divisions is maintained constant, but the parameters of the cell cycle are changed selectively. A number of workers analyzed groups of cells enclosed in a common primary wall. In doing this, they were following Sanio's idea that, at cell division, a new primary cell wall is formed not only between two daughter cells, but around the whole perimeter of both

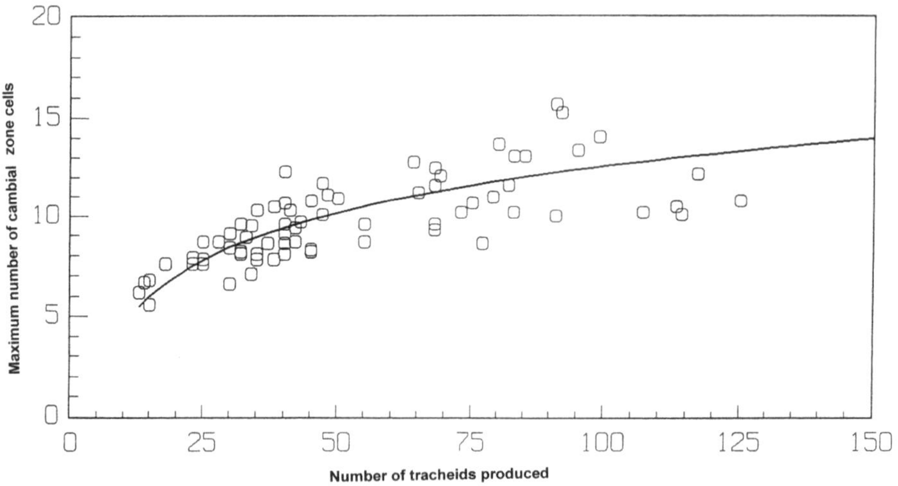

Fig. 3.10. Dependence between maximum number of cells in the cambial zone and total number of tracheids produced in pine (*Pinus sylvestris*) tree rings from the Siberian taiga

daughter cells. It is possible to determine an average number of divisions of xylem mother cells by the size of such groups, assuming that each group arises from the division of an initial. From these data, the number of divisions of xylem mother cells can be from one up to three (Newman 1956; Murmanis 1970, 1971; Mahmood 1990). Bannan (1955, 1957) has offered a more elegant method of evaluation of this value. It is well known that not only can initials divide anticlinally, but also mother cells, though more rarely. Thus, there is a temporary cell file only in xylem or in phloem, but not in both. The number of divisions of mother cells is evaluated by the number of cells in such a file. It varies depending on the growth rate of the xylem and can reach five (Bannan 1957; Wilson 1964; Imagawa and Ishida 1981a, b).

Few evaluations of the duration of the cell cycle and its separate phases exist for the cambial zone (Larson 1994). For example, in *Thuja occidentalis* the duration of the cell cycle changes from four to 14 days during a season (Bannan 1962b). In *Pinus strobus*, it is ten days and remains practically constant during the period of active growth in both fast- and slow-growing trees (Bannan 1962a; Wilson 1964). In *Larix laricina*, the cell cycle is estimated at ten days (Kennedy and Farrar 1965). In contrast to Gregory and Wilson (1968), who considered the rate of division to remain constant, Skene (1972) observed a change in the cell cycle from 28 days at the beginning of a season to ten days during the active growth period. Moreover, it turned out that the cell cycle was more stable in slow-growing trees. Typical seasonal dynamics of the mitotic index in the cambial zone are shown in Fig. 3.11. All the available data on the duration of the cell cycle and the separate phases of it show significant variability between samples taken in different parts of a tree during a growing season, especially for the size of the cambial zone. This essential variability is determined by the weather conditions of the season and other factors. Thus it is clear that the length of the cell cycle in the cambial zone changes during the growth season. Combining this statement with the observed curvilinear relationship between the number of cells in the cambial zone and the annual xylem increment, we are led to the following conclusion: the regulation of cell production by the cambial zone can be achieved by

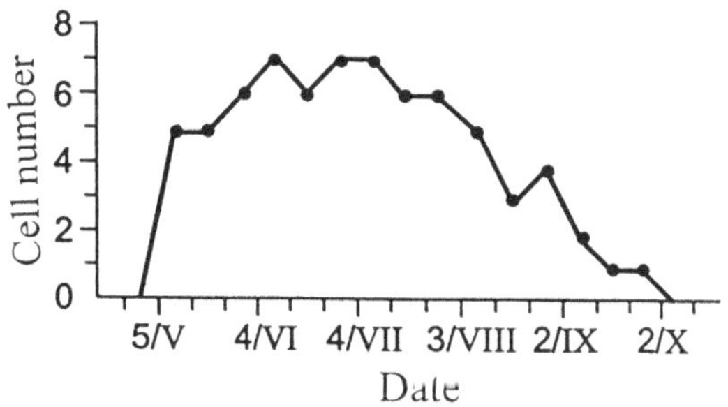

Fig. 3.11. Seasonal change in cell division frequency in cambial zone of *Larix leptolepis* Gord., calculated on the basis of cell production for 10-day intervals (Imagawa and Ishida 1970)

increasing the number of cells in the cambial zone as well as by increasing the rate of cell division in the cambial zone.

The most convincing data are obtained from records of the seasonal dynamics of distribution of the mitotic index within the cambial zone (Imagawa and Ishida 1972; Fig. 3.12). Detailed research (Bannan 1955, 1962b; Wil-

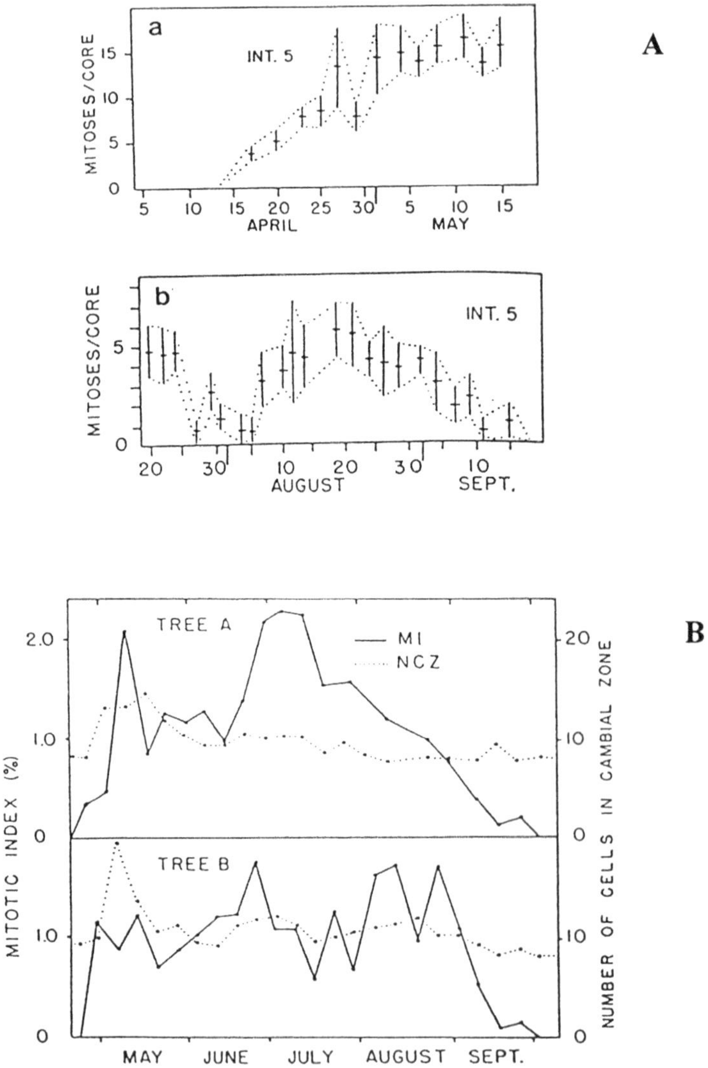

Fig. 3.12. Seasonal variations in some parameters of the cambial zone. **a** Seasonal change in number of mitoses per increment core in the fifth internode of five *Pinus strobus* trees (*a* at the initiation of cambial activity in spring, *b* at the cessation of cambial activity in the fall, *dotted lines* indicate the range of sampling error; Wilson 1966). **b** Seasonal change in the mitotic index (*MI*) and radial number of cells in the cambial zone (*NCZ*) of two *Pinus strobus* trees (Wilson and Howard 1968)

son 1964) has shown that the mitotic index and, therefore, the duration of the cell cycle changes within the cambial zone, i.e. the cells in the cambial zone are heterogeneous in their rate of division. Most frequently, cells in the mitosis phase are experimentally observed in the central part of the cambial zone or the maximum in the mitotic index is shifted close to the phloem (Wilson 1964; Fig. 3.13b). If these data on the distribution of the mitotic index are transfered to the rate of cells passing through the cell cycle, the outcome will not be so obvious. For example, Wilson (1964) supposed that all xylem mother cells could have an identical cell cycle duration and that the reduction of the mitotic

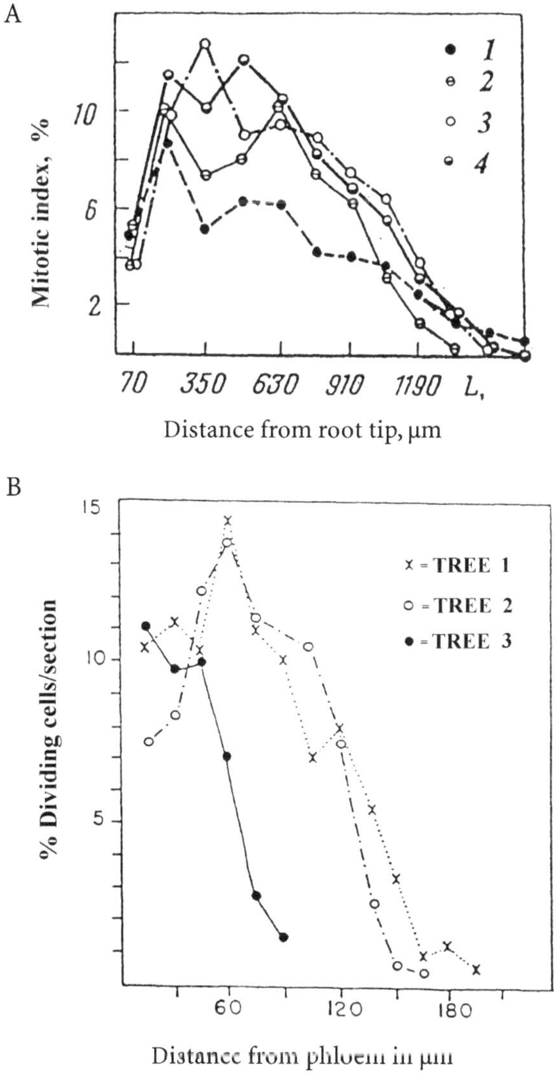

Fig. 3.13. Within meristem variations: **a** change in mitotic index up the maize root (average data on ten roots of seedlings: *1* epidermis, *2* exoderm, *3* bark, *4* pericycle; Balodis 1968), **b** percentage of cells in division at different distances from the phloem in three white pine trees (Wilson 1964)

index towards the cambial zone edge was connected with the variation in the size of the cambial zone in different cell files. So, the resultant calculation of the mitotic index included undivided cells. But even if we consider that all cells are divided, it is possible to get completely opposite conclusions concerning the distribution of the cell cycle within a cambial zone, using the measured distribution of mitotic index. This issue will be considered in more detail later.

We summarize the results concerning the kinetic parameters of cambial activity in xylem cell production observed in different conifer species:

1. The number of cambial cells in dormant cambium and active cambium is rather different. There is a significant relationship between the number of dormant cells in the cambial zone and subsequent annual xylem increment (Skene 1972; Sviderskaya 1999).
2. The number of cells in the cambial zone varies during a season due to internal and external factors. The average duration of the cell cycle in the cambial zone varies during a season. Usually the cycle is shorter when earlywood cells form and longer during the formation of latewood cells, especially at the end of the growing season.
3. The total annual xylem cell production is closely related to the number of cells in the cambial zone. There is, however, a curvilinear relationship between the annual xylem cell number increment and the average cambial cell number (Skene 1972) or the maximum cell number in the cambial zone (Vaganov et al. 1985; Sviderskaya 1999). This relationship indicates that a low rate of cell production during a season is supported mainly by an increase in the cambial zone (more xylem mother cells), but a higher rate of cell production during a season must also be associated with an increase in the rate of cell division or a faster cycling of xylem mother cells.
4. There is a typical distribution in the measured mitotic index along the cambial zone: moving from the last phloem cells produced, the mitotic index increases to a maximum approximately one-third of the way across the cambial zone and then decreases slowly or rapidly, depending on whether the cambial zone is wide or narrow (depending on the number of cells in the cambial zone; Bannan 1957; Wilson 1964).
5. There are differences in the rates of cell production even if the relationship between the number of cells produced during a season and the number of cells in the cambial zone is the same (Gregory and Wilson 1968). These differences are associated with a higher or lower mitotic index.
6. In conifers, the first cell divisions (in the growing season) are evident in cambial cells near to or often adjoining the last differentiated tracheids of the previous year's growth ring (Bannan 1955, 1962b; Grillos and Smith 1959; Zimmermann and Brown 1971; Savidge 1993).

Some of the assumptions made are in conflict with the evidence:

1. The assumption that the distribution of the mitotic index along the cambial zone indicates the rate of cell division (or rate of xylem mother cell

cycling) cannot be accepted. The small radial diameter of cells in the cambial zone results from cell divisions when a mother cell achieves nearly double the normal size. The formation of a primary cell wall between two daughter cells in the final stage of division does not stop the permanent linear extension of the tangential cell wall. If we presume that data on the distribution of cell division (mitotic index) across the cambial zone are in accordance with the rate of linear cell growth, then the frequency of cell division has to decrease and vice versa as cell cycle duration increases near the boundary of the cambial zone. In that case, the linear rate of cell growth has to decrease in the same direction, but then to increase sharply when a cell moves to the zone of radial expansion. In the case of the absence of sliding growth between adjoining radial files, this must automatically lead to significant differences in the radial growth of cells in different files and to big differences in radial cell dimensions (up to 2–3) in neighboring files, because as cells enlarge they pull on their neighbors. Such cases are observed very rarely experimentally, for example, if xylem mother cells die under strong radiation (Ivanov 1974; Musaev 1996).

2. The second assumption is also related to variation of the cell cycle across the cambial zone. Based on experimental data (first of all, on data for the number of cell divisions during the growing season in trees of different growth classes) and the linear relation between the rate of cell production and annual xylem cell increment, Wilson (1964) assumes a constant rate of cell division across the cambial zone. But, given this assumption, it is impossible to explain the curvilinear relation between the number of cells in the cambial zone and annual xylem cell production, especially for wide tree rings with large numbers of cells in the files. The only way to reconcile this assumption with the experimental records is to assume that the whole population of cambial cells increases its rate of cell division when it achieves a near-maximal size (14–16 cells per file). There is no known mechanism for this. Moreover, if we use the mitotic index as an indicator of the rate of cell division, Wilson's assumption disagrees with the observed independence of mitotic index from tree vigor (the same as production of wider tree-rings; Gregory 1971).

3.6 A Phenomenological Approach to the Description of the Observed Patterns of Cambial Activity

As one can see from the data presented above, the main aspects of cambial activity are closely connected to: (a) distribution of the rate of cell division (or cell cycle duration) across the cambial zone, and (b) seasonal variation in the size of the cambial zone and the average rate of cell production. To produce a model of cambial activity appropriate for quantitative description, we must integrate these parameters and check with available observations. Unfortunately, there are few data on cambial activity suitable for this, but we can go to numerous data recorded for another meristem, namely the root meristem, whose structure is very similar to that of the lateral meristem. Clowes (1961a, b) pointed out that experimental data for the mitotic index or distribution of the mitotic index in the meristem cannot provide a direct evaluation of cell cycle duration if the duration of the mitosis phase varies with the cell cycle duration. He showed such a relationship for the root meristem in the cap initial, where the central stele was just above the quiescent center and the central stele was 200 μm from the quiescent center. This statement was supported by the results of Hejnowicz (1959), who brought out the interesting point that the mitotic rate is largely constant through an appreciable length of root, but the frequency of cell division is then directly proportional to linear cell density. In summary: "This is an effective warning against reliance on subjective impressions: the mitotic rate is not necessarily greater in those parts of a tissue where mitotic figures appear to be most abundant" (p. 71 in Dormer 1972). Several fine publications by Grif and Ivanov (1975, 1980, 1995) provide a data base of the temporal characteristics of cell cycles in the meristems of different plants. This will allow us to explore in detail the relationship between mitosis and the cell cycle.

The relationship between the duration of mitosis and the duration of the cell cycle for the root meristem of three plants growing under different temperatures is shown in Fig. 3.14. Uniform methods of experimental measurement of the duration of mitosis were applied (Grif and Ivanov 1975, 1980, 1995). There is a good linear relation between these two characteristics, both decreasing when temperature increases. The same relationship occurs in different tissues (stele, epidermis, bark) at different distances from the tip in roots growing under fixed temperature (Fig. 3.15). That is, there is a linear proportional dependence between the duration of mitosis and cell cycle duration in the root meristem.

The observed experimental dependence is described quite well by the linear equation:

$$M = a(C - C_0) \tag{3.1}$$

where M is the duration of mitosis, C is the cell cycle duration, C_0 is the minimal duration of the cell cycle when theoretically the duration of mitosis is equal to zero, and a is a coefficient. The values a and C_0 are different for the data shown (Table 3.2), but the form of the equation is constant. The graphs in Fig. 3.16, based on data from Grif and Ivanov (1975, 1980) show that the linear relationship between mitosis duration and cell cycle duration is characteristic of many plant meristems. In Fig. 3.16, the data are shown not only for different plants but also for different ranges of cell cycle variations (panel a for C values from 7 h to 30 h, panel b for C values from 10 h to 210 h). The range of

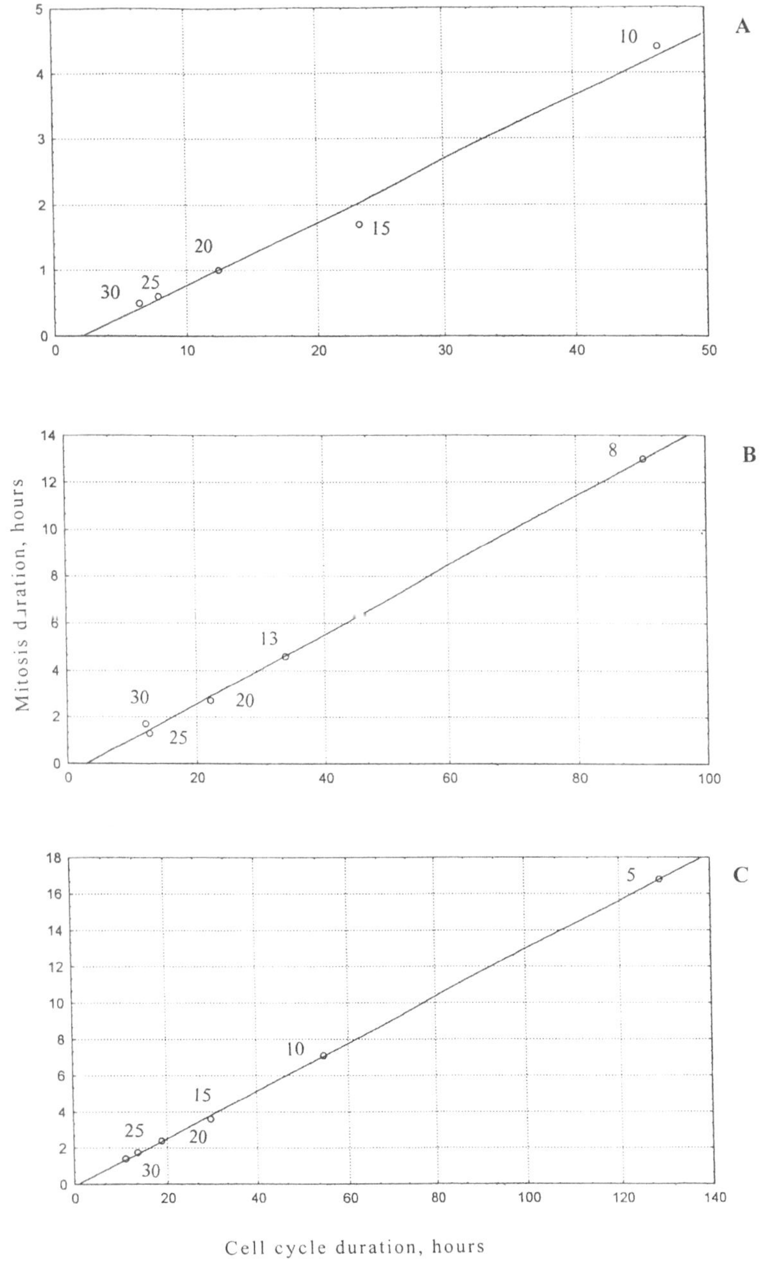

Fig. 3.14. Dependence of duration of mitosis on cell cycle duration in roots of various plants at different temperatures: **a** roots of *Helianthus annuus*, **b** roots of *Vicia faba*, **c** roots of *Allium cepa* (based on data from Murin 1964, 1967; Lopez-Saez et al. 1966; Burholt and Van't Hof 1971). Numbers near data points show growth temperature (°C)

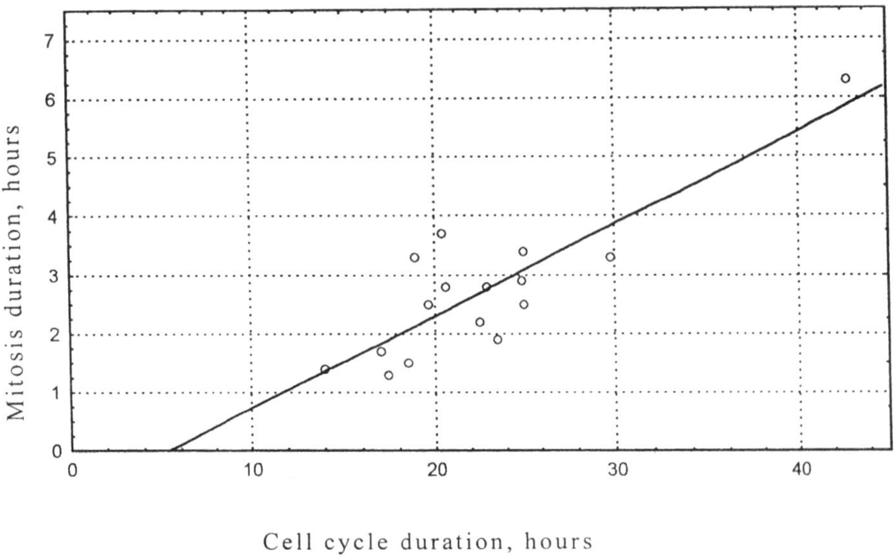

Fig. 3.15. Dependence of mitosis duration on cell cycle duration in maize roots for the cells of various tissues and at different distances from the apex (based on data from Ivanov 1974)

Table 3.2. Values characterizing the relationship between mitosis duration and cell cycle duration in the roots of different plants (Vaganov and Djanseitov 2000)

Species	From Eq. 3.1		Correlation coefficient
	a	C_0	
Allium cepa L.	0.131	0.67	0.99
Helianthus annuus L.	0.096	1.98	0.98
Vicia faba L.	0.148	2.71	0.97
Data on many plants	0.154	4.20	0.96
Zea mays L.	0.156	5.30	0.85

cell cycle durations shown in Fig. 3.16b is close to the observed cell cycle duration in the cambial zone of conifers – more than 200 h or 7–8 days (Larson 1994). Perhaps the slope (coefficient a in Eq. 3.1) depends on the species of plant and the sensitivity of the cell cycle duration to temperature changes and maybe to changes in growth conditions in a broader sense.

An important peculiarity of Eq. 3.1 is that the line crosses the abscissa at a positive value. Experimental study of the rate of cell division and the distribution of the cell cycle along the meristem (and through the cambial zone) is based on evaluation of the mitotic index or the relative number of cycling cells

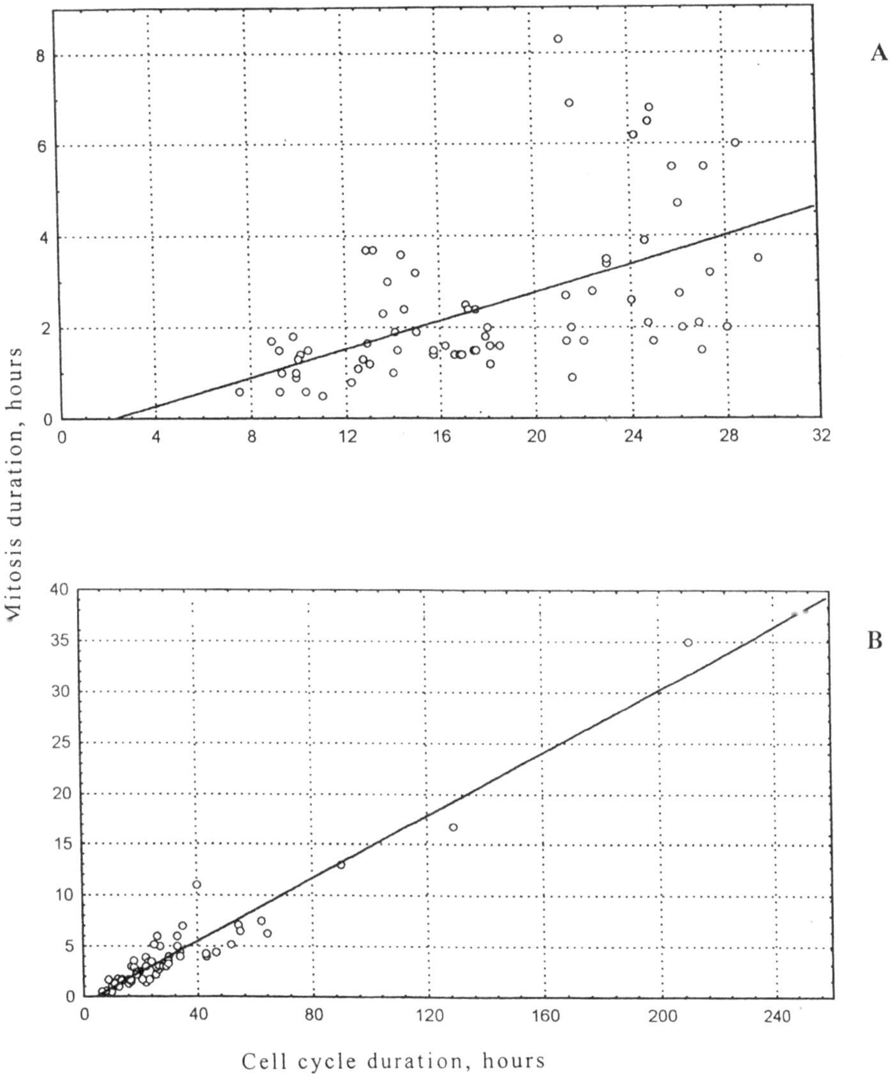

Fig. 3.16. Dependence of mitosis duration on cell cycle duration in roots of different plants: **a** data from Grif and Ivanov (1980), **b** data from Grif and Ivanov (1975)

in mitosis. If we consider the proliferation pool to be equal to 1 (potentially all cells in the meristem may divide), then according to Ivanov (1974):

$$M_I = \ln 2 * (M / C) \qquad (3.2)$$

or taking into account the Eq. 3.1:

$$M_I = a * \ln 2 * (C - C_0) / C = b - b * C_0 / C \qquad (3.3)$$

where b is equal to $a \times \ln 2$. From this it follows that, when the cell cycle duration decreases (and the rate of cell division increases as well as the total production of meristem), the mitotic index (M_I) also decreases and vice versa. For example, the mitotic index values decrease in roots growing under increasing temperature (Fig. 3.17, calculations from data presented in Murin 1964, 1967; Lopez-Saez et al. 1966).

According to Eq. 3.3, the data on mitotic index distribution across the meristem have to be interpreted contrary to the generally accepted understanding (see, for example, Fig. 3.13). So, for the Balodis (1968) data, the max-

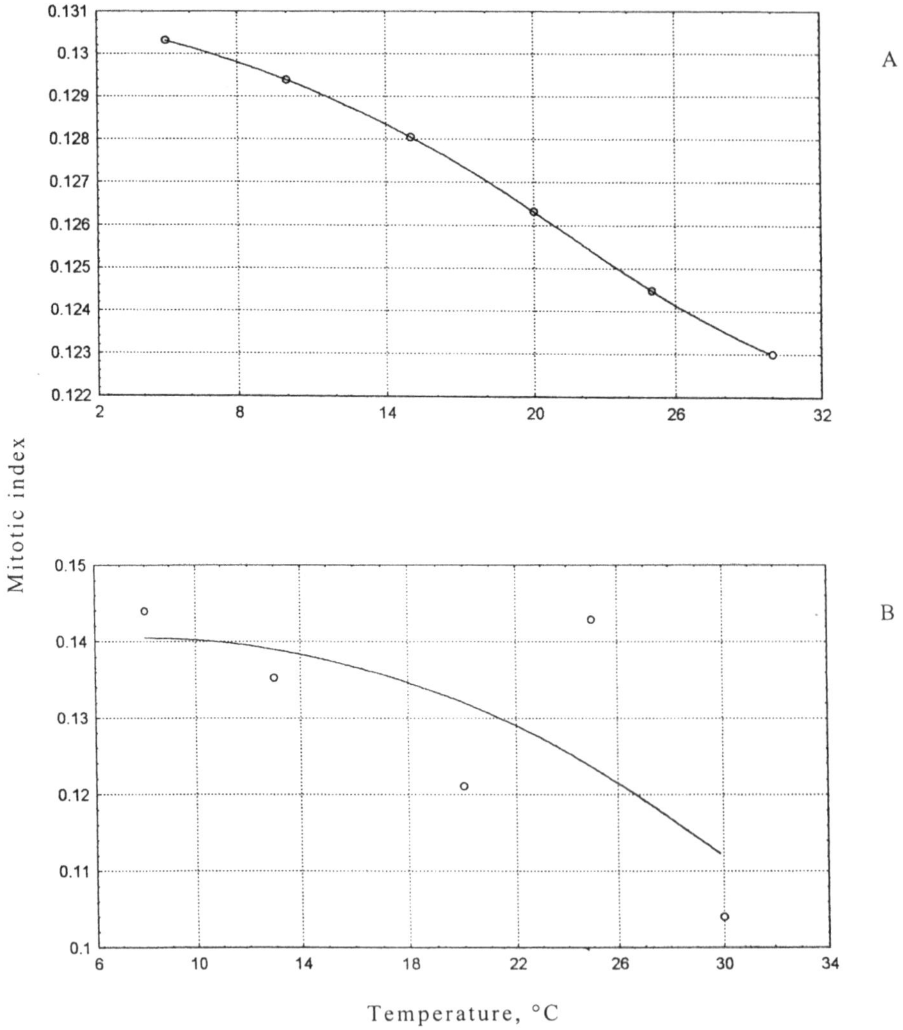

Fig. 3.17. Dependence of mitotic index on temperature for roots of different plants: **a** *Allium cepa*, **b** *Vicia faba* (based on data from Murin 1964, 1967; Lopez-Saez et al. 1966)

imal cell cycle duration occurs near the root apex and the mitotic index has maximal values there. For the cambial zone in conifers, according to Eq. 3.3, the observed mitotic index will be maximal in the central part of the cambial zone close to the initial and will decrease towards the boundary with the enlargement zone. In the same direction, the duration of the cell cycle decreases (i.e. the linear rate of cell radial growth increases). The calculated theoretical curves of specific growth rate and M_I distribution across the cambial zone are shown in Fig. 3.18 according to the position of the xylem mother cell relative to the initial. These calculations were made for a typical conifer cambial zone. They show that the mean duration of the cell cycle when the number of cambial cells is 14 is close to 10 days, the initial (cell in zero position) has a cell cycle duration of 23 days, and xylem mother cells near the boundary with the enlargement zone have a cell cycle duration of about 4 days. The theoretical calculations are within the experimental range of these characteristics.

Based on the previous considerations, we can explain the observed curvilinear relationship between the number of cells in the cambial zone and annual xylem increment (see Figs. 3.9, 3.10). From Eq. 3.1, it follows that: (a) there is a linear relation between the duration of mitosis and cell cycle duration, and (b) this relationship is described by lines with different slopes (coefficient a) and different values of C_0. If we assume that there are two species (or one species grown in different conditions) for which such linear relationships could be described by the curves presented in Fig. 3.19a, then the mitotic index in the cambium will be higher in the case with the steeper line (curve 5

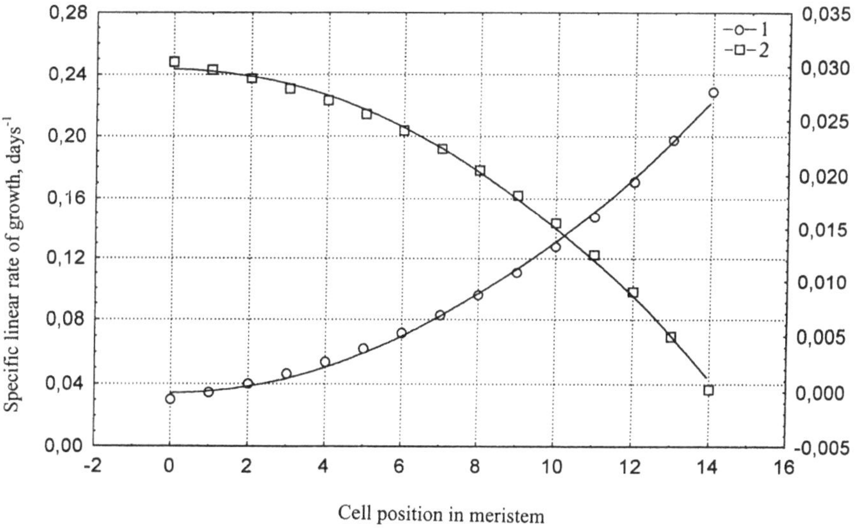

Fig. 3.18. Theoretical example of change in specific rate of linear cell growth (*1*) and mitotic index (*2*) across a meristem according to the equation: $M = a(C - C_0)$

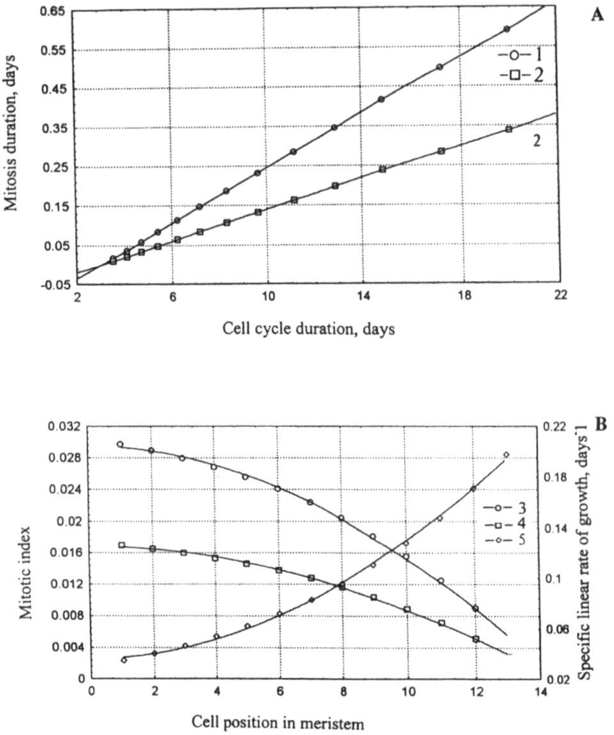

Fig. 3.19. Theoretical example of two dependencies (*1, 2*) of mitosis duration on cell cycle duration (**a**) and mitotic index changes (**b**) corresponding to these variants across the cambial zone (*3, 4*) with the same specific linear cell growth rate (*5*)

in Fig. 3.19b) than for the case with the less steep line, even though the specific growth rate across the cambial zone is equal. This explains Gregory and Wilson's (1968) results from a comparative analysis of mitotic index in the cambial zone of white spruce from Alaska and New England.

This assumption is not, however, sufficient to explain another observed relationship, namely the greater production of xylem derivatives with an equal number of cells in the cambial zone. It is necessary to assume that the steeper linear relationship between *M* and *C* coincides with a steeper increase of the linear growth rate of cells within the cambial zone (Fig. 3.20a). Then, the mitotic index value within the cambial zone differs by a factor of two from the experimental data (Fig. 3.20b). Finally, accepting this assumption we can calculate the relationship between the cell number in the cambial zone and the annual xylem increment (Fig. 3.20 c). It completely agrees with the observations. That is, the higher sensitivity of *M* changes to changes of *C* (higher coefficient *a* in Eq. 3.1) has to combine with the higher linear growth of cells in the cambial zone or higher L_{SPL} and L_{SPN}. This is more or less obvious because the same mechanisms are responsible.

The phenomenological aspects of these mechanisms are the following. The steeper characteristics of *M* as a function of *C* indicate that the relative portion of *M* in the cell cycle increases (in time) and the sum of the durations of

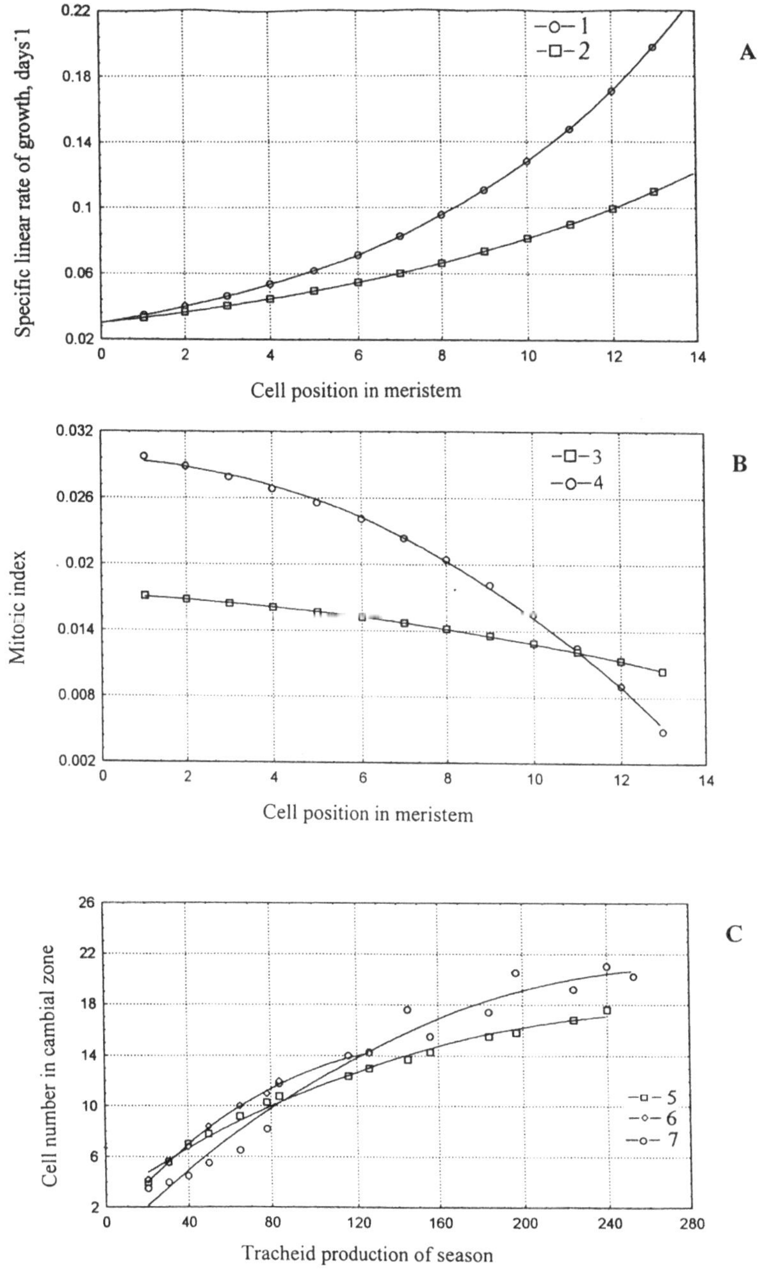

Fig. 3.20. Dependence of mitosis duration on cell cycle duration, with distribution of specific rate of linear cell growth across the cambial zone: **a** two curves, characterizing variability of specific growth rate across the meristem (*1* fast, *2* slow), **b** corresponding curves of mitotic index changes across the meristem (*3* fast, *4* slow), **c** dependence of cell number in cambial zone on total cell production for the season [*5* calculated theoretically, *6* experimentally observed for black spruce (Gregory and Wilson 1968), *7* experimentally observed for Scots pine (Vaganov et al. 1985)]

other phases in the cell cycle (G_1, S, G_2) decreases.This means that, in cell cycles of the same duration, the cell goes through the other phases faster. Such an increase must exist across the cambial zone. Then curves 3 and 4 in Fig. 3.20b have to coincide with curves 1 and 2 in Fig. 3.20a.

Finally, the assumptions mentioned above are sufficient to explain and to describe quantitatively the observed relationships in annual xylem production and the number of cells in the cambial zone, the differences in mitotic index across the cambial zone, and those between annual rings of trees growing in different conditions. When earlywood cells form, the size of the cambial zone is maximal and the average duration of a cell cycle in the cambial zone decreases because the latest cells on the cambial edge next to the enlargement zone, which have the shortest cell cycle, are involved in division. Such cells are in positions 10–14 according to Fig. 3.18. During the second part of the season when the latewood cells form, the size of the cambial zone decreases, mainly at the expense of short cycling of the last cells in the cambial zone. The average duration of the cell cycle increases and the rate of cell production decreases. Thus it is clear that, in indirect measurements of cell cycle duration (as done by Skene 1972; Kennedy and Farrar 1965), it is necessary to take into account the kinetics of two variables: cambial zone size and the portion of cycling cells with the shortest cycle. The assumption of a stable cell cycle duration throughout the cambial zone, as Wilson (1964) suggested, led to unrealistic values of the length of the cell cycle (see also Larson 1994).

There are some important consequences. The first is the increasing rate of cell division towards the cambial edge. The linear rate of tangential wall growth increases in the same direction and reaches its maximum in the enlargement zone. The experimentally determined frequency of dividing cells (the number of cells in mitosis as a proportion of the total number of cells) across the cambial zone is inversely related to cell cycle duration and the rate of cell growth. The initial has the longest cell cycle; and the xylem mother cell of the first generation has a cell cycle duration similar to that of an initial. In such a case, the initial looks like the cell of the quiescent center of a root meristem (Ivanov 1974, 1987). The difference in cycle duration between the initial and the latest xylem mother cells at the cambial edge may be as much as 5–6 (in the calculation presented above, C of the initial is 23 days, C of the short cycling xylem mother cells is about four days). There is no doubt that the xylem mother cell passes each position on its way to the edge but divides less than the number of positions available, because the next xylem mother cell in the previous position also divides and pushes it out of the cambial zone. So, each xylem mother cell can make fewer divisions within the cambial zone than the positions available. This "movement between subsequent divisions" is considered by Ivanov (1974) in detail. We are describing a more stable picture that can be observed with measurements and averaged for many radial files of cells in the meristem.

Second, for relatively slow-growing trees in the temperate zone with a growing season of about 90–100 days (annual cell production up to 60 cells, tree-ring width up to 2.0 mm), the cambial cells involved in proliferation are situated in positions 1–7 and the annual production linearly increases with increasing size of the cambial zone. If, however, more cells are involved in division (for example, if the cambial zone adds cells in positions 8–14), then the rate of production will be higher than predicted by the increasing size of the cambial zone, because of the shorter cycling of cells. In this case, the annual production will increase to a level higher than that indicated by the number of cells in the cambial zone.

Third, perhaps the most important mechanism of control of cambial activity is realized through the relationship between *M* and *C*. As shown above, it leads to a higher linear growth of cells in the same position within the cambial zone (see Fig. 3.19). There are mechanisms to control this if it is connected with relative shortening of the G_1, S, and G_2 phases. Together with physical–chemical gradients across the cambial zone, there can be specific growth factors affecting the level of transcription–translation (Savidge 1996) because this level of regulation includes the biosynthetic phases of the cell cycle. A search for such factors may be more likely to succeed in conditions where the effect is largest. In the case of conifers, these might be trees growing near the northern timber line where relatively high production during the shortest growing season is made possible by a steeper linear dependence of *M* on *C*. According to our data, annual production in *Larix* spp during the very short growing season (about 30 days) in some tree-rings is about 40 cells; i.e. more than one cell per day. This production is even higher than that observed by Gregory and Wilson (1968) in Alaska.

Another aspect of this consideration is a possible mechanism of a cessation of cell division in the cambial zone before the cell starts to enlarge. The rate of cycling increases towards the cambial edge, and as hypothesized, it is mainly associated with shortening of the interphases (G_1–S–G_2). Perhaps the kinetic mechanism of a cessation of division occurs if cells pass through the G_1 phase faster than the permitted interval. Rapid movement through this phase might lead to the reduced production of substances required for DNA synthesis and the change in direction from division to enlargement. It is shown that, according to the DNA content in cambial cells before and in the period of maximal activity, most of the cells in interphases are in phase G_1 (Mellerowicz and Riding 1992). Also, summarizing the division activity in the context of its regulation, Roberts (1976) especially noted: "Cells in G_1 show visible signs of tracheary element cytodifferentiation, but evidence is not conclusive that the appearance of cytodifferentiation is strictly confined to G_1, G_2 remains a possibility..." (p. 53 in Roberts 1976). The indirect data on the shortening of the G_1 phase in the edge of the root meristem are widely presented and discussed in Ivanov's review (1987). The importance of the G_1 phase for xylem differentiation is extensively discussed by Dodds (1981) and Aloni (1987).

The variations of cell cycle duration in meristems are still an issue under discussion (Barlow 1976; Clowes 1976; Webster and MacLeod 1980; Baskin 2000; Evans 2000; Evans et al. 2001; Ivanov et al. 2002). In shoot meristems, there is heterogeneity of growth rates within the apex. It has been clearly shown that the cell-doubling time (mean cell cycle duration) decreases from the summit cells to the flanks of the shoot apical meristem by a factor of two or more (Denne 1966; Nougarede et al. 1990; Lyndon 1998). For instance, the cell cycle within the *Pisum* apex varies almost five-fold, from 15 h to 69 h; and all cells are cycling but the shortest cell cycles (15 h) are in the region where the next primordium will emerge (Lyndon 1973, 1998). In the moss *Hookeria*, the mitotic index was inversely proportional to the length of the cell cycle, which means that a higher mitotic index was observed in the lower-growing cells (Hallet 1976). An increase of temperature will shorten the cell cycle (Francis and Barlow 1988). Lyndon (1979) reported that in *Silene* plants the cell cycle in the vegetative apical dome recorded at 13 °C was 57 h, twice as long as at 20 °C. Interestingly, at 27 °C, close to the upper limit of temperature tolerance, the cell cycle was much longer (93 h) than at the lower temperature. The cell cycle in shoot apices of *Chrysanthemum* plants grown at higher light intensity had mean cell cycles in all regions of the apex which were shorter than those at lower light intensity (Nougarede et al. 1990).

In the root meristem, the more common opinion is that the duration of the cell cycle does not change along the meristem (Ivanov and Dubrovski 1997; Baskin 2000; Ivanov et al. 2002). However, there are some results that contradict this opinion. The kinetics of growth show the bell-shaped distribution of the elemental linear growth rate along the meristem and the zone of cell elongation, which indicates that the specific growth rate changes. It accelerates at/after the meristem, reaches its maximum in the most elongated cells, and declines towards the edge of the elongation zone (Erickson and Sax 1956; Silk and Erickson 1978; Ishikawa and Evans 1995; Mullen et al. 1998). If the acceleration of growth rate includes most of the meristem, in parallel with measured sizes of cells that are supposed unchanged or equal along the root meristem (Ivanov et al. 2002), the cell cycle must be shorter toward the end of the meristem.

In the cambium, most observations related to cell cycle duration are based on measurements of mitotic index across the cambial zone and on the evaluation of cell production in xylem tree rings during a season (Wilson 1964; Skene 1972; Larson 1994). There are fewer possibilities to do special experiments than in root or shoot meristems, but the main issues of internal and external control seem to be the same (Savidge 2000). Recent analysis of poplar wood tissues revealed a strong variation in gene expression across the cambial zone: "Approximately two-thirds of the cell cycle-related genes (cyclins, cyclin-dependent kinases, histones, PttXET, etc.) showed highly similar profiles across the cambial zone with a steep increase in expression toward the xylem side" (p. 2,283 in Schrader et al. 2004). These results not only support our view of variation in cell cycle duration across the cambial zone but also support the results of our phenomenological analysis concerning the increasing rate of cycling towards the enlargement zone.

The quantitative description of the main kinetics of the cambial zone in some cases combines the main features of shoot and root meristems and some primary assumptions which are based on our estimations and speculations. We may be mistaken in this, but we believe it valid to fill the gaps in knowledge of processes and patterns in cambium with observations from shoot and root meristems.

3.7 Control of the Important Kinetic Parameters of the Cambial Zone for Cell Production

For the kinetic description of the production of cells, we note that two main kinetic parameters determine seasonal change in the production of N cells in an annual ring: N_C (the number of cells in the cambial zone) and SGR (the specific growth rate), which is inversely proportional to the mean duration of the cell cycle of xylem mother cells (C). The specific growth rate in this case is the specific rate of cell production and the specific rate of their linear growth in the cambial zone. For the specific rate of the radial growth of cambial cells it can be written that:

$$R_{SPL} = (1/D_R)(dD_R/dt) \tag{3.4}$$

or:

$$R_{SPL} = \ln 2/C \tag{3.5}$$

where R_{SPL} is the specific growth rate, D_R is the radial size of a cell, C is the mean time of cell cycle in the cambial zone.

The specific rate of cell production is calculated as:

$$R_{SPN} = (1/N_C)(dN/dt) \tag{3.6}$$

During an average cell cycle, C production will make N_C, therefore the specific rate of production will be equal to:

$$R_{SPN} = 1/C \tag{3.7}$$

Thus, the specific rate of radial growth of cells in the cambial zone and the specific rate of cell production from the cambial zone are linearly connected and are determined by the length of the average cell cycle.

Total production for a season is an integral:

$$N = \int R_{SPL} * N_C * dt \tag{3.8}$$

i.e. the production of cells (and the width of the tree-ring) is determined by two time-dependent kinetic parameters: the specific growth rate of cells in the cambial zone and the number of cells in the cambial zone. It is possible to evaluate the specific growth rate R_{SPL} in two ways:

1. Using the known dynamics of cell production rate, d$N(t)$/dt, and the number of xylem mother cells, $N_C(t)$. In this case, $R_{SPL}(t) = 1 / N_c)*(dN(t) / dt)$.
2. Using the seasonal change of mitotic index in the cambial zone $M_I(t)$ and the known relation between the duration of mitosis and the cell cycle.

Unfortunately, when tested experimentally, both methods give significant errors. Long-term observations of the seasonal formation of annual rings reveal common changes in the kinetic parameters of cell production (Vaganov et al. 1985, 1992). The specific rate of cell production reaches a maximum early in the season, then the absolute rate of the production of cells and the number of cells in the cambial zone reach their maxima. Particularly careful measurements of R_{SPL} were made by Sviderskaya (1999), whose results we consider in the next chapter.

The size of the cambial zone (N_C) and the specific growth rate (L_{SPL}) change during the growing season. Is there any connection between the changes in these parameters? There are not many relevant data available. For example, for white spruce there is a weak correlation between the number of cells in the cambial zone and the mitotic index during a season (this is based on measurements; Gregory 1971). The variability in the rate of production and the specific growth rate of cells in the cambial zone for tree rings of different width are calculated using the relationship between the number of cells in the cambial zone and the total production of cells in a season (Vaganov et al. 1985; Table 3.3). The number of cells in the cambial zone varies much more than the specific growth rate. In experiments with trees of different vigor, most authors find a linear relationship between the maximum sizes of the cambial zone and the size of the annual ring, though the mitotic index varies little during the season and does not depend much on the size of the ring (Gregory and Wilson 1968; Sisson 1968; Gregory 1971). This is in good agreement with the results of the phenomenological approach described in Sect. 3.5, where the rate of tracheid production changes non-linearly with the change in size of the cambial zone because of the non-linear distribution of the cell cycle within the cambial zone.

Table 3.3. Calculation of kinetic characteristics of cambial zone for wide pine tree rings

Cell number in tree ring	Cell number in cambial zone	Rate of tracheid production (cells/day)	Specific rate of production (1/day)	Average number of divisions ($\ln N_C/\ln 2$)	Average of duration cell cycle (days)
260	16.2	3.3	0.20	4.02	5.0
230	15.5	2.9	0.19	3.95	5.3
190	14.0	2.4	0.17	3.81	5.8
150	13.5	1.9	0.14	3.75	7.1
120	11.9	1.5	0.12	3.57	7.9
105	11.0	1.3	0.12	3.46	8.5
80	9.6	1.0	0.10	3.26	9.6
62	8.7	0.8	0.10	3.12	10.8

3.8 Conclusions and Discussion

The basic terms needed for the study of cambial activity and the production of new xylem cells by examining the seasonal course of these phenomena were introduced. The methods needed to study the seasonal kinetics of tree-ring growth and structure were described. The cellular organization of the cambial zone and seasonal patterns of cambial activity were considered before the introduction of a phenomenological approach to patterns in the kinetics of cambial activities. Based on all this, it has been possible to draw some basic conclusions on the control of the kinetic parameters of the cambial zone relevant to cell production.

So, the following important kinetic parameters of the cambial zone are responsible for cell production and tree-ring width:

1. The dormant and starting size of the cambial zone (the number of cells)
2. The specific growth rate distribution across the cambial zone, which is the same as the distribution of the cell cycle duration (related to the dependence of M versus C)
3. Average (maximal) size of the cambial zone in the main growing period
4. Average specific growth rate or cell cycle dependent on cambial zone size

What can we say about the possible control of each characteristic? Perhaps the specific growth rate distribution within the cambial zone depends on both the tree species and the geographical zone of growth (regional climate). There is some evidence for this from observations (Gregory and Wilson 1968; Sviderskaya 1999). From the dendrochronological point of view, these characteristics are more or less stable through a long period of growth and do not affect inter-annual variations of cell production. The dormant and especially the starting size of the cambial zone may be closely related to the previous growth and tree vigor (Dodd and Fox 1990; Sviderskaya 1999). Our dendroclimatic analysis based on "cell chronologies" of larch and spruce near the northern timber line shows the significant effect of starting conditions (temperature, soil melting) on the production of cells and tree-ring width (Hughes et al. 1999; Vaganov et al. 1999). These characteristics affect the inter-annual variability of tree-ring width and cell production. However, the starting number of cells in the cambial zone may also be affected by growth conditions in the previous year and, so, might be responsible for autocorrelation in tree-ring series that appear, at first sight, to have no climate cause (Fritts 1976). The main possible environmental control is related to infra-seasonal variations in size of the cambial zone and the specific growth rate of xylem mother cells. This thesis is supported by practically all experimental data, which show a close correlation between the average and maximal size of the cambial zone and the annual xylem increment (see Wilson 1964; Gregory 1971; Skene 1972; Vaganov et al. 1985). In combination with cell cycle distribution across the cambial zone, these seasonal variations may explain much of inter-annual deviation in total cell production and tree-ring width.

Analysis of the data shows that there are two determinants of variability in the sizes of the cambial zone: (1) more or less constant over many years, determined by the condition of the tree as a whole (growth class, vigor, energy of growth, age, position in stand, etc.), (2) intra-seasonal, which is determined by a change in the specific growth rate of xylem mother cells and on which their number also simultaneously depends (Vaganov 1996a). The first determinant can be considered as a constant on the long-term scale of the life-span of a tree (some years, some decades), the second depends on current environments (climatic conditions within a season).

The assumption that the cell cycle is equal along the cambial zone throughout the season leads to the following: to control division it is necessary to control the width of the cambial zone (the number of dividing cells) and their specific growth rate. This means that two mechanisms are involved in control. One is clearly positional and the other may have a dependence on concentration or have some other nature. Formally, the unequal (in the case of this hypothesis = increased) cell growth rate across the cambial zone needs only one control: positional, which is easily described in mathematical terms and the corresponding equations. Of course, nature does not always conform to the simplicity of its mathematical description.

4 Radial Cell Enlargement

4.1 Introduction

Cell enlargement after division is a basic process that is studied relatively easily and seems to be very sensitive to manipulation by chemical reagents or environment. This is why numerous investigations have been made on seedlings of conifers with the application of different hormones to change the process of enlargement or its final result – radial tracheid dimension. Few investigations have been made in natural conditions with older trees. In this chapter, as elsewhere in this volume, we do not deal with the topic of hormonal control of tracheid enlargement. It is well covered in monographs and review papers (Fukuda 1996; Savidge 1996; Kozlowski and Pallardy 1997). We focus on the kinetic characteristics of cell enlargement (Sect. 4.2), especially those observed in "normal" growth conditions (without intervention) and discuss the methods used to do this (Sect. 4.3). We also consider observations made from the point of view of the relationship between two related processes: cell division and cell enlargement (Sect. 4.4). Are these processes completely independent or is there a strong influence of the division process on the kinetics of enlargement and its final result: radial tracheid size? If a relationship exists, how strong is it and what can violate it during seasonal growth in natural (not artificial) conditions? We consider some consequences of the relationship we postulate between the rate of cell division and cell expansion (Sect. 4.5). Our basic conclusions are summarized and discussed (Sect. 4.6) in the context of some recent work on root development and the distribution of plant hormones, enzymes, and carbohydrates.

4.2 Cell Expansion and Associated Processes

Expansion of the radial size of cells after they leave the cambial zone is the next main stage of the cytodifferentiation of xylem (Gamaley 1972; Roberts 1976). The latter (see pp 36–37 in Roberts 1976) noted that at this stage: "the cells exhibit high variability in the extent and regulation of expansion. The deposition of primary wall material during expansion requires the synthesis of primary wall monomers. Protein synthesis occurs. DNA replication involving endoreplication and gene amplification may occur." The visible result of the enlargement is a greatly increased radial cell size. In earlywood, the radial dimension of tracheids reaches 50–60 μm; in latewood it is about 15–25 μm. So, during the formation of earlywood, the radial size of tracheids increases up to 7–8 times, in latewood up to 2–3 times in comparison with the starting size of a cell in a cambial zone, which is about 7–8 μm on average.

4.3 Methods to Study the Kinetics of Cell Enlargement and the Main Results.

There are two main experimental approaches to the study of the kinetic characteristics of radial cell enlargement during tree-ring formation. The first is the quantitative analysis of the number of cells in the different zones in the developing tree ring, based on measurements of wood samples taken periodically within a growing season (Whitmore and Zahner 1966; Skene 1969; Wodzicki 1971; Antonova et al. 1995). This method allows one to quantify the time needed for a cell to complete enlargement. The second method uses a "instantaneous tracheidogram" and is discussed later in this section.

In earlier research it was found that: (1) there was little correlation between the duration or average daily rate of radial enlargement (V) and the final radial tracheid size (Wodzicki 1971), and (2) from the analysis of radial cell expansion in tree rings forming in trees of different classes and in different positions in the stem. Dodd and Fox (1990) showed that the duration of tracheid expansion is closely correlated to the radial dimension of tracheids, because the rate of expansion was approximately constant and the same value was found for early- and latewood tracheids during a growing season. They estimated the rate as 2.0 μm/day. In most observations, the rate of radial tracheid expansion changes insignificantly among earlywood tracheids and between early- and latewood tracheids. The total period to complete enlargement for earlywood tracheids is about 3–4 weeks, and for latewood about 1–2 weeks (Whitmore and Zahner 1966; Skene 1969; Wodzicki 1971; Vaganov et al. 1985). Shorter cell enlargement durations (<1 week for earlywood, 5–10 days for latewood) have been reported for *Abies balsamea* in Quebec boreal forest (Deslauriers et al. 2003). It has also been clearly shown that, in slowly growing trees (where a narrow tree ring was formed), the expansion of tracheids was approximately one week late compared to fast growing trees, in spite of the radial size of tracheids being the same (Vaganov et al. 1985; Sviderskaya 1999).

The second method requires more precise measurements of radial cell sizes on thin cross-sections periodically taken during a growing season, compiled into so-called "instantaneous tracheidograms" (Vaganov et al. 1985; Vaganov 1990; Sviderskaya 1999). This allows one to use the relative position of a cell within the completed tree ring to obtain the curves of time versus increasing cell size in the same relative position. This method gives two kinetic characteristics of enlargement: the time-period of enlargement and the approximate rate of radial cell growth for different intervals of enlargement.

Using such an approach, Sviderskaya (1999) made measurements in growing tree rings of four conifer species (*Pinus sylvestris*, *P. sibirica*, *Picea obovata*, *Abies sibirica*) and found that the average rate of tracheid expansion did not differ significantly for early- and latewood tracheids (Fig. 4.1). Small differences occurred in the average seasonal rate of radial tracheid expansion between different species (e.g. for *Pinus sylvestris* it was about 1.4 μm/day with a maximum of 2.0 μm/day, for *P. sibirica* it was about 1.6 μm/day with a maximum of 2.2 μm/day). She also estimated the duration of the radial enlargement of early- and latewood tracheids. It was 25–27 days for earlywood and about 10 days for latewood tracheids. She noted that the rate of radial expansion of a single tracheid changed across the zone of enlargement. The highest rate was observed in the first part of the zone and then the rate of expansion gradually decreased towards to the edge of the zone.

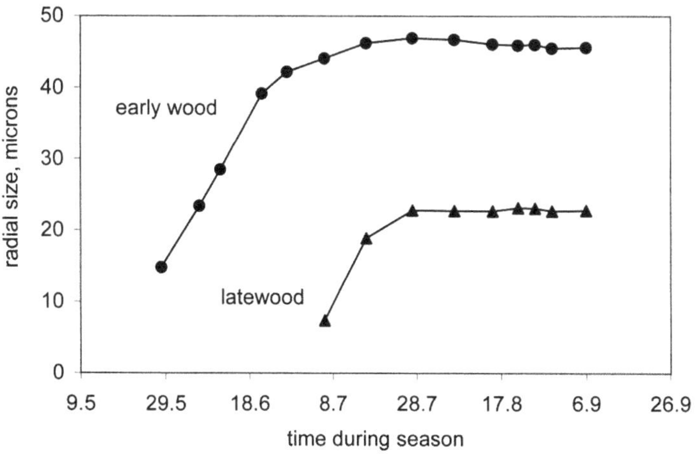

Fig. 4.1. Typical curves of tracheid radial size by date of early- and latewood cells of *Pinus sylvestris* in the southern taiga

An original approach was applied in work on the growing root apex (the standard object of research on processes of cell production and differentiation of plant meristems; Erickson and Sax 1956; Ivanov 1974; Silk and Erickson 1979; Gandar 1983; Silk et al. 1989). Estimates of growth rates of cells were based on measurements of the sizes of cells in separate radial files. The essence of the approach is based on a calculation of specific rates of increment in the linear sizes of cells, depending on their sequential positions in the file. The results of these histometric methods of calculation of the rates of linear growth has shown good correspondence with data obtained by direct measurements of the growth of apices, with the help of small marks (Ivanov 1974; Silk et al. 1989). It is also possible to use a histometric approach for the calculation of cell radial growth rates in an active period of growth (as a rule, dur-

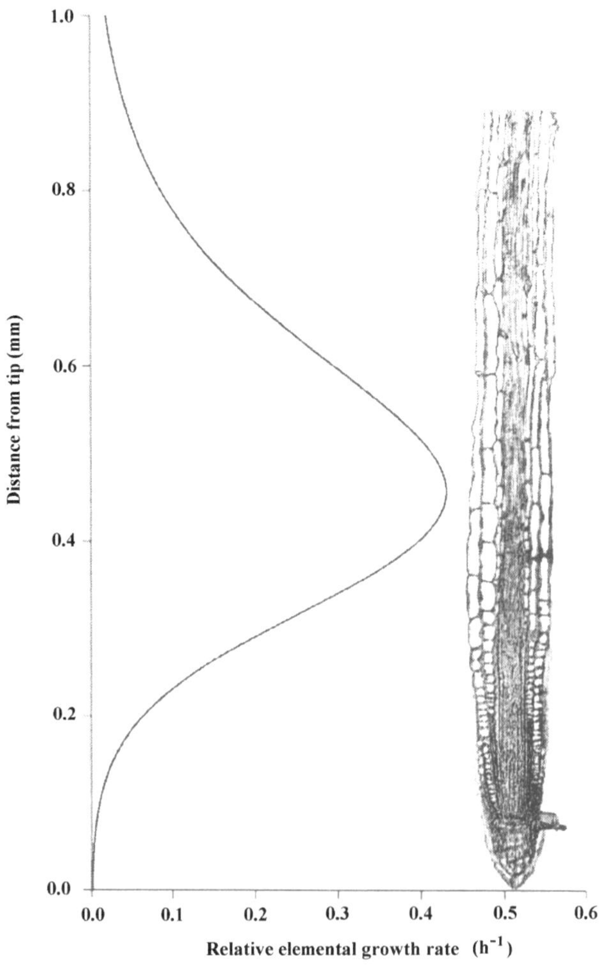

Fig. 4.2. Relative elemental growth rate profile for a vertically growing *Arabidopsis* root. The *inset* is a longitudinal section of the root and is scaled to the graph (Mullen et al. 1998)

ing the formation of earlywood) on cross-sections of complete tree rings, measuring their radial sizes in separate files. The outcomes of such measurements confirm the histometric data from the analysis of linear growth of cells in a root apex (Fig. 4.2). The specific rate of radial growth of cells begins to increase in approximately the middle part of the cambial zone, reaches a maximum at the beginning of the zone of enlargement, and then gradually decreases. This approach is based on the assumption that the tissue growth is steady. Seasonal formation of the tree ring seems not to be constant throughout the season, but is close to constant during a short interval when earlywood tracheids are being produced. Vaganov et al. (1992) tried to apply the histometric approach to evaluate the radial rate of tracheid growth, using measurements of the "instantaneous (immediate)" tracheidogram (Fig. 4.3). The results completely confirm the previous results and indicate that the maximal radial growth rate of tracheids occurs in the first part of the enlargement zone. Moreover, if we consider the part of the curve relating to the cambial zone, the gradual increase in the cell specific growth rate completely agrees with the data on the duration of the cell cycle and the rate of passing through the cell cycle illustrated in Fig. 3.18. Unfortunately, it would be wrong to apply this approach to the cambial zone, because of cell division. The only possibility to estimate the absolute linear growth rate (not the specific growth rate) indirectly would be to determine the average cell production. Thus, in a case with six or seven dividing cells in the cambial zone and a total absolute production of 40–50 cells, cambial cells were produced at the rate of 0.5 cells/day during a season of 90 days. If we take into account the radial size of a cell leaving the cambial zone (8 μm) and the average cell size within the cambial zone (10–12 μm), then the absolute growth rate of the whole cambial zone is approximately 4 μm/day and the specific growth rate is about 0.06 per day. It seems to be higher during the production of earlywood cells because of the higher production rate and appears lower in the cambial zone than in the enlargement zone. The outcomes of indirect histometric analysis of the growth rates of cells in the cambial zone confirm the essential heterogeneity of cells in the cambial zone so far as the main kinetic parameters, cell cycle and linear growth rate, are concerned.

Most experimental results indicate that final tracheid size is not determined during radial expansion excepting, perhaps, observations made by Antonova and co-authors (Antonova and Stasova 1993; Antonova et al. 1995). Thus we must look elsewhere for the factor controlling radial tracheid dimension. It is obvious that such a search must focus on the period before that cell radial expansion, namely on cell production. So the question is: what kinetic parameter(s) of tracheid production and, naturally, what in the cambial zone may be responsible for the determination of radial tracheid size?

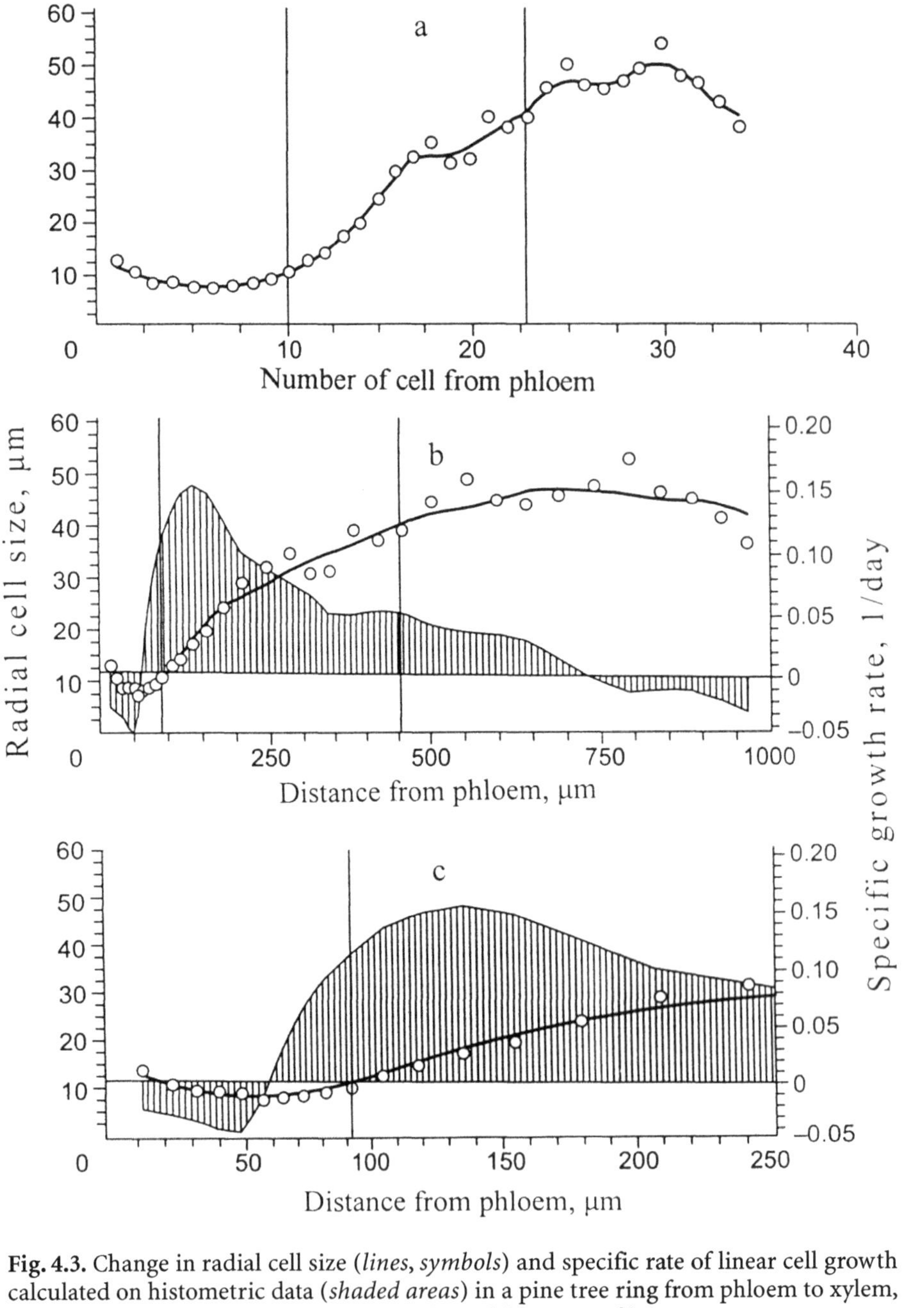

Fig. 4.3. Change in radial cell size (*lines, symbols*) and specific rate of linear cell growth calculated on histometric data (*shaded areas*) in a pine tree ring from phloem to xylem, by number (**a**) and by distance (**b, c**). c Enlarged fragment of **b**

4.4 Results Confirming the Relationship Between the Rate of Division and Tracheid Expansion

Roberts (1976; see p. 52), summarizing published results on xylogenesis, pointed out that: "The progressive development of the different stages of the cytodifferentiation sequence occurs at varying rates, under different in-vitro conditions, and results in a variety of sizes and shapes of differentiated cells. The possible relationship of cell expansion to the cell cycle has not been studied." We will consider the results of further research that indicates the existence of such a relationship and supports the hypothesis of the primacy of fluctuations in the rate of cambial activity (i.e. the rate of division) in controlling tracheid expansion and the final tracheid dimension.

4.4.1 Direct Comparison of Radial Growth Rate and Radial Tracheid Dimension

Seasonal observations on the number of cells in different zones of the developing tree ring allow one to make some important comparisons. It is necessary to compare the final tracheid dimensions with the main kinetic characteristics of cambial growth: the specific growth rate of cell production or the specific radial cell growth within the cambial zone. Using the basic equation (Eq. 3.6), Sviderskaya (1999) carefully estimated the specific growth rate from cell production data and the number of cells in the cambial zone for precise times in the growing season. She then compared these values with the final radial dimension of tracheids which at that time had just left the cambial zone. To match the tracheids on the final tracheidogram, she used the scheme shown in Fig. 4.4.

To estimate the specific growth rate of tracheid numbers it is necessary to obtain two curves: one is the seasonal curve of total number of cells in the growing tree ring (including the cambial zone) and the other is the number of cells in the enlargement, the maturing and mature zones (excluding the cells in the cambial zone). This is because:

$$R_{SPN} = (dN / dt) / N_C$$

or for practical use:

$$R_{SPN} = (\Delta N / \Delta t) / N_C$$

where ΔN is the difference in the total number of cells in the growing tree ring during the chosen time interval Δt.

If the two curves needed are obtained, then the estimation of seasonal changes in specific growth rate may be extracted from Fig. 4.4 according to:

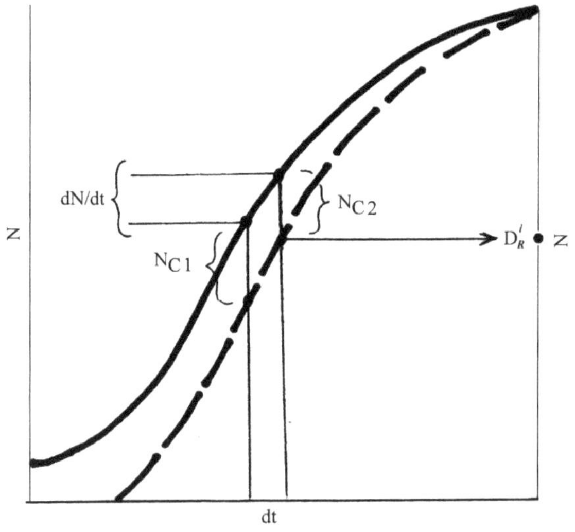

Fig. 4.4. Scheme illustrating the calculation of specific growth rate from observed seasonal data in order to compare it with radial tracheid diameter

$$R_{SPN} = (\Delta N / \Delta t) * (2 / (N_C^1 + N_C^2)$$

in order to include the possible changes in the number of cells in the cambial zone during the time interval chosen for calculation. The calculated value is then compared with the radial dimension of tracheids that left the cambial zone at the same time during the season.

The results of a comparison within a growing season are shown in Fig. 4.5. Obviously, the accuracy of the specific growth rate calculations is rather less than the accuracy of determination of the position of tracheids on the timescale. Nevertheless, there is a good relationship between intra-seasonal changes in specific growth rate and intra-seasonal changes in the size of tracheids which leave the cambial zone with a certain growth rate. This means that the radial tracheid dimension is mainly determined either by the average radial growth in the cambial zone or (because it is the same when expressed as specific growth rate) by the average rate of cell division within the cambial zone. In other words, radial tracheid dimension is a replica of the intensity of new cell production by the cambial zone. We will consider how this statement is supported by other indirect data.

4.4.2 High Frequency Variations of Radial Tracheid Dimension in Conifers

In several publications where measurements of radial tracheid sizes were presented, the authors pointed out that it is necessary to measure multiple radial

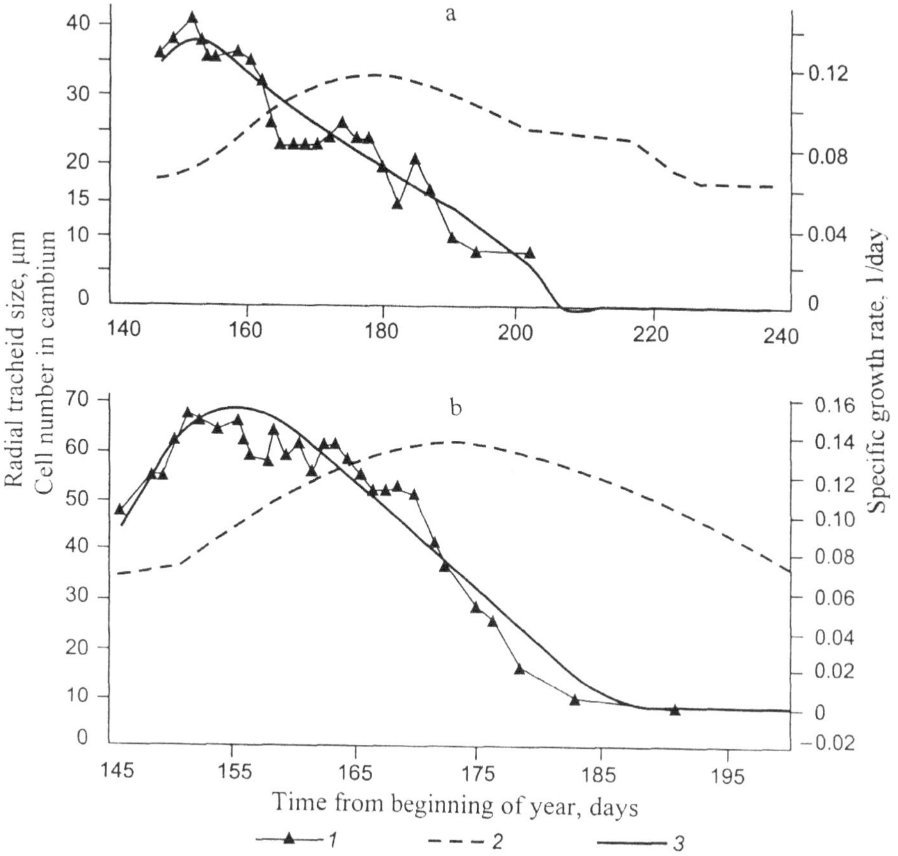

Fig. 4.5. Seasonal dynamics of radial tracheid size, specific growth rate and cell number in the cambial zone of trees of Siberian fir (*Abies sibirica*; **a**) and Scots pine (*Pinus sylvestris*; **b**): *1* radial cell size, *2* cell number in cambial zone (5), *3* specific growth rate

files of tracheids to get stable results, owing to the random variability of tracheid dimensions (Wilson 1963; Wodzicki 1971). A rather large variability in radial tracheid sizes, especially in earlywood, affects even densitometric profiles of tree rings, in spite of the large number of files of tracheids integrated (Schweingruber 1988). Three components of variability are suggested from an analysis of the variability of radial tracheid sizes in larch and pine trees (Vysotskaya et al. 1985; Vysotskaya and Vaganov 1989): (1) a common seasonal trend, (2) variations due to climate, (3) high frequency variations (Fig. 4.6). To evaluate each of the components of variability, Djanseitov and co-authors (2000) used the transformation of initial measurements to index values.

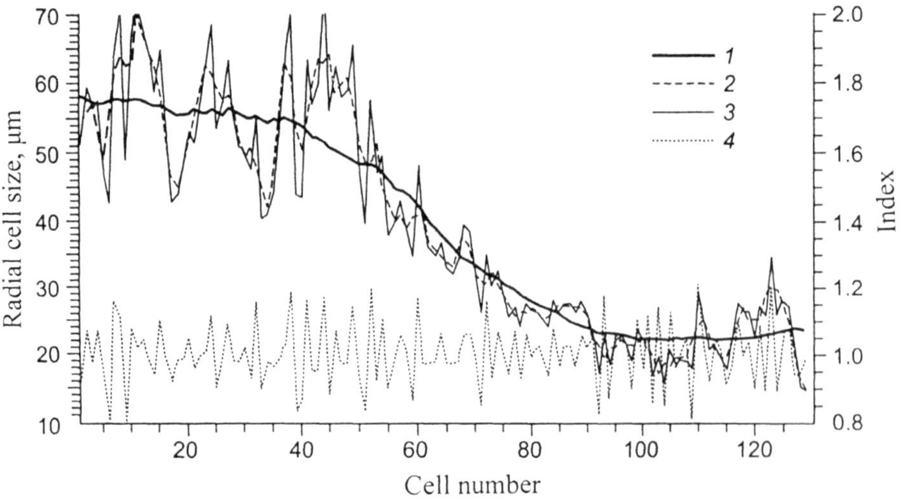

Fig. 4.6. Variability of radial cell size in typical individual file of pine tree ring: *1* seasonal trend, *2* climatically determined variation, *3* initial file of measurements, *4* calculated high-frequency component of radial size variability

(1) For common variability:

$$I_{COM} = D_R^i / D_R^{MEAN} \tag{4.1}$$

(2) For evaluation of variability due to climatic factors in combination with high frequency variations:

$$I_{SEASON} = D_R^i / D_R^{SPLINE} \tag{4.2}$$

(3) For evaluation of high frequency variations:

$$I_{HIGH} = D_R^i / D_R^{SMOOTH} \tag{4.3}$$

D_R^i is the radial tracheid size of i-th tracheid in the file, D_R^{MEAN} is the average radial tracheid size for the whole file measured, D_R^{SPLINE} is the seasonal trend in tracheid sizes approximated by spline, D_R^{SMOOTH} is the curve of radial tracheid sizes approximated by 3-point or 5-point smoothing, and I_{COM}, I_{SEASON}, I_{HIGH} are the corresponding index values. The dispersion or standard deviation of I_{COM} evaluates the common variation of radial cell size within a tree ring, the dispersion or standard deviation of I_{SEASON} evaluates the component of variability caused by current weather conditions in a combination with high frequency variation, and I_{HIGH} evaluates the portion caused by high frequency variability. Table 4.1 shows the relative values of each component of

Table 4.1. Estimation of total and high-frequency variability (see text) of radial cell sizes in conifer tree rings

Species	Site conditions	Total variability (standard deviation)	High-frequency variability	
			Standard deviation	% of total dispersion
Larix dahurica	Northern timber line	0.389	0.131	11.2
Picea obovata Ldb.	Northern limit of species	0.308	0.092	8.8
L. sibirica Ldb.	Mesic conditions of south taiga	0.306	0.092	8.5
L. sibirica Ldb.	Forest steppe	0.352	0.101	8.1
Pinus sylvestris L.	Mesic conditions of south taiga	0.342	0.077	5.1
P. sylvestris L.	Forest steppe	0.372	0.088	5.8
P. densiflora Sieb.	Dry sites, South Korea	0.348	0.103	8.8
P. rigida	Dry sites, South Korea	0.326	0.075	5.6
L. leptolepis Gord.	Dry sites, South Korea	0.409	0.091	4.8

variability within tree rings of different species growing in different conditions. The common variability in tracheid dimensions increases in tree rings from poor growth (temperature limits or soil moisture deficit) and relates to tree species. For example, in larch trees, the standard deviation and dispersion are larger than in pine trees. At the same time, the percentage of common variability caused by high frequency changes varies from 5% to 11% (Table 4.2).

There are two hypotheses on the causes of this high frequency variation. If it has a clearly random source, i.e. there is no correlation between the radial dimensions of two subsequent tracheids in a file, its source might be in the process of cell enlargement. In the absence of a relationship between two subsequent tracheids in a file, the idea of complete independence of radial enlargement and production of cells in cambial zone is supported because the extension of each tracheid ceased at a different time and under different conditions. If the high frequency variation has a regular character (for instance, if it is cyclic), its source might be in the cambial zone, because subsequent tracheids in a single file are the daughter cells of the last-divided xylem mother cell before it loses the ability to divide.

Table 4.2. Estimation of high-frequency component of standard deviation in cell size variability of tracheid files in *Larix leptolepis* Gord. at different smoothing windows (averaged for 50 files in ten tree rings)

Smoothing	Cell number		
	7	5	3
Standard deviation	0.118±0.0068	0.112±0.0060	0.091±0.0053
Relative percentage from 7-point smoothing	100	94	78

To test the hypotheses, Djanseitov et al. (2000) used a widely representative selection of tracheid radial size measurements within tree rings of different species (*Larix sibirica*, *L. gmelinii*, *L. leptolepis*, *Pinus sylvestris*, *P. rigida*, *P. densiflora*, *Picea obovata*) and with a large range of numbers of cells in files (from 30 to 230). Sequential measurements of tracheid radial sizes may be considered as pseudo-time-series, because each subsequent tracheid appeared in the enlargement zone behind the previous one; and the Fourier mathematical approach is suitable for such data (Box and Jenkins 1970; Bolshov and Smirnov 1983; Mazepa 1990). The authors used Fourier analysis to calculate the following characteristics for each series of I_{HIGH}: (1) the function of spectral density as smoothed periodograms, indicating significant frequencies in the series, (2) the autocorrelation function, which selects the significant and insignificant relationships between cell sizes of adjoining cells in a file with different lags, (3) a comparison of the I_{HIGH} distribution curve with a Gaussian curve, (4) a comparison of the cumulative probability curve with Kolmogorov–Smirnov's criterion which reveals the frequency for which the variations exceed the "white noise" level with a probability of 0.95.

A typical example of these characteristics is shown in Fig. 4.7 for a single radial file of tracheid radial sizes within a tree ring of *Pinus densiflora*. There are 145 cells in this ring. All these characteristics and criteria show that there are cycles in the high frequency variations of tracheid dimensions. The peak of spectral density indicates that the average and highly significant cycle is equal to 2.14 cells, autocorrelation shows the high negative relation between the radial dimensions of two subsequent cells in a file, and the Kolmogorov–Smirnov criteria confirm the high probability of the cyclic component in tracheid size variations. The frequency histogram of the index has no significant difference from a Gaussian curve but indicates a slight excess of cells with a low index. The first-order autocorrelation has an especially high value that clearly establishes that the high-frequency variations in tracheid dimensions come from relationships between successive values and hence their source lies in the cambial zone rather than in the zone of enlargement. One can see the cyclic variation in I_{HIGH}, especially clearly in Fig. 4.8, where these varia-

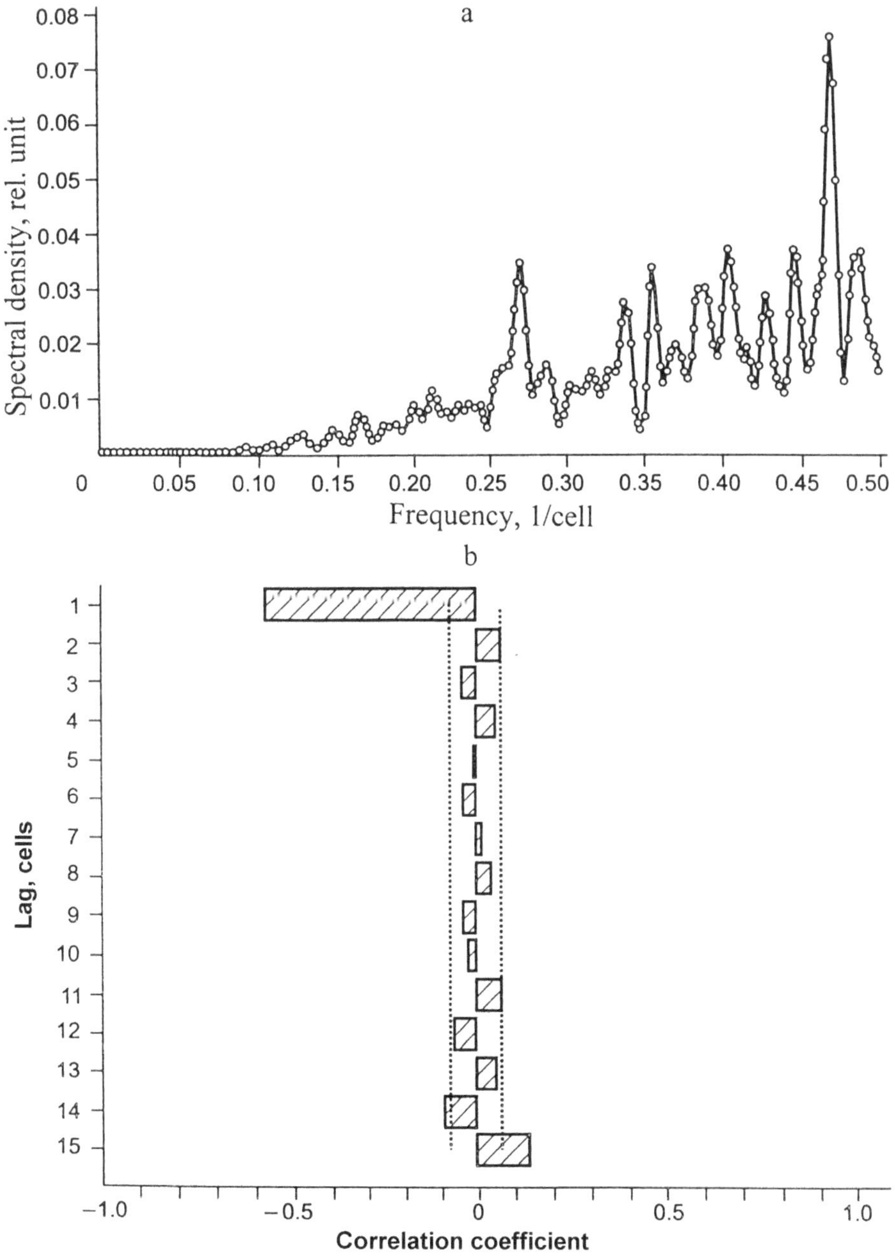

Fig. 4.7. Fourier analysis data of the high-frequency component of variability of radial tracheid sizes in *Pinus densiflora* Sieb. tree rings: **a** spectral density function, **b** autocorrelation function, **c** (*see Page 118*) histogram of index size distribution compared to the curve of Gaussian distribution

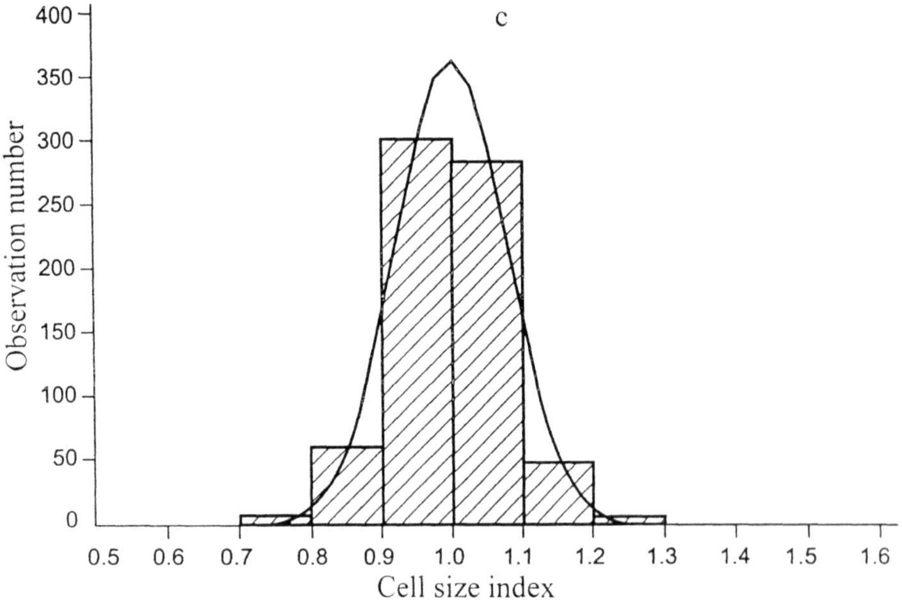

Fig. 4.7. (*Continued*)

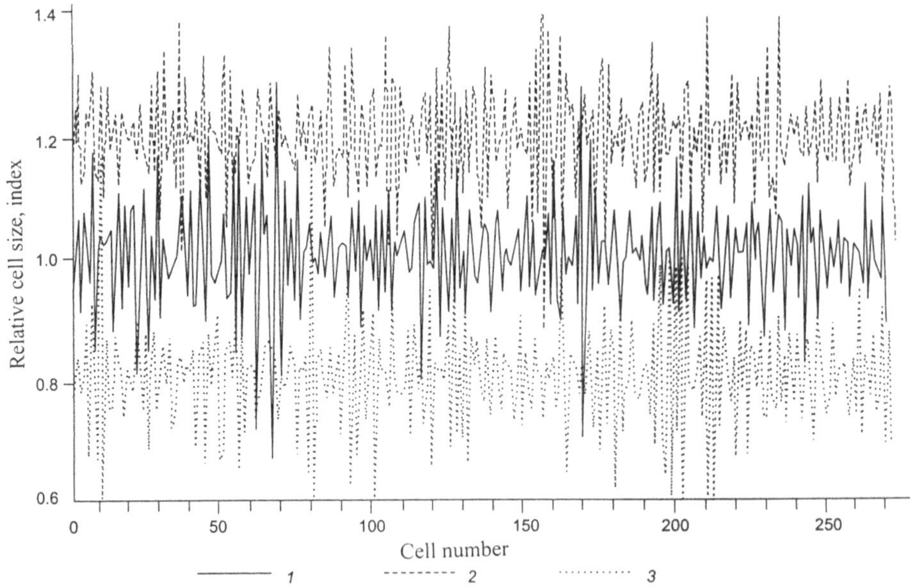

Fig. 4.8. High-frequency variability of radial cell size in individual files (*1–3*) of a *Larix leptolepis* Gord. tree ring (the curves are shifted in the *y*-axis ordinates for a better comparison)

tions are shown for a wide tree ring with a large number of tracheids in its files. Note that the amplitude of the variations is not stable through the season.

Figure 4.9 summarizes the basic statistical characteristics of the Fourier analysis of tracheid files in a wide range of tree-ring width and annual xylem increments, namely the average significant cycle, the first-order autocorrelation, and the standard deviation. It is obvious that the duration of the basic cycle, the standard deviation, and the value of the first-order autocorrelation

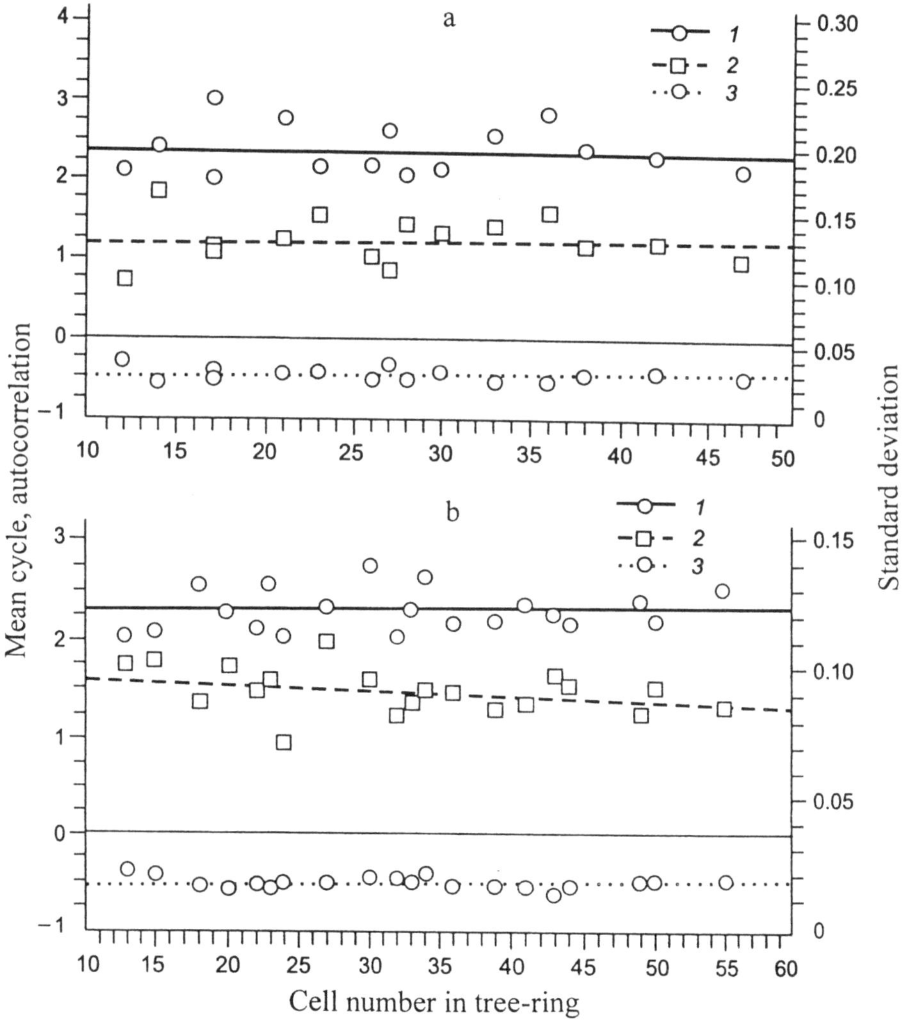

Fig. 4.9. Stability of statistical characteristic of high-frequency cell size changes in annual rings with different width of *Larix dahurica* (**a**), *Picea obovata* Ldb. (**b**), *Pinus densiflora* Sieb. (**c**), and *P. rigida* (**d**): *1* mean cycle length in cells, *2* standard deviation, *3* first-order autocorrelation (**c** and **d** see page 120)

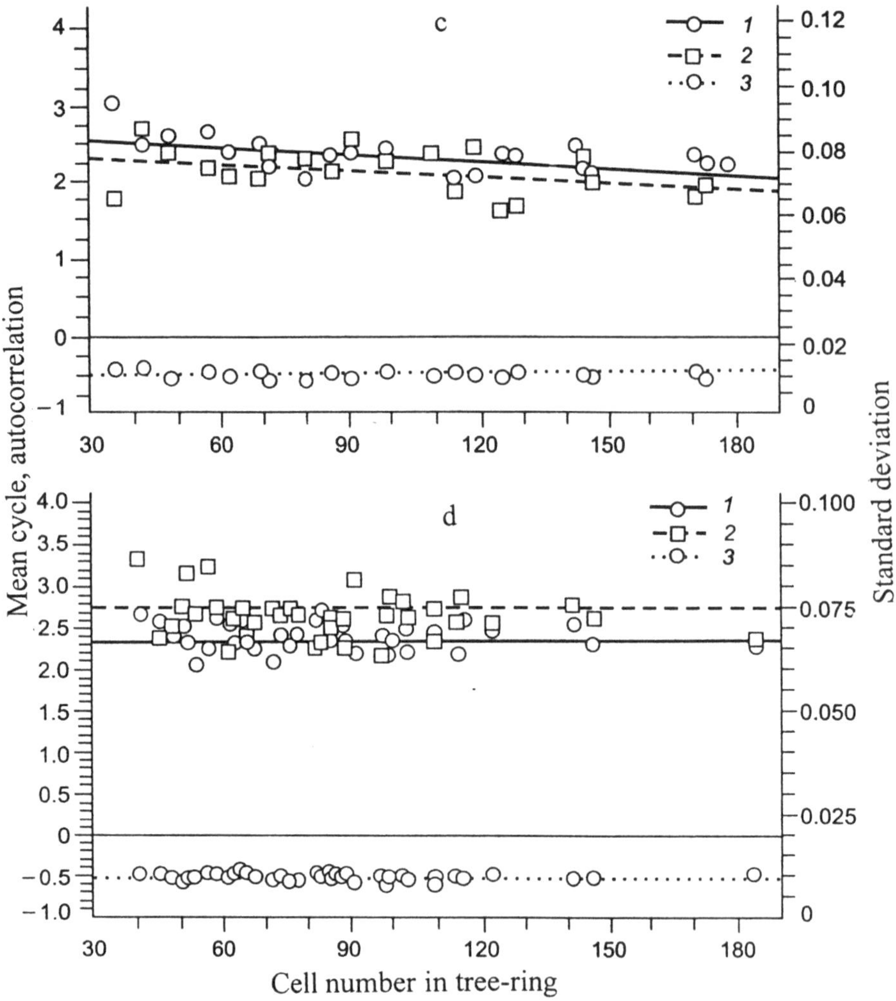

Fig. 4.9. (*Continued*)

are practically independent of annual xylem increment (tree-ring width; an insignificant tendency can be seen for tree-rings of red pine, *P. densiflora*). The average cycle varies between 2.2 and 2.4 cells and the first-order autocorrelation is also very stable (from –0.520 up to –0.540) in tree rings of different species. The high frequency variations are caused by differences in the dimensions of two neighboring tracheids; and the data in Table 4.2 confirm this. The main input to high frequency variations is the input of differences between two neighboring tracheids, which is shown by using different smoothing windows (3, 5 and 7 cells).

According to the evaluations presented in Table 4.1, the cyclic component of tracheid dimension variations plays a significant role in their common variations. We can expect that, in some cases, this component may mask the small variation in tracheid dimension caused by seasonal weather conditions or even intra-seasonal trends, especially in tree rings where differences between early- and latewood tracheids are not so great (for example, in tree rings formed in seedlings). Three-point smoothing of the raw measurements of tracheid radial size will decrease this variability.

It is most probable that this cyclic variation is due to the last division of the xylem mother cell, which produced the two neighboring tracheids. The asymmetry might be created by differences in the rate of linear growth near the cambial edge. The outer daughter cell expands a little faster than the inner daughter cell. This is in accordance with the distribution of the rate of linear growth across the cambial zone and the beginning of the zone of enlargement (see Fig. 4.3). Undoubtedly those differences in radial cell dimension that originated in the last portion of the cambial zone remain in the final tracheid size if the process of expansion does not have a significant influence on the final tracheid size and is not halted by trauma.

There are important consequences of these results for the relationship between cambial growth and the subsequent cell expansion:

1. After the last division of the xylem mother cell at the edge of the cambial zone, both daughter cells in most cases go on to enlarge. This is confirmed by the average value of the peak in spectral density of the high frequency variation, which is close to 2.0 cells (range 2.2–2.4).
2. The hypothetical boundary between the cambial zone and the enlargement zone may be defined as a boundary formed by the asymmetric division of xylem mother cells. Asymmetry of the last division of the xylem mother cell can be supported by physical and chemical gradients as well as the transport of hormones and assimilates through the cambial zone, combined with water potential (Denne and Dodd 1981; Savidge 1991, 1993, 1996).
3. It is possible to calculate the average rate of cell expansion at the boundary between the cambial zone and the enlargement zone, because of the asymmetry of cell tracheid dimensions that is maintained during cell expansion. Differences of 5–11 % may appear during the growth of the cell plate between two daughter cells (called also fragmoplast movement; Wilson 1964; Brett and Wardron 1996), even if a mother cell is divided into two equal parts at mitosis. The formation of a new primary wall between new cells takes about 20 h (Wilson 1964). So, during this time, one daughter cell expands 5–10 % faster than the other, or this difference appears at a distance of 1.6 μm (because each daughter is about 8 μm in radial size just after the mitosis). This means that the rate of linear growth of cells near the cambial boundary is equal to 0.8–1.6 μm over 20 h or approximately 1.0–1.9 μm/day. This value coincides closely with the reported rates of tracheid expansion in the enlargement zone (Wodzicki 1971; Dodd and Fox 1990; Antonova et al. 1995; Sviderskaya 1999). Note that Webster (1979)

pointed out the difference in cell cycle duration between sister cells in root meristem was about 14%.

4. The observed asymmetry of cell division indirectly supports the hypothesis of an increased rate of cell growth towards the edge of the cambial zone. If we believe with Wilson (1964) that the rate of cell division across the cambial zone is constant (taking into account that a xylem mother cell can divide 3–5 times in its cambial life) it is very likely that the last division occurs in xylem mother cells at 20–30% of the width of the cambial zone from the cambial edge. Hence we can expect those cells to have a very large radial size because of the absence of sliding growth between files. Such cells do not usually exist in the mature xylem structure but appear after the cambial population suffers severe damage by nuclear radiation (Musaev 1996) or strong frosts. If however we assume that the rates of division and linear growth increase towards the cambial zone edge, then most of the divisions occur in the second part of the cambial zone. In this case, the high probability of the xylem mother cell moving to the enlargement zone does not lead to its radial dimension increasing in time.

4.4.3 Indirect Evidence for a Relationship Between the Rate of Cell Division in the Cambial Zone and Cell Expansion

Summarizing published data on the variability of the radial tracheid dimension within conifer tree rings, we can say (Vaganov 1996a):

1. The radial tracheid dimension shows a clear seasonal trend (from early- to latewood), except in some subtropical and tropical trees (Vaganov et al. 1985).
2. The radial tracheid dimension shows a variability due to climatic factors operating within a growing season that combines with the seasonal trend (a typical example of this is the formation of "false" rings caused by intraseasonal drought; Fritts 1976; Schweingruber 1988, 1996).
3. The average radial tracheid dimension is more or less constant over a long period of tree growth. This results in the close linear relationship between tree-ring width and the number of cells produced annually (Gregory 1971; Vaganov et al. 1985, 1992). The small variability in radial tracheid dimension in comparison to tree-ring width can also be seen in Tables 2.2, 2.3, where statistical data from some "cell chronologies" are shown.
4. The range of variation in radial tracheid dimension in tree rings of different conifer species is usually limited – from 8 μm to 70 μm (Vaganov et al. 1985).

What indirect data confirm the suggestion that the final radial tracheid dimension cannot be effectively controlled by external influences during enlargement? They come mainly from measurements of tracheidograms

(Vaganov et al. 1985, 1992). Narrow layers including only two or three tracheids with a small radial dimension and thin cell walls can often be observed in wide tree rings from dry conditions (Fritts 1976; Shashkin and Vaganov 1993). The appearance of those cells in the earlywood zone may, in most cases, be the result of periods of moisture deficit only lasting from several days to a very few weeks. Obviously, the existence of such a layer of small, thin-walled cells cannot be due to the effect of water stress on cells that were enlarging at that time. There are several reasons to exclude this. First, all observations confirm that earlywood tracheid enlargement takes at least three or four weeks (see Fig. 3.7), i.e. longer than the duration of the water stress. In wide rings (60–80 cells or more in the file, or 2.0–2.5 mm or more in width) during a growing season of about 100 days, the mean cell production must be around 0.6–0.8 cells/day (during the formation of earlywood it will be higher). So, there must be 12–16 cells in the enlargement zone at any point in time. The layer of small cells cannot result from a water stress-induced cessation of cell expansion near the end of enlargement, because they will already be close to fully expanded. But if water stress affects the cells that have only just started to enlarge, then why are the other cells in the enlargement zone not affected? Consider, for example, the formation of small cells within the earlywood of *P. rigida* growing in dry sandy soil in Korea. Every year they form one to three "false rings" caused by short droughts, especially in the eight weeks between the beginning of growth and the onset of monsoons. The most interesting detail of a tracheidogram of rings with two "false rings" (Fig. 4.10) is the sharp transition from small cells to typical large earlywood cells. This coincides with the onset of the rainy season. There are, approximately, at least five to six cells in the enlargement zone at this time and water supply by rain would have to differentially affect the radial tracheid dimensions. The greatest influence of the recovery of turgor would, of course be felt by cells at the beginning of the enlargement zone. Less influence would be seen in cells at the end of the enlargement zone, because they have already done most of their expanding. As a result, if the effect was felt in expanding cells, the transition in size must be gradual, but it is not. The same pattern of sharp transitions was also observed within tree rings of *P. ponderosa* growing in semi-arid conditions in the south-west United States (Fritts 1976; Fritts et al. 1991).

Wilpert (1991) carefully analyzed the infra-seasonal changes in the radial tracheid dimension of *Picea abies* corresponding to increasing soil-water deficit. To assign each tracheid in a radial file to the time of the season when it was formed, he used a seasonal growth curve obtained by the pinning or point-damage method (see Chap. 3). According to this method, the water deficit values corresponded to the cells that were in the enlargement zone at this time. A comparison gave him the lag in time between the recorded water deficit and the radial tracheid dimension (Fig. 2.22). The pattern of changing radial tracheid dimension followed the changing soil-water potential by about 12 days, when averaged for all ten years and all trees. He explains this

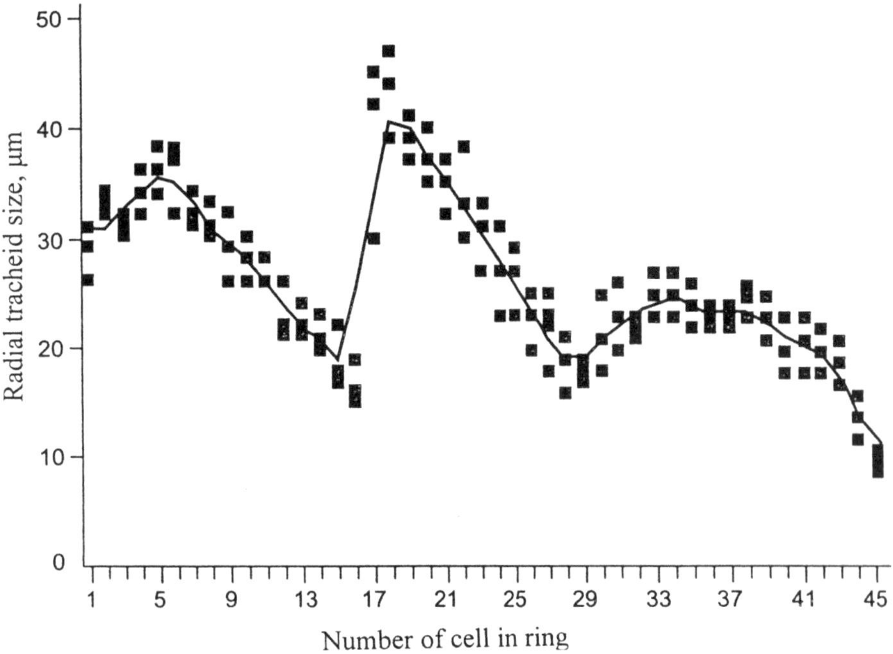

Fig. 4.10. Tracheidogram of pitch pine (*Pinus rigida*) tree ring with two intra-seasonal droughts. See text for further explanation

lag as typical for a biological system which shows a delayed reaction to an environmental stimulus. However, if we assume that tracheids accept the external signal of water deficit in the cambial zone, this explanation is not needed.

More evidence for the role of the cambial zone as the site where final radial tracheid size is determined was obtained from tracheidograms measured on cross-sections of developing tree rings or "instantaneous" tracheidograms (Vaganov et al. 1985). The tracheidograms of fast- and slow-growing trees were compared at the same time of the season (Fig. 4.11). The first-formed tracheids already achieved their final size in the middle of June in the ring of a fast-growing tree, but the equivalent tracheids were still in the enlargement zone in a slow-growing tree. Later in the season, the first-formed tracheids in both fast- and slow-growing trees had the same radial size and their tracheidograms were very similar (see also Terskov et al. 1981).

If we accept the statement that radial tracheid dimension is predetermined by the kinetic characteristics of cell growth within the cambial zone, it is necessary to define those characteristics. During cell "life" in the cambial zone, the major kinetic process is its radial growth between two consequent divisions inversely related to the duration of the cell cycle. It is possible to consider that the duration of the cell cycle (or the specific growth rate of xylem

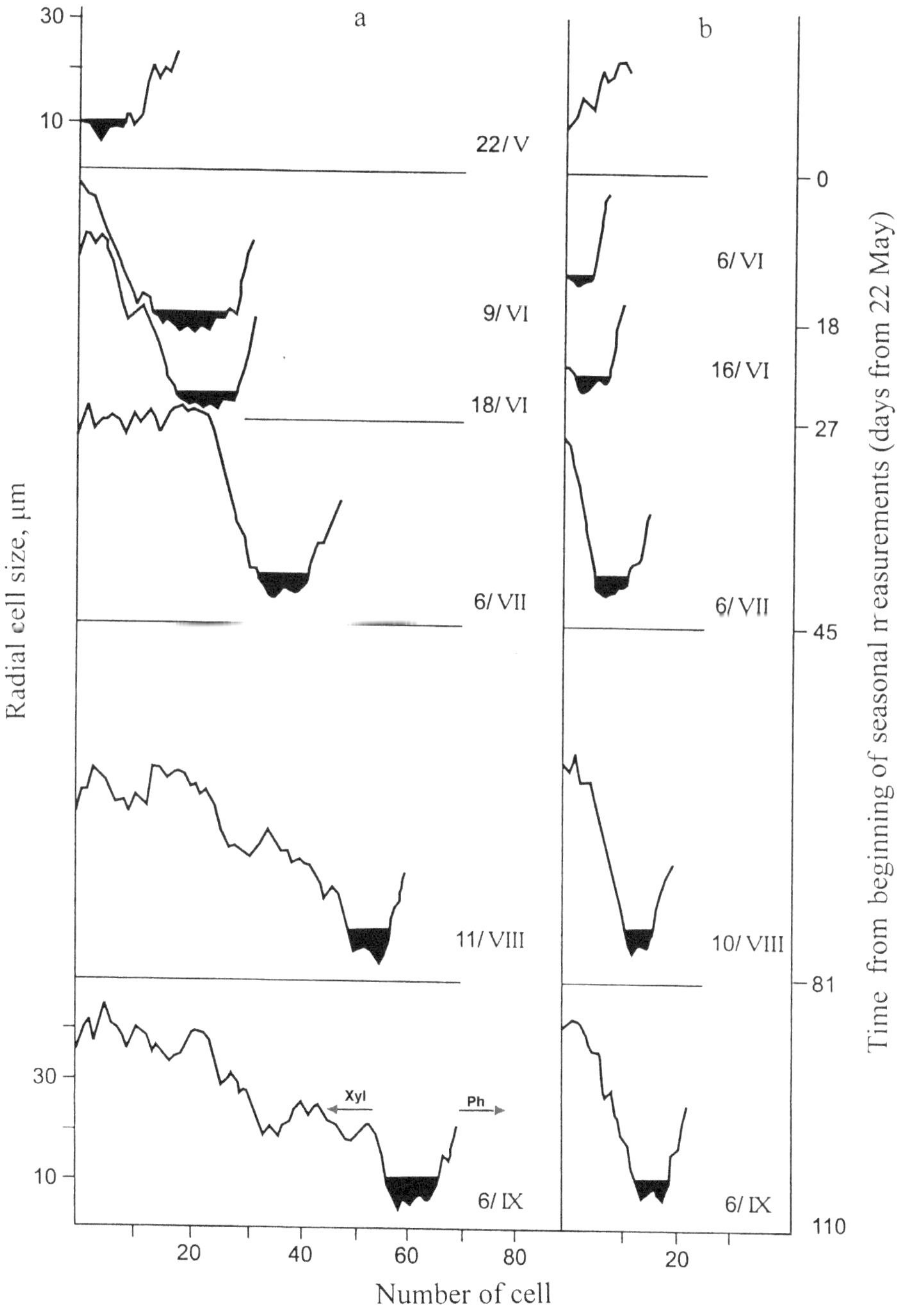

Fig. 4.11. Tracheidograms of pine tree rings being formed at different times in a growth season: **a** rapidly growing tree, **b** slowly growing tree (*shaded area* shows the cambial zone). Sample dates are shown to *right* of each tracheidogram. *Ph*, *Xyl* indicate the position of new phloem and xylem cells produced, respectively

mother cells in the cambial zone) predetermines the future final radial tracheid dimension. Many years of observations of seasonal growth of tree rings in several conifer species (*Pinus sylvestris, P. sibirica, Picea obovata, Abies sibirica*) in the southern taiga and taiga–steppe zones in Siberia (Vaganov et al. 1985, 1992; Sviderskaya 1999) provided material for direct and indirect comparisons of several kinetic characteristics of tree-ring formation and the radial dimension of tracheids formed. The kinetic approach developed and used by Whitmore and Zahner (1966) and Skene (1969) was applied. The specific growth rate is calculated from the rate of cell production and the number of cells in the cambial zone. The position of cells at the point of leaving the cambial zone to commence enlargement is determined in the tracheidogram according to the curve of the increasing number of cells in the maturation and enlargement zones (see explanations in Fig. 4.4). A typical example (Fig. 4.5) shows the correspondence between the kinetics of the specific growth rate and the final radial tracheid dimension for cells produced by the cambial zone at a particular time in the season, with the corresponding measured specific growth rate. This temporal correspondence leads to a close statistical relationship between the specific growth rate and the radial tracheid dimension (Fig. 4.12). It is interesting that the minimal specific growth rate that corresponds to the minimal cell size is about 0.038 per day, which is equal to the duration of a cell cycle of approximately 26 days. This value is close to the approximate duration of the cell cycle for a cambial initial and the first xylem mother cell (see Chap. 3).

The main indirect evidence for relationships between the kinetic characteristics of cell production (in the cambial zone) and radial tracheid dimension comes from measurements of tree-ring width (and the number of cells) and tracheidograms of tree rings growing in extreme climatic conditions. In all our measurements, we clearly observed the influence of annual xylem production (both in the ring width and the number of cells) on average values of radial tracheid dimension for rings narrower than 0.5 mm (see Fig. 2.36; Vysotskaya and Vaganov 1989). The narrow rings were produced under unfavorable climatic conditions in certain years, and if these conditions suppressed production of tracheids, then suppression of their radial dimension immediately followed. For narrow rings, the specific growth rate of tracheid production is low because of the narrower cambial zone and active xylem mother cells have a longer cell cycle. If the climatic suppression is very high, then the cambial zone produces only one or two cells during the growing season. These cells have a radial dimension close to that of typical latewood cells but cell walls with the same thickness as earlywood tracheids. Such rings (known to dendrochronologists as "micro-rings") are often formed in trees near latitudinal or elevational timber lines and their recognition can be difficult because they are in contact with typical latewood cells formed in the tree ring of the previous year. According to the calculation of the specific growth rate and average cell cycle during the production of these tracheids, it seems

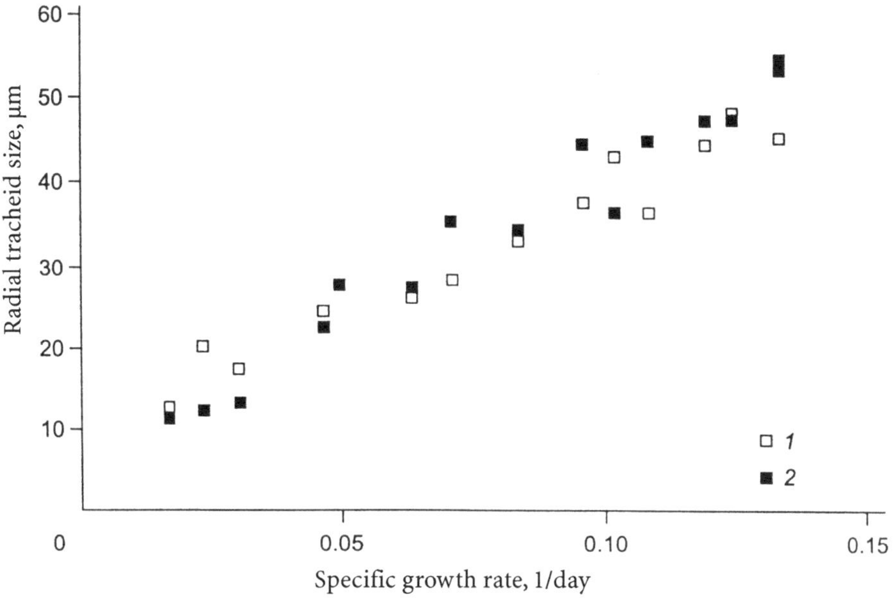

Fig. 4.12. Relationship between specific growth rate and radial size of tracheid being formed for Scots pine (*1*) and Siberian fir (*2*) tree rings

to be very close to the cycle of the initial because the radial dimension is close to the minimal value: 7–8 µm. This indicates that the production of new cells in those extreme years results from the division of only a few cells in a narrow cambial zone, probably by one or two divisions of the first xylem mother cell in a position close to the initial.

The arguments we have made for the primary role of the rate of cell division in the cambial zone and the ultimate radial dimension of the resulting tracheids are apparently contradicted by work based on the study of seedlings, as opposed to mature trees (Larson 1962, 1964; Richardson 1964; Gordon and Larson 1968). We suggest that the comparison may well be invalid, for three primary reasons. First, the range of tracheid dimensions in conifer seedlings is often much narrower than in the wood of mature trees (see, for example, our Table 2.2). This makes it much harder to detect the relationships we discuss here. Second, the tissues used in many studies are from close to the apical meristem, and so are not analogous to the situation on the main stem. Third, a number of these analyses refer to primary, not secondary, xylem.

4.5 Some Consequences of the Relationship Between Growth Rate and Radial Tracheid Dimension

The relationship between the radial tracheid size and the specific growth rate in the cambial zone permits, subject to some simple assumptions, the reconstruction of the kinetics of the seasonal course of tracheid production in a developing tree ring.

Because:

$$R_{SPL} = (1/N_C) * (dN/dt) \quad (4.4)$$

then:

$$dN/dt = R_{SPL} * N_C \quad (4.5)$$

According to the results shown in Fig. 4.12, we may assume that $D^i_R = a^* R_{SPL}$. For the n-th tracheid in the file, it is possible to define R_{SPL} as:

$$R_{SPN} = (1/(t_n - t_{n-1}))/N_C = D^n_R / a \quad (4.6)$$

where t_n and t_{n-1} are the dates when the n-th and $(n-1)$-th tracheids left the cambial zone. Assuming that the average number of cells in the cambial zone changes little during the season (a strong assumption corresponding to the model of Wilson 1964), for coefficient a we may write:

$$t_P * N_C = a * \sum_{n=1}^{N} 1/D^n_R \quad (4.7)$$

where t_P is the period of new cell production by the cambial zone.

Thus, to reconstruct the specific growth rate and rate of tracheid production, we have to define the radial tracheid diameter, the number of tracheids produced, and the average size of the cambial zone. As there exists a strong relationship between the average size of the cambial zone and the number of cells produced ($N_C = k^* N$; see Chap. 3), the next equation may be written:

$$t_P * N = a_1 * \sum_{n=1}^{N} 1/D^n_R \quad (4.8)$$

where $a_1 = a/k$.

The calculations which were made for data from the seasonal formation of tree rings in the southern taiga (Vaganov et al. 1985) show clearly that a_1 is a constant value which does not depend on the duration of growth and the total annual xylem production (the number of cells in tree rings). Using Eq. 4.8 and the values of a_1 and t_P, it is possible to define the date of each tracheid's production by the cambial zone from the measured tracheidogram (Vaganov et al. 1985, 1990). For each subsequent tracheid in the file, we may calculate the date accordingly:

$$t_n = a_1 /(N * \sum_{i=1}^{n} 1/D_R^i) \quad (4.9)$$

$$n = 1,\ldots,N$$

or combine with Eq. 4.8:

$$t_n = t_P * (\sum_{i=1}^{n} 1/D_R^i)/(\sum_{i=1}^{N} 1/D_R^i) \quad (4.10)$$

Tracheidograms of two tree rings with different internal structures are shown in Fig. 4.13: the first tracheidogram contains a small portion of earlywood tracheids and increased zones of transition and latewood tracheids (which is characteristic of larch tree rings from moderate–moist conditions) and the second tracheidogram contains a wide earlywood zone and a small number of transitional and latewood tracheids (which is characteristic of larch tree rings from moist conditions). Figure 4.13b shows the calculated seasonal growth curves which correspond to these tracheidograms (in the case of a 150-day season of tracheid production). One can see that, in the first case,

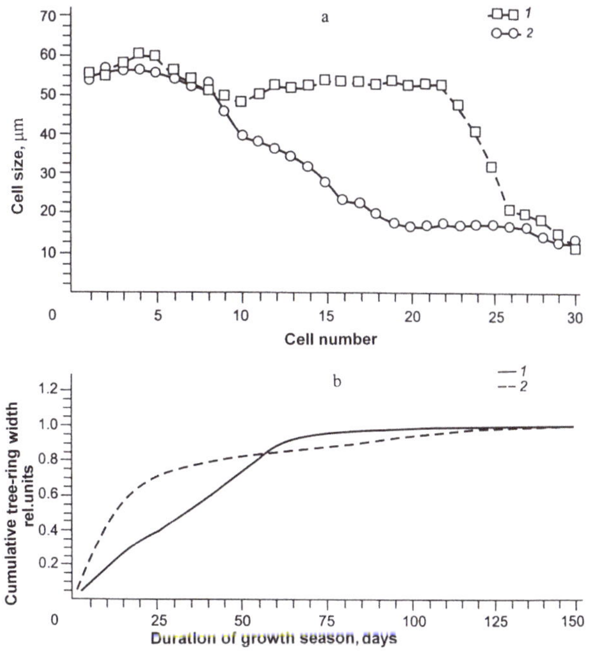

Fig. 4.13. Tracheidograms of tree rings with different late- (*1*) and earlywood (*2*) widths (**a**) and seasonal growth curves of tree rings (**b**)

about 70 % of tracheids in the tree ring were produced by the cambial zone during the first month of the season, whereas in the second case the rate of the tracheid production was approximately constant during 2.5 months. The features of the tracheidograms are repeated in the seasonal kinetics of cell production.

The constancy of the number of cells in the cambial zone is a strong assumption, which can be observed in certain limited cases. The argument made in Chap. 3 indicates that the size of the cambial zone changes during a season and these changes are related to the variations of specific growth rate in the cambial zone. In the simplest cases, the variations in the number of cells in the cambial zone may be described as its dependence on the growth rate:

$$N_C = b * R_{SPL} \tag{4.11}$$

or:

$$Nc = f(R_{SPL}(t + \tau)) \tag{4.12}$$

where is the time lag which corresponds to the average "lifetime" of a cell in the cambial zone. In such a case, the calculated seasonal growth curve is significantly different from those calculated when it is assumed that average N_C is constant. Figure 4.14a,b shows the growth curves for these two options, based on the same tracheidograms. If the number of cells in the cambial zone is linearly related to the specific growth rate, it leads to accelerated tracheid production during the second part of the season (Fig. 4.14a). If the number of cells in the cambial zone lags, the specific growth rate changes and there will be slower production of tracheids in the first part of the season.

Several examples of the use of tracheidograms to reconstruct seasonal growth curves and the rate of cell production have already been published (Vaganov et al. 1985, 1992; Shashkin and Vaganov 1993; Vaganov and Park 1995). All these examples not only showed good agreement between the calculated and measured or otherwise estimated growth rates, but also showed that the anatomical structure fixed in the radial tracheid sizes could be a replica of the seasonal variations in the intensity of the growth processes of xylem differentiation and thus they record information about the changing environment.

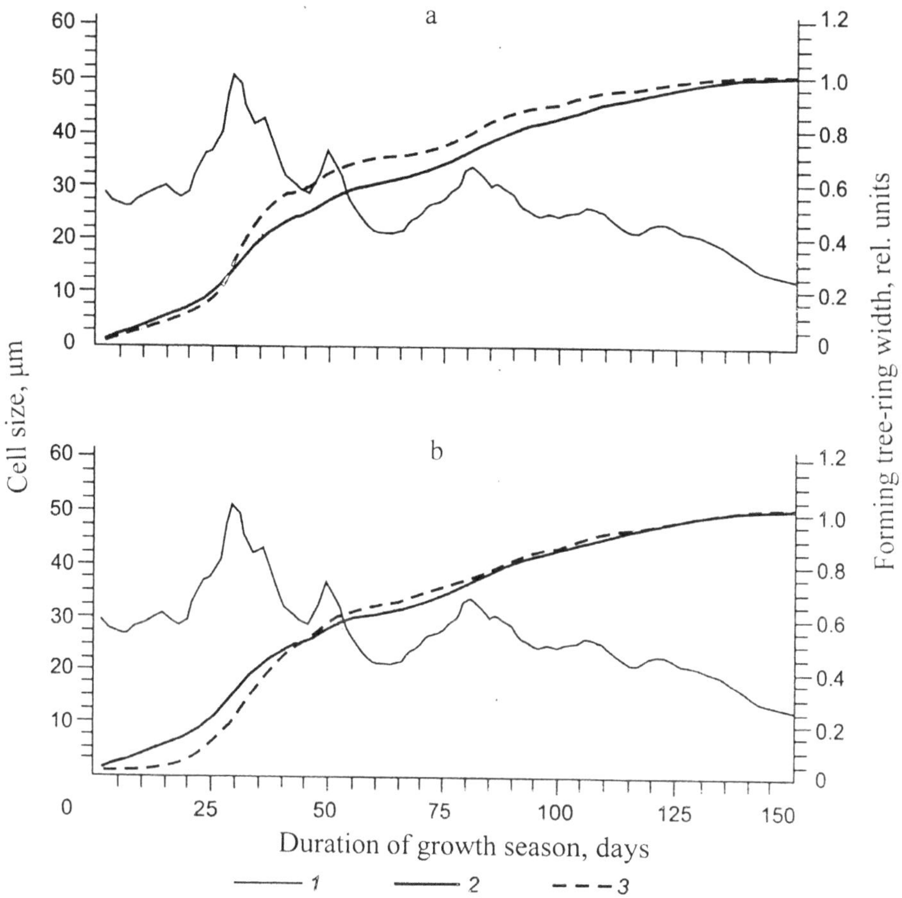

Fig. 4.14. Tracheidogram of annual ring (*1*) and calculations of seasonal growth curves (*2*, *3*) for different variants of specific growth rate and cell number in cambial zone. See text for further explanation

4.6. Conclusions and Discussion

Much may be learned from the study of the kinetics of radial cell enlargement in non-experimental conditions. Methods for doing this have been discussed, along with observations of the relationship between cell division and cell enlargement. Evidence derived in these ways has been applied to the following question: are these processes completely independent or is there a strong influence of the division process on the kinetics of enlargement and its final result, radial tracheid size? Evidence has been presented for such a relationship and concerning its intensity and robustness.

Several lines of evidence conflict with the idea that the environmental control of the final radial dimension achieved by a tracheid acts on the enlargement stage. It is however possible to explain this evidence if this control acts on cell production, that is, it is effective in the cambial zone, not in the enlargement zone.

The evidence for the relationship between the growth rate of cambial cells and the final radial dimension of the tracheids they produce is obtained by means of a kinetic approach. It needs to be tested by other direct and indirect sets of data, because not all the questions that arise have been answered. Most questions come from the enormous volume of research on the hormonal control of wood formation (see Zimmermann 1964; Barnett 1981; Creber and Chaloner 1984; Savidge 1996; Kozlowski and Pallardy 1997). There are many examples of the specific and non-specific effects of existing hormones [auxin, indole-3-acetic acid (IAA), gibberellin, ethylene, and others] on the production rate of tracheids and their size. As discussed earlier in this chapter, some of this work is based on saplings and may be of limited applicability to mature trees in natural stands.

We do not believe these data are necessarily contradictory to the statements made above. The first main argument is that most of the hormone research was made using the hormones as artificial additives which can change the "normal" mechanism of the relationships of the two basic processes of tracheid differentiation. The next argument is that hormonal control of wood formation has not yet produced a quantitative theory allowing the description of the observed effects of external factors on wood formation. The kinetic approach, with obvious restrictions, allows us to do this. Another argument (but not the last) is that, for a better understanding of the processes of wood formation in relation to changing environment, we urgently need to design such a quantitative theory, even if we are not able to include all known mechanisms. This is because it is essential to study the response of wood formation in different tree species to the natural range of growth conditions on a quantitative basis. The quantitative approach may be used as a tool to understand dendroclimatic and dendroecological results, because these results are completely quantitative.

We do not know the precise mechanism by which the kinetics of cell production and growth in the cambial zone determine the ultimate radial dimension of a tracheid. Recent work on root development may, however, indicate the kinds of mechanisms that may be involved. Baluska et al. (1994, 1996, 2001) hypothesized and substantiated the existence in growing roots of a so-called "transition zone" between the root meristem and the elongation zone. The main significance of this "transition zone" is as a sensory zone of the root that monitors diverse environmental parameters and effects appropriate responses (Baluska et al. 1996, 2001). For example, gravistimulation significantly changes the distribution of the relative elemental growth rate pattern along the growing root (Mullen et al. 1998).

The real mechanisms are still unclear. Perhaps they operate through interactions of the gene expression responsible for the control of coordinated growth processes with hormones and growth regulators (Savidge 2000; Baluska et al. 2001). For our purposes, the definition of such a special zone is very likely. This is because the results we have presented here lead us to expect the existence of some special mechanisms of external control of cell production and enlargement at the edge of the cambial zone. We can only hypothesize that such mechanisms are related to growth rate and "movement" of cells through the cell cycle, especially through the G_1 phase. These mechanisms would provide the link between growth rate near the edge of the cambial zone and further radial enlargement. Using a simplified scheme (i.e. that the division rate at the edge of the cambial zone controls the ultimate radial diameter of the tracheid) clarifies the positional control of cell growth within the cambial and enlargement zones. Both zones (dividing cells and enlargement) are characterized by a high rate of primary wall expansion (elongation), although the second stage has a higher rate of linear growth.

The importance of auxin in this process is clear (Rayle and Cleland 1992; Casgrove 1993; Brett and Waldron 1996). Precise determinations of IAA distribution within growing xylem and phloem show that IAA can be the hormonal signal for the positional control of cell growth (Uggla et al. 1996, 1998, 2001). The results clearly indicate that IAA concentration is higher within the cambial zone with dividing cells and decreases to zero at the end of the zone of enlargement (Uggla et al. 1996, 1998, 2001). We have not entered the discussion on the origin and maintenance of a constant level and even the total value of IAA (is its control external, from the shoot or internal – from the cambial zone itself?). Such a picture of the distribution of one of the main hormones that is closely related to cell wall growth and cell growth regulation supports some of the assumptions we have made in a way we cannot get from direct evidence.

Uggla et al. (2001) describe graphically the generalized distribution pattern of IAA, carbohydrates, and sucrose-metabolizing enzyme (SuSy) activities across the cambial zone and the zones of enlargement and maturation. If we compare the measured concentration of IAA from experiments by Uggla et al. (1996, 1998, 2001) and the specific growth rate from our evaluations, we see that the maximal linear extension growth rate of cells in the enlargement zone coincides with decreasing IAA concentration (Fig. 4.15). The sucrose concentration decreases in the same direction; and SuSy activity increases in the enlargement zone and falls during cell wall thickening. All the main processes involved in the radial growth of developing tracheids and the formation of the secondary wall are coordinated by the position of the cell in the file. The simplest explanation for the IAA pattern is that the expanding cells that have left the cambial zone use up the amount of IAA produced within the cambial zone by the dividing cells.

This implies that IAA mediates the positioning signal from the cambial zone to the enlargement zone. IAA is also under the higher level control of the

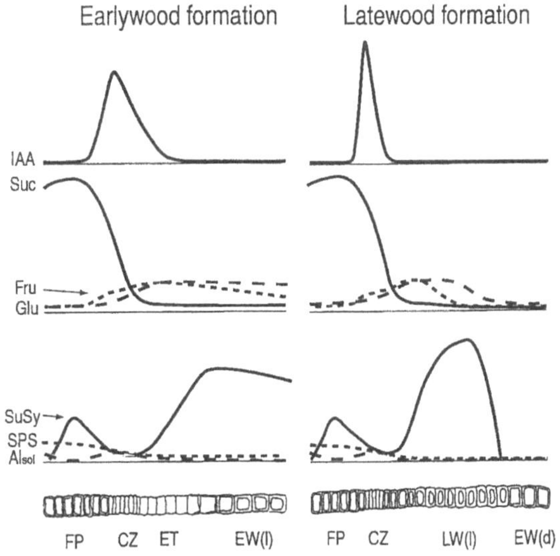

Fig. 4.15. Schematic drawing describing the generalized distribution patterns of IAA, carbohydrates, and sucrose-metabolizing enzyme activities across the cambial meristem and its differentiating and mature derivatives in early- and latewood formation. *FP* Functional phloem, *CZ* cambial zone, *ET* expanding tracheids, *EW(l)* maturing (living) earlywood tracheids, *EW(d)* mature (dead) earlywood tracheids, *LW(l)*, maturing (living) latewood tracheids (reprinted with permission from Uggla et al. 2001; copyright American Society of Plant Biologists)

shoot (and root) meristems and so it can be the mediator of the external control of cell growth within the cambial and enlargement zones. The main differences between early- and latewood formation are related to seasonal changes in illumination, temperature, and water supply, so the right and left patterns on the diagram (Fig. 4.15) indirectly show the results of environmental control of wood formation. According to the diagram, the main pattern remains very similar through the season. Quantitative variations are mostly related to the width of the cambial zone and the position of cells. This indirectly supports the hypothesis defining xylem formation as a partially independent system after the external signal from higher levels of regulation has been accepted.

5 Cell Wall Thickening

5.1 Introduction

The last stage of differentiation of the xylem elements that form the water-lifting system of the plant is characterized by completion of a rigid secondary wall with the consequent autolysis of the protoplasm. The secondary wall contains cellulose microfibrils, xylan, protein, and lignin. They provide a strictly ordered structure on the exterior of the cell membrane. The three main layers in the secondary wall are distinguished by the orientation of cellulose microfibrils (Preston 1974). First there is S1, then a main, much thicker layer, S2, in which cellulose microfibrils are oriented along the axis of the cell and frequently display spiral structures. Then comes S3, a layer that is absent in compression wood. The completion of the secondary wall involves a complex of intracellular processes and systems: the endomembrane system for transport, specialization of certain areas of the cytoplasmic diaphragm and elements of the cytoskeleton, expression of new genes, activation of numerous enzymes, and biophysical processes connected to between-cell gradients, properties of membranes, and organization of the cell wall (Catesson 1994; Demura and Fukuda 1994; Fukuda 1994; Savidge 1996).

We introduce the seasonal course of cell wall thickening (Sect. 5.2) and give particular attention to the experimental investigation of compression wood formation (Sect. 5.3). Such experiments have the special advantage, for our purposes, of revealing the timing of responses in cell wall formation to specific environmental changes. The relationship between radial tracheid dimension and the final cell wall thickness that a tracheid achieves is explored (Sect. 5.4) and some basic conclusions are reported and discussed (Sect. 5.5).

5.2 Seasonal Course of Cell Wall Thickening (Process and Basic Results)

The formation of a secondary wall is observed just before the termination of the radial growth of tracheids (Gamaley 1972). Lignification of the cell wall happens a little later, beginning with the middle lamella and primary wall and then reaching the internal layers (Barnett 1981; Timell 1986). It is not yet clear whether the termination of growth of the tracheids is an outcome of the reduction of elasticity of the primary wall and beginning of completion of the secondary wall or if it is dictated by other signals.

The completion of the secondary wall with its consequent lignification can be considered as the final stage in the biogenesis of the cell wall, which happens continuously during the closing stage of tracheid differentiation. Actually, the tangential and radial walls of cambial cells represent two levels in the process of maturation of primary cell walls (Catesson 1990, 1994). Tangential walls have a more rigid polysaccharide matrix in comparison with radial walls, the chemical composition and ultrastructure of which arises from the mechanical properties radial walls need for subsequent radial growth (Roland 1978; Catesson and Roland 1981). The initial heterogeneity of cambial cell walls disappears during the first stage of tracheid maturation, when the radial growth of cells is completed. It is supposed that the ratio of synthetic rates of different types of polysaccharide and their selective inclusion in radial and tangential walls predetermines the fate of the cells (Catesson 1990, 1994). For example, the earlier beginning and fast rate of xylan synthesis, mainly included in radial walls, can increase radial growth and result in the formation of large tracheids or vessels. In contrast, the synthesis of cellulose reduces an initial non-uniformity of the wall and so hinders radial growth (Catesson 1990, 1994).

It is known that the angle of microfibril orientation in the secondary wall depends on the size of the cell (Preston 1974). The correlation between the shape of the cell and the biosynthesis of microfibrils is common for many types of cells, including vessels and phloem elements. In thin and long cells of any type, microfibrils are placed at a smaller angle to the vertical axis of the cell. Therefore, the structure of the secondary wall can be pre-determined by the development of the cell wall (Barnett 1981).

All mature xylem elements (except parenchyma cells) have a thick secondary cell wall. However, thickness can noticeably differ in tracheids formed at different times in the growth season. In earlywood tracheids, the cell wall is noticeably thinner (1.5–3.0 μm). In latewood tracheids, the cell wall thickness can reach 7.0–8.0 μm. As well as tracheid sizes, the thickness of a cell wall can vary greatly, especially in latewood cells in various tree rings and in various parts of the tree (Zahner 1968b; Larson 1969; Creber and Chaloner 1984; Vaganov et al. 1985). The greater cell wall thickness compared with tracheids differentiated at the beginning of the growing season is one of the main criteria that define latewood. Until now, the issue has not been solved as to how the transition from the formation of earlywood to the formation of latewood is regulated. In any case, such a transition is connected with the control of secondary wall synthesis. External and internal factors can influence the formation of the secondary wall. Experimental research shows that seasonal change of external factors, such as light exposure, photoperiod, water deficit, nutrients, or temperature influence both the quantity of latewood cells and the thickness of their cell walls (see Brown and Sax 1962; Wilson 1964; Zahner 1968b; Larson 1969; Denne and Dodd 1981; Creber and Chaloner 1984; Vaganov et al. 1985; Carlquist 1988a, b; Downes and Turvey 1990; Antonova and Stasova 1993; Lev-Yadun and Aloni 1995; Savidge 1996). The ratio between the number, the size, and the thickness of xylem cell walls formed at the same time in different parts of a tree changes also: in young shoots, in branches, in the main stem, in the roots. Thus, the system of regulation at the level of the whole tree can be modified to a certain degree by local conditions in different parts of the tree.

In early works conducted by Denne (1971a), Wodzicki (1971), and Skene (1972), it was pointed out that the rate of cell wall deposition varied relatively little within a growing season. For example, in spite of differences in the growth rates of individual trees (different tree vigor), the actual rate of deposition of cell wall material was about 0.1–0.2 μm^2/day and seemed to show lit-

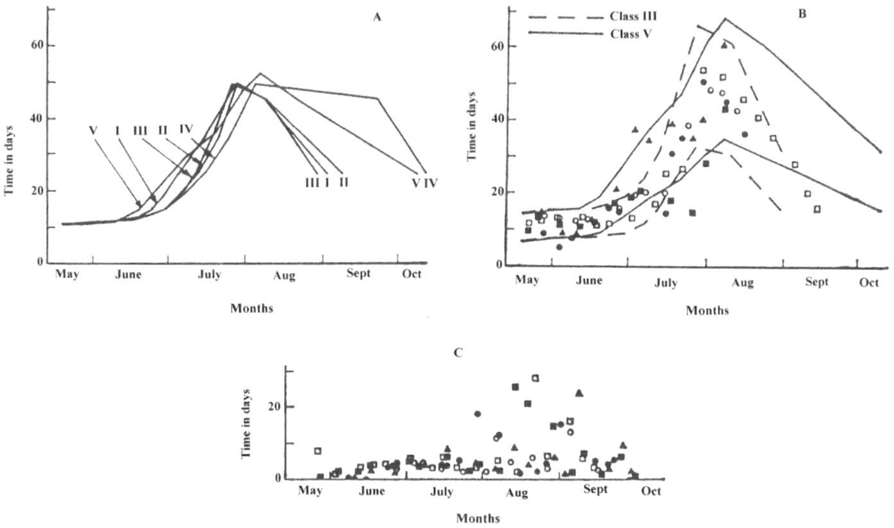

Fig. 5.1. Calculated (**a**) and observed (**b, c**) time periods for completion of secondary wall deposition (**a, b**) and for lysis of cytoplasm (**c**) against time of the season in trees of *Tsuga canadensis* of different vigor classes (I–V). The curves represent calculated times, assuming constant daily deposition rates of 0.1 $\mu m^2/\mu m$ and 0.2 $\mu m^2/\mu m$ (Skene 1972)

tle change during the course of the season in *Tsuga canadensis* (Fig. 5.1; Skene 1972). The time-period required for lysis of the cytoplasm was about 4 days, with no evidence of any changes with tree vigor (Skene 1972). For tree rings of Douglas-fir (*Pseudotsuga menziesii*), it was 4–5 μm^2/day and in trees with a well developed crown at the beginning of a season 6–7 μm^2/day (Dodd and Fox 1990). These authors compared the rates of radial growth and cell wall formation in young trees distinguished by development of the crown, and for different heights in the stem. The differences were not marked, except in the case of the growth rate in the second half of a season in trees with a well developed crown (Fig. 5.2).

The experimental data of Wodzicki (1971) showed that the rate of maturation has no clear influence on the radial size and thickness of the cell wall, and the main role is played by the duration of these stages of differentiation. The average value of the rate of cell wall deposition in *Pinus sylvestris* was a little

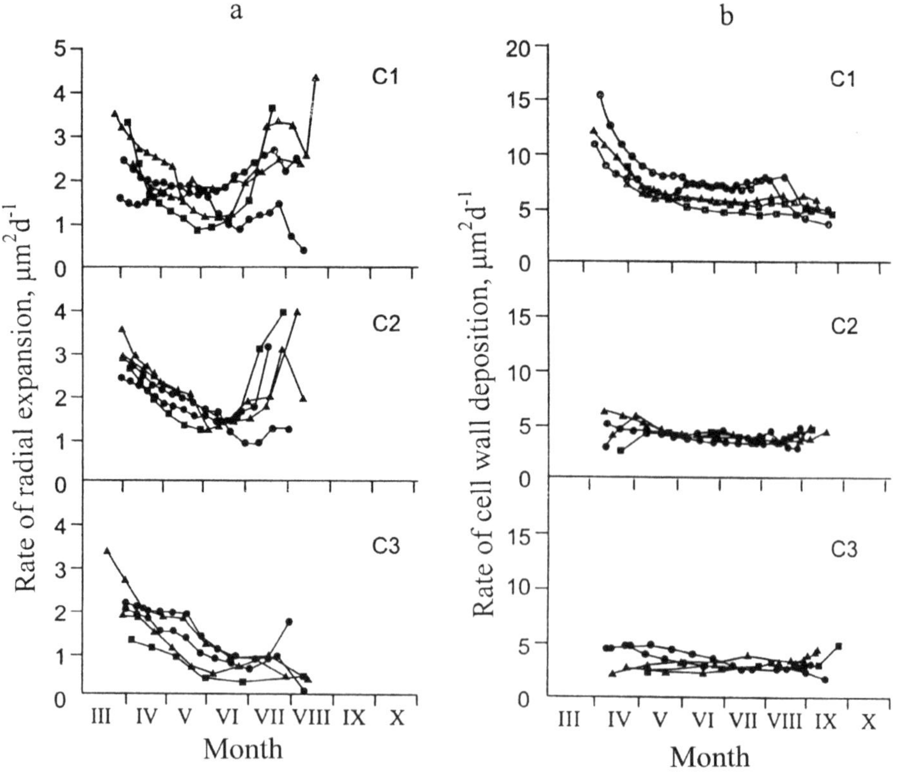

Fig. 5.2. Rate of radial expansion (**a**) and secondary wall deposition (**b**) of tracheids produced at different time in the season in annual rings of *Pseudotsuga* in trees with a different vigor and crown development (*C1*, *C2*, *C3*; Dodd and Fox 1990)

more than that in *Tsuga* and *Pseudotsuga*, reaching 5–10 μm²/day. Similarly, Sviderskaya (1999) obtained a rate of cell wall deposition of 5–7 μm²/day in observations of seasonal tree-ring formation in three coniferous species (*Pinus sylvestris, Picea obovata, Abies sibirica*). So, as in the case of the radial cell dimension of the tracheid, the experimental data on the kinetics of cell wall deposition indicate that the average rate of deposition of cell wall does not differ much during the growing season; and this supports the statement that the leading role in determining the final cell wall thickness is played by the duration of this process and by the radial size of the tracheid in which the cell wall was deposited.

5.3 Formation of Compression Wood in Experiments with Inclination

Yoshizawa and co-authors (1985a–c, 1986a, b, 1987; Yoshizawa 1987) made experiments on the internal mechanisms of tracheid development and the formation of compression wood in xylem, as affected by artificial tilting of young trees of several conifer species. They focused on the analysis of several anatomical characteristics of tracheids formed under the inclination effect: radial tracheid dimension, cell wall thickness, tracheid length, the number of pits, microfibril angle, and the position and angle of spiral perforations on the inner side of tracheids. Especially interesting results were obtained from periodic tilting of trees.

Figure 5.3 shows microphotographs of thin wood cross-sections of young trees of *Taxus cupridata* after periodic tilting. The lighter zones of wood correspond to tracheids of compression wood with thicker cell walls, the darker zones correspond to normal xylem. Data on cell production in normal and compression wood and the corresponding time intervals are presented in Tables 5.1 and 5.2. Tilting significantly increases the rate of tracheid production; and the production of compression wood is up to 2.0–2.5 times higher than normal in the same intervals of time. This is a common feature for compression wood production (Timell 1986). One important result is a rather sharp increase in radial dimension and cell wall thickness in the transition between the formation of normal wood and compression wood. Such sharp increases are most detectable in the measurements associated with short intervals (5 days) of tilting (Fig. 5.4). Under the influence of inclination in tracheids, a clear transition is formed from S-type to Z-type perforations in the inner cell wall sides. These are systematic changes in the orientation of the structural features of the inner layers of the tracheid cell wall. In compression wood, the S-type of perforation forms and in normal tracheids, the Z-type (Fig. 5.5). The transition zone from one type to the other caused by treatment or normal growth contains only three or four tracheids. Bearing in mind the

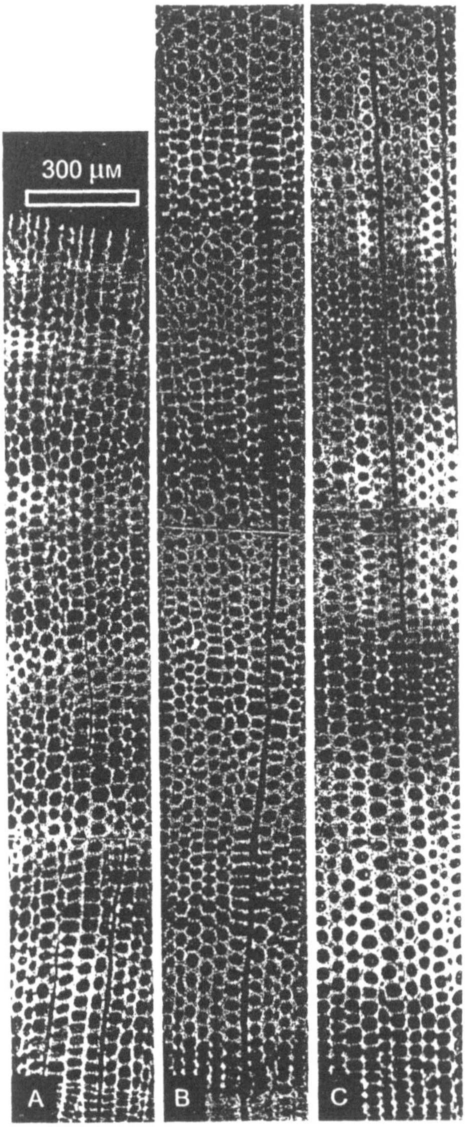

Fig. 5.3. Several compression wood arcs formed in the lower side of stems subjected to alternating positioning. **a** Sample tree 1, **b** sample tree 2, **c** sample tree 3. Note: compression arcs are recognized as bands of heavily lignified cells under a fluorescence microscope (Yoshizawa et al. 1985b)

peculiarities of cell wall formation in the tracheids of compression wood in relation to the timing of the tilting, Yoshizawa pointed out that:

1. The different stages of differentiation of tracheids have a different sensitivity to tilting. Only cambial cells accept the stimulus completely and develop typical compression wood tracheids. Only small modifications in cell wall structure occur in tracheids which are already in the enlargement or thickening zones when the treatment starts.

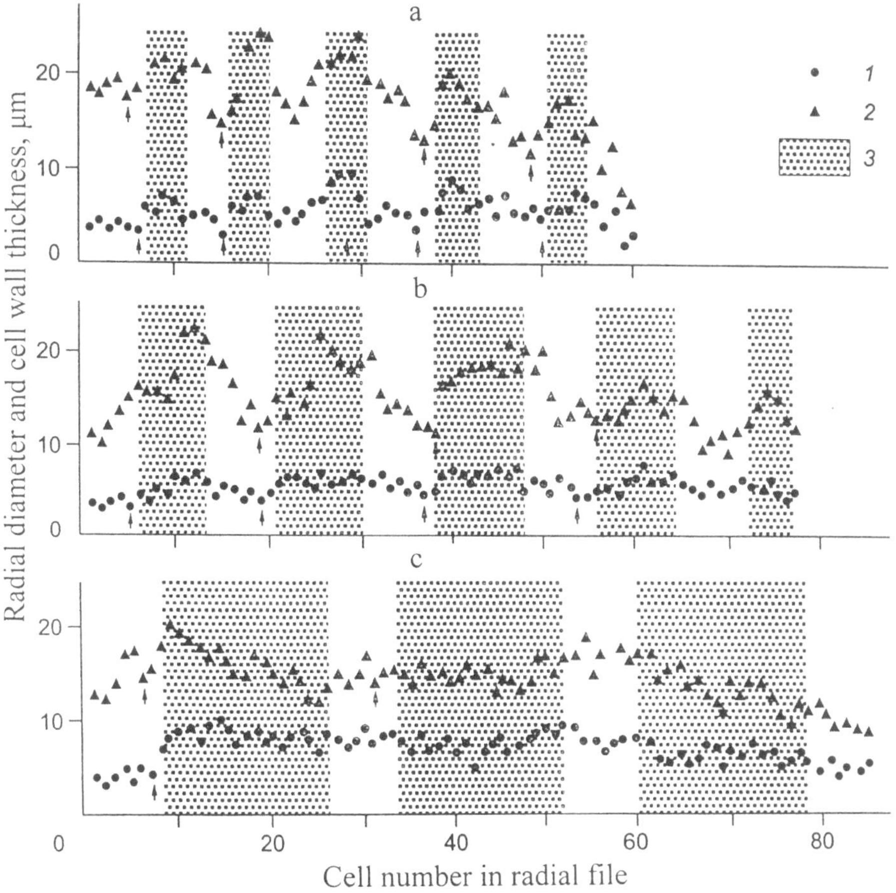

Fig. 5.4. Change in cell wall thickness (*1*) and radial diameter (*2*) within a growth ring. **a** Sample tree 1, **b** sample tree 2, **c** sample tree 3. *Single arrowheads* indicate temporal decreases in radial diameter and wall thickness, respectively, *shaded areas* are intervals of inclination (*3*) (Yoshizawa et al. 1985b)

2. There is a lag in time between the tilting that initiates compression wood formation and its effect appearing in the anatomical characteristics of the tracheids. This lag is caused by the need of tracheids to complete the last stages of cytodifferentiation (for example to pass through the formation of the S_1 and S_2 secondary wall stages).
3. After the cessation of tilting, only cells that are at that moment in the cambial zone develop into normal tracheids.

Thus, these results from the application of a stimulus which has a clear anatomical response (the increase in radial tracheid dimension and cell wall thickness) confirm the hypothesis of the existence of a key process which is affected at one stage of development, with consequences during the subse-

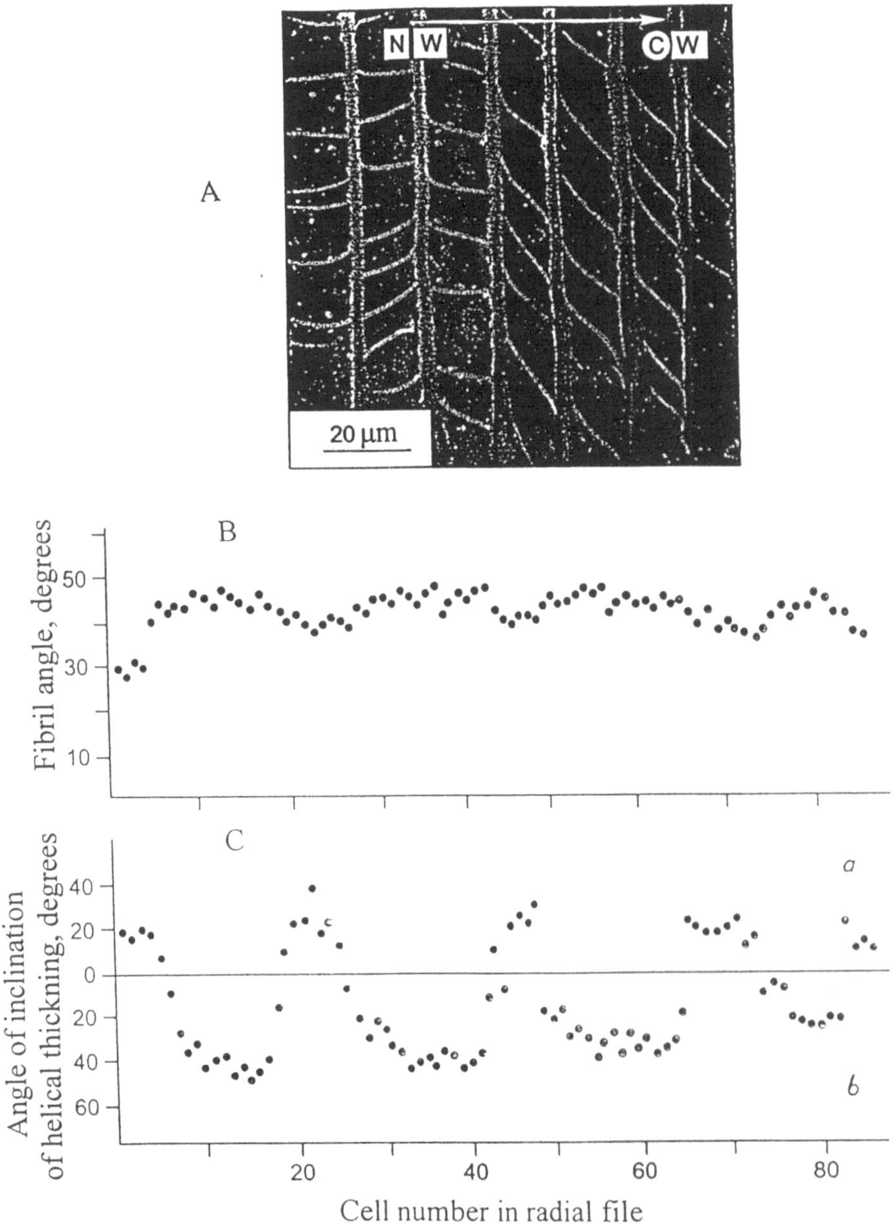

Fig. 5.5. a Scanning electron micrograph of the inner surfaces of transitioning from normal (*NW*) to compression (*CW*) wood. Note: *arrow* indicates the branched thickening crossing in innermost microfibrils where the S3 layer is absent. **b, c** Changes in the orientation angle of helical thickenings and the fibrillar angle of the S2 layer within an annual ring. Note: the turns in the direction of helical thickenings correspond to the changes of the stimulus of inclination (Yoshizawa et al. 1985b)

Table 5.1. Number of compression (*C*) and normal (*N*) tracheids formed at a height of 20 cm in stems subjected to alternating positioning (after Yoshizawa et al. 1985b). Note: inclination angle was 60 degrees, vertical periods were 20 days in all samples

Sample tree	Inclination period (days)	Cell type	Cell number				Total	Average
			1	2	3	4		
A	5	C	5	4	5	4	18	4.5
		N	4	6	6	6	22	5.5
B	10	C	8	9	11	9	37	9.3
		N	6	6	8	7	27	6.8
C	20	C	18	22	21	–	61	20.3
		N	6	8	6	–	20	6.7

Table 5.2. Number of tracheids formed per day (after Yoshizawa et al. 1985b). *C* Compression tracheids, *N* normal tracheids

Sample tree	Cell type	Compression zone number			
		1	2	3	4
A	C	1.0	0.8	1.0	0.8
	N	0.2	0.3	0.3	0.3
B	C	0.8	0.9	1.1	0.9
	N	0.3	0.3	0.4	0.4
C	C	0.9	1.1	1.1	–
	N	0.3	0.4	0.3	–

quent phases of tracheid cytodifferentiation. The cambial zone was revealed as the primary target of the influence. Note also that the increased radial tracheid dimension and the increased cell wall thickness in compression wood coincide with the increased rate of cell production in the cambial zone.

5.4 Relationship Between Radial Tracheid Dimension and Cell Wall Thickness

Usually the relationship between radial tracheid dimension and cell wall thickness is described as inverse. There are several rules to discriminate earlywood and latewood tracheids, based on the relationship of radial dimension (and lumen) to double wall thickness (Denne 1988). But more detailed measurements of single files of tracheids within tree rings show that the relationship is more complicated than this. The typical within-tree ring variations in radial tracheid dimension and cell wall thickness are presented in Fig. 2.30. In earlywood, variations in tracheid dimension correspond to weak negative changes in cell wall thickness. Ford and Robards (1976), analyzing short-term variations in tracheid dimension and cell wall thickness in earlywood of *Picea sitchensis* found that (pp 220–221): "the cross correlation function for cell diameter and cell wall thickness along the radial file illustrates that there is a general correlation between the two parameters: large cells tend to have thick walls and vice versa". Moving further along a file we see the transition zone, where the radial tracheid dimension decreases until the cell wall thickness increases. Then approximately in the middle of the latewood zone, the cell wall thickness reaches its maximum, decreasing after that towards the ring boundary as the radial tracheid dimension decreases. The changes in the last-formed tracheids are observed in the anatomy of tree rings in many conifers. A graph of the relationship between radial tracheid dimension and cell wall thickness shows two big and relatively separate groups of tracheids (see Fig. 2.38). One of these groups has slightly negative relations between the compared features, while the other shows a strongly positive relationship. The first group includes most of the earlywood tracheids and tracheids from the transition zone; and the second includes most of the latewood tracheids.

According to these relationships, we can write the following equations that statistically describe them:

1. for earlywood and the transition zone

$$L_W = L_W^{MAX} * \exp(-a * (D_R - D_R^{MIN})) + L_W^{MIN} \tag{5.1}$$

2. for the latewood zone

$$L_W = b * (D_R - D_R^0) + L_W^{MIN} \tag{5.2}$$

where D_R is the radial tracheid dimension and L_W^{MIN} is the minimal value of the cell wall thickness. From numerous data, the approximate value of L_W is close to 1.2–1.5 μm. This value seems to be associated with the minimal wall thickness of a complete secondary wall. D_R^{MIN} is close to the minimal tracheid size (7–8 μm). L_W^{MAX} and D_R^{MIN} are the theoretical values of the cell wall thickness and the diameter of the last transition zone tracheids if the slightly inverse relationship between the diameter and the cell wall thickness continues to the latewood zone. Note that there are strong limitations on the components of this equation. In the range where D becomes less than D_R^{MIN}, L_W may be more than half of the diameter, which is impossible. So, the next ratio must be:

$$L_W^{MAX} + L_W^{MIN} \leq D_R^{MIN} / 2 \tag{5.3}$$

In fact, the sum of these two components is less than half of D_R^{MIN}. Only for compression wood may the sum be close to 1/half of D_R^{MIN} because, for example, for tracheids with a diameter of 20 μm, L_W may reach 7–8 μm (Timell 1979).

The weak negative relationship between the tracheid diameter and cell wall thickness in earlywood and transition zones leads to linear relations between the tracheid diameter (radial cell size) and the cell wall area (Fig. 5.6). Cell wall area increases practically linearly with the tracheid diameter in both early- and latewood. Both approximation lines cross the abscissa at close to 8 μm and the ordinate close to 100 μm², which corresponds to data for tracheids with the smallest diameter and the thinnest cell wall. For a quantitative description of density variations according to these two relations we can write:

$$\rho_X = \rho_W * (A_W / D_R \times D_T) \tag{5.4}$$

in the case of earlywood and taking into account the relation in Fig. 5.6, it is:

$$\rho_X = \rho_W (9D_R + 28) / D_R * D_T \tag{5.5}$$

Assuming the average size for D_T is 35 μm (Vaganov et al. 1985; Vysotskaya and Vaganov 1989):

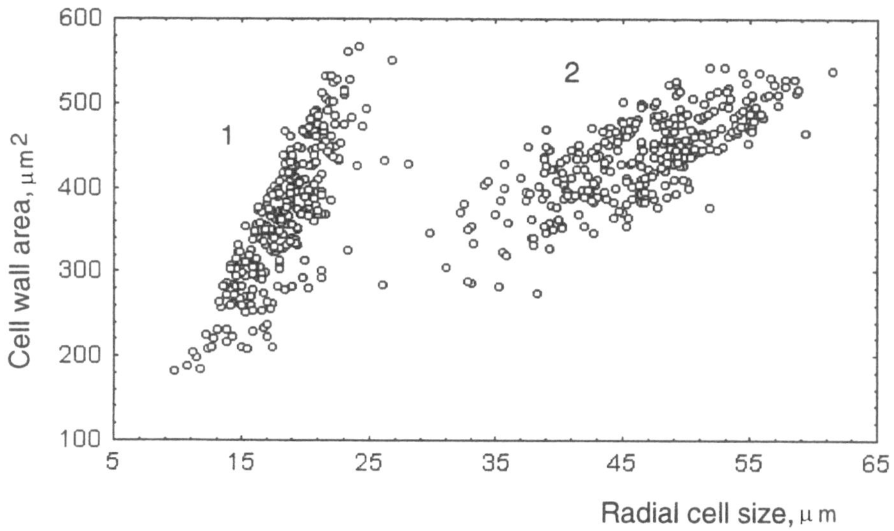

Fig. 5.6. Radial tracheid dimension versus cell wall area for earlywood (*2*) and latewood (*1*) tracheids in larch (*Larix Gmelinnii*, Silkin and Kirdyanov 1999)

$$\rho_X = \rho_W(0.8/D_R + 0.26) \tag{5.6}$$

In the case of latewood:

$$\rho_X = \rho_W(116 - 27D_R)/D_R * D_T \tag{5.7}$$

Assuming that the average size of D_T is about 35 μm:

$$\rho_X = \rho_W(0.64 - 2.43/D_R) \tag{5.8}$$

ρ_W is the density of wood material which is constant and equals 1.4 g/cm^3 (Nikitin 1962; Schweingruber 1988). The relationship between density and tracheid diameter for early- and latewood (Eqs. 5.6, 5.8) is shown in Fig. 2.38b. The density in the latewood zone increases with increasing tracheid diameter. The density in the earlywood zone decreases with increasing tracheid diameter. The calculated curves are very similar to real measurements of wood samples from sites near the northern timber line. The analysis described above shows the strength of the relationships between the main anatomical characteristics of wood in conifers.

It is also interesting to note that the average values of cell wall areas for earlywood are the same as for latewood. Benthel (1964) was perhaps the first to publish precise measurements of cell wall area changes in single files of tree rings of *Abies grandis* (Dougl.) Lindl. and *Pinus contorta* Dougl.: "It is suggested [that] the amount of cell-wall material that can be developed in maturation of a specific cell is a function of the parentage of the cell and is fixed for that cell at the time of cell division. It is further suggested that [the] quantity of cell wall is essentially the same for all daughter cells originating from a particular cambial initial within an annual ring" (p. 89 in Benthel 1964). Skene (1969) obtained the same results by measuring the volume of wall material in early- and latewood tracheids. This volume ranged from 8.2×10^{-3} mm^3 to 6.2×10^{-3} mm^3 in earlywood tracheids and from 8.9×10^{-3} mm^3 to 6.9×10^{-3} mm^3 in latewood tracheids. These data seem very similar to those presented in Fig. 5.6. This means that the quantity of "building material" for a cell wall does not change when the cambial zone transfers from production of earlywood tracheids to latewood tracheids, but is controlled by the tracheid diameter and initially by the rate of tracheid production in the cambial zone. These data, if applied to the concept of internal control of wood formation, can be interpreted as control of cell wall thickness through the control of tracheid diameter.

We agree with Ford and Robards (1976) who pointed out (p. 221): "From the point of view of cambial physiology, a most intriguing feature of the analysis is that variation in radial tracheid diameter is closely related to variation in cell wall thickness. The two developmental processes are physically separated in the growing ring, there is a lag in time as a cell passes through one stage to the next and yet the results are correlated. This suggests that the environment which influences wall thickening is that which exists early in the life of the tracheid and before the process of thickening has started, i.e., that cells move out of the process of expansion with the course of their wall thickening process largely determined".

5.5 Conclusions and Discussion

As with cell production and enlargement in the previous two chapters, the seasonal course of cell wall thickening has been explored, with a special emphasis on the experimental investigation of compression wood formation. This has the great advantage of permitting precise investigations into the timing of response to environmental stimuli. It reveals the cambial zone as the primary target of environmental control of cell wall thickening. We argue that final tracheid diameter is controlled through the rate of cell production, as is cell wall thickening, because variations in radial cell diameter and cell wall thickness are closely related.

Mechanisms for the control of tracheid radial expansion and hence, through the predetermination of final tracheid radial dimension, cell wall thickness may be mediated through control of the rates of synthesis of different components of tangential and radial walls.

The average rate of deposition of cell wall does not differ much during the growing season and supports the statement that the leading role in determining the final cell wall thickness is played by the duration of this process and the radial size of the tracheid in which the cell wall is deposited. As discussed in Chap. 4, this radial size is determined in the cambial zone.

Experiments on the effect of tilting on compression wood production indicate that the cambial cells are the main target of environmental influence on cell wall thickening.

The results shown on Fig. 2.38 may also be interpreted in terms of the influence of growth rate (that is, specific growth rate in the cambial zone) on cell wall thickness, if we take into account that tracheid diameter is closely correlated to specific growth rate (Chap. 4). According to this hypothesis, tracheids that are produced by the cambial zone with a high specific growth rate obtain more material to build secondary cell walls than those produced with a low specific growth rate. The next sequence of events is realized from a kinetic point of view: a high specific growth rate (high rate of cambial cell division) provides a high radial tracheid diameter after enlargement that, in its turn, provides the thicker cell wall because more material is supplied. The same sequence can be applied to the formation of a latewood tracheid because there is a similar linear relationship between the tracheid diameter and the cell wall area or biomass of latewood tracheids. The pathway of tracheid differentiation starts in the cambial zone, whose kinetic characteristics play a key role in subsequent stages of tracheid maturation. During the growing season, the combination of external factors changes and this affects the rate of tracheid production. So, it is easy to imagine that, in the first part of a season, the conditions are not favorable and the specific growth rate is less, which leads to the production of smaller earlywood tracheids with a smaller biomass. However, if conditions become favorable to a high specific growth rate during the second part of the season, then larger latewood tracheids with thicker walls are produced. This is the simplest path of environmental influence on wood formation. We will consider this topic in more detail.

The process of tracheid differentiation is, as a matter of fact, a process of implementation of the genetic program of differentiation, starting at the level of cambial cells and finishing with secondary cell wall formation (Fukuda 1996; Graham 1996; Hertzberg et al. 2001; Chaffey et al. 2002; Ito and Fukuda 2002; Goujon et al. 2003; Kirst et al. 2003; Nieminen et al. 2004). Schrader et al. (2003), for example, showed that the expression of specific members of the auxin transport genes are associated with different stages of vascular cambium development and demonstrated that trees have developed mechanisms to modulate auxin transport in the meristem in response to developmental and environmental cues. A variety of anatomical parameters of tracheids among trees and years indicates that the eventual result is not absolutely determined and that it depends on the local conditions where the differentiation occurs. The process of differentiation can be presented as a series of events, in which the duration and intensity of each stage depends on the previous one. Then, in changed conditions, if two events are carried on further in space and time, their deterministic relationship may change. If we identify processes most sensitive to the influence of external factors, we see that the external signal should be most clearly perceived by the cambial zone. These external effects will leave their mark on further processes of differentiation and, ultimately, the anatomical characteristics of tracheids. As a result, the direct influence of environmental conditions on the process of tracheid enlargement, as recorded in their final anatomical characteristics, will be significantly smaller. There are a number of indirect data that indicate that the events occurring during cell division in the cambial zone can have a strong influence on the ultimate sizes of tracheids. Already, at this early stage of differentiation, biochemical changes in the primary cell wall are necessary for the radial growth of cells and determination of its rate (Taiz 1984; Catesson 1990, 1994; Pritchard 1994).

The sensitivity of separate stages of differentiation to changing local conditions may vary. Larson (1964) supposed that it is possible to get practically any ratio between the tracheid sizes and their cell wall thickness in conifer tree rings by manipulating environmental conditions. The hypothesis that each stage of tracheid differentiation is affected by environment independently is strongly supported in several publications of Antonova and Stasova

(1993, 1997; Antonova et al. 1995). They ignored any relations between consecutive stages of tracheid differentiation reported by others (Ford 1981; Vaganov et al. 1985; Vaganov 1996a). In the next chapter, we consider the environmental control of wood formation and the results of the effect of environmental changes on the anatomical characteristics of tree rings.

6 Environmental Control of Xylem Differentiation

6.1 Introduction

The possibility of external influences operating independently on each stage of tracheid differentiation received considerable support in a number of papers published in the 1960s and 1970s (Larson 1962, 1963, 1964, 1994; Zimmermann 1964; Wodzicki 1962, 1971). This was mainly based on experiments with young trees (seedlings) under controlled conditions of temperature, light, and moisture (Richardson 1964; Denne 1971a, b, 1979; Barnett 1981). Larson (1964), based on the results of his experiments and a review of others, adopted the hypothesis that it is possible to obtain tracheids with different combinations of radial diameter and cell wall thickness by manipulating external factors like temperature, light, and moisture. Unfortunately, this hypothesis was not supported further by quantitative measurements, especially those of relations between growth kinetic parameters and anatomical features. The apparent independence of external control at each stage during tracheid differentiation could be a result of the known relationship between these processes, especially as the relationship between tracheid radial size and cell wall thickness has quite a complicated character (Chap. 5).

The hypothesis of independent control at each stage of tracheid differentiation (division, enlargement, wall-thickening) requires the operation of a complicated system for the remote control of differentiation, involving a combination of assimilates, growth hormones, and other growth factors (Savidge 1996). Contrast this with an arrangement in which each process in a tissue is regulated by a higher center in a hierarchy of control. Hierarchical systems in which a stimulus from a higher level manages a key process on a lower level, and then the subsequent processes at the lower level enact this stimulus, are more effective. The enactment of the genetic program is also simplified in such cases. So, in the case of xylem differentiation, the higher level of control is at the level of the whole tree and the lower level is that of tissue differentiation. The key process managed at the lower level is the process of cell growth in the cambial zone, as we have tried to show in previous chapters.

Here we consider a conceptual scheme of the environmental control of xylem formation (Sect. 6.2). We examine its relevance to an analysis of tree-ring formation under strong temperature-limitation (Sect. 6.3) and to the use of "differential tracheidograms" in the study of the effects of weather conditions on tree-ring formation within a season. The applicability of the scheme in the very different case of tree-ring growth in a monsoon region is explored (Sect. 6.5) and some basic conclusions are reported (Sect. 6.6).

6.2 Conceptual Scheme of the Environmental Control of Xylem Differentiation

A basic scheme for a physiological rationale for interpreting environmental effects on xylem differentiation was proposed by Denne and Dodd (1981). As a working hypothesis, they assumed that the effects of environmental factors may be either direct, or indirect through substrate availability or growth regulator balance (Fig. 6.1). Figure 6.1a shows a simplified version of the main effects of environmental changes; and Fig. 6.1b summarizes the complex variety of interacting physiological processes that would be expected to affect xylem differentiation. The central arrow in Fig. 6.1a shows all the direct effects; and the other arrows show different indirect effects on transpiration, translocation of assimilators, rate of photosynthesis, rate of growth regulator production, etc. According to the authors, "the model shown is greatly oversimplified; for example the box labeled 'shoot activity' includes a wide range of developmental processes (such as apical activity, production of leaf primordia, flowers and fruits, and stem elongation), each of which might be affected by environmental factors" (p. 239 in Denne and Dodd 1981).

This can be accepted only as a qualitative model. Even if it were desirable to transform this model into a scheme of calculation, it would be impossible to verify using experimental data, because: (1) many of the processes have no adequate quantitative description, (2) there are so many unknown coefficients that too few degrees of freedom are available for the model, and (3) the processes have different characteristic times (for example, the rate of respiration and photosynthesis changes within hours, but leaf area or nutrient uptake may be considered constant throughout a large part of the growing season).

Such a model might serve as a first step in constructing a quantitative model, based on some theoretical assumptions, which reduces the main uncertainties and highlights the main targets of environmental control. Concerning the environmental control of tree-ring formation, dendrochronological data give enough material to reveal the effects of climate on inter-annual variations of tree-ring width, density, and tracheid dimension, and to fit statistical relations between them. We illustrate this by several typical examples

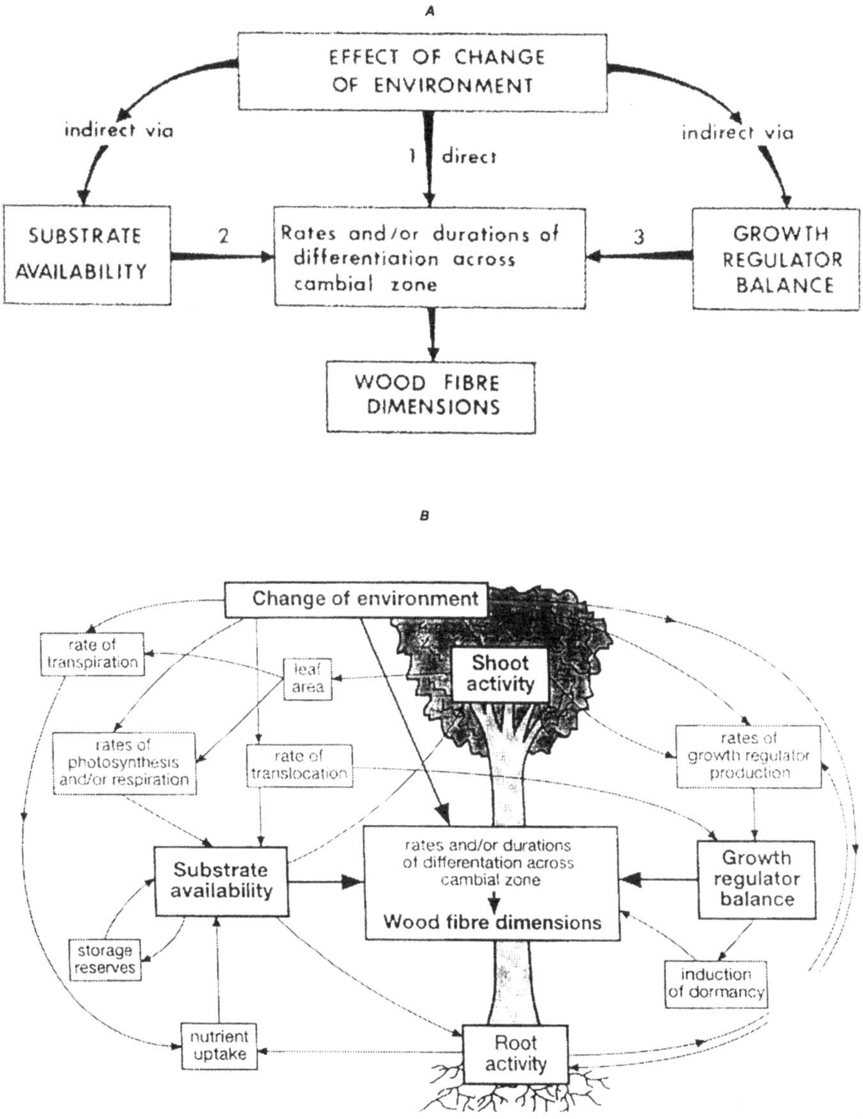

Fig. 6.1. Schematic diagram of the main effects of environmental factors on tracheid dimensions. **a** Interacting physiological processes affecting xylem differentiation. **b** At present, it is almost impossible to quantify individual parameters (Denne and Dodd 1981)

that lead to important conclusions for the further design of a simulation model.

6.3 Tree-Ring Formation Under Strong Temperature Limitation (Northern Timberline)

At high latitudes, all growth processes are under the control of the main limiting factor – temperature (Shiyatov 1986; Briffa et al. 1990, 1998; D'Arrigo et al. 1992; Vaganov et al. 1996a–c). Temperature is responsible for up to 70 % of tree-ring width inter-annual variability. The seasonal course of temperature has a strong effect on the seasonal kinetics of tree-ring formation and tracheid production (Kandelaki 1979; Kandelaki and Dem'yanov 1982; Vaganov et al. 1994, 1996a, c; Liang et al. 1997). It is interesting to consider the temperature effect on seasonal tree growth and the formation of tree-ring structure at high latitude sites, using measurements of width, density, and anatomical characteristics of annual rings.

Tree-ring measurements of conifers from the Siberian northern timberline were used (Vaganov et al. 1999; Kirdyanov et al. 2003) to define relationships between tree-ring growth and temperatures at different times in the growth season. In addition to tree-ring width, measurements of cell and density structure were used to characterize within-season details of ring growth. Cell size variations in tree rings characterize changes in tree growth rate during the season. Cell size, cell wall thickness, and density connected with these two parameters are characteristics of mass accumulation in the stem (Fig. 2.11). Annual mass accumulated in a tree stem can be defined as the product of two parameters: mean tree-ring density and tree-ring width.

Tree-ring width is connected with the number of cells produced during a season. Consequently tree-ring width variability characterizes variation in cell production. Cumulative cell production in a tree ring depends on the cell number in the cambial zone (Bannan 1955; Wilson 1964; Vaganov et al. 1985).

Material from six sites near the northern timberline was analyzed (Fig. 6.2). The choice of these sites was guided by the relative proximity of long records of daily temperature. Dendroclimatic analysis of standardized tree-ring width chronologies was carried out for four sites. For two other sites, the following chronologies of tree-ring structure parameters were used in addition to tree ring width: cell size, maximum latewood density, and cell wall thickness. Maximum latewood density was obtained from density profiles measured by means of X-ray microdensitometry, as described in Sect. 2.2.2. Chronologies of cell wall thickness variability were calculated according to maximum density and cell size in latewood chronologies. Cell chronologies were obtained from averaged individual cell files standardized to the same

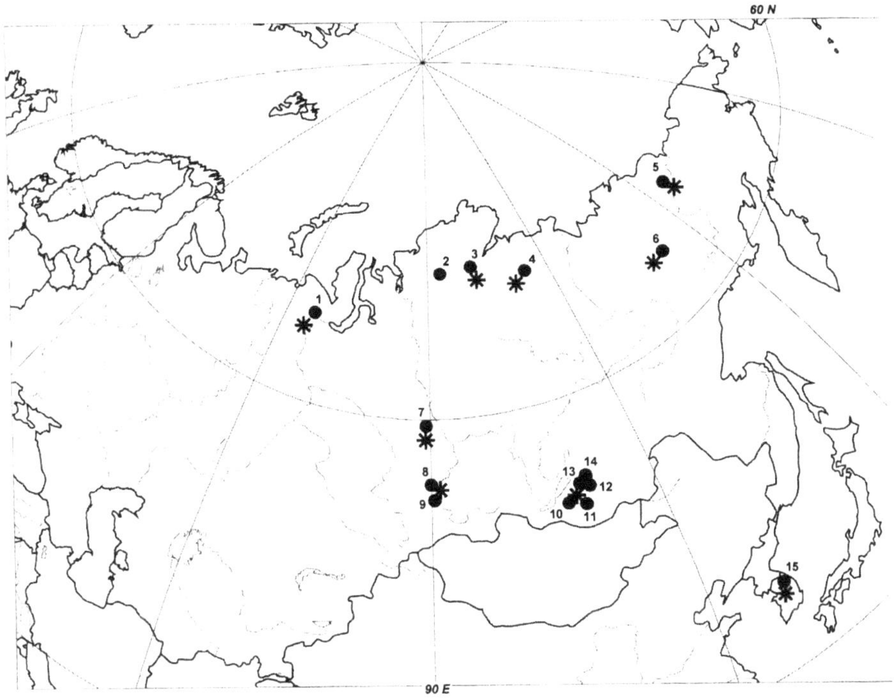

● Location of sites

✱ Location of nearest meteorological stations

1-SOB, *Larix sibirica*
2-KAM, *Larix gmelinii*
3-NOV, *Larix gmelinii*
4-ARY, *Larix gmelinii*
5-INP, *Larix cajanderi*
6-Verkhoyansk, *Larix cajanderi*
7-BOR, *Pinus sylvestris*
8, 9-Khakasia, *Pinus sylvestris, Larix sibirica*
10-14-Buryatia, *Pinus sylvestris*
15-Korea, *Pinus densiflora, Pinus rigida*

Fig. 6.2. Map of weather stations and site locations for which the tree-ring chronologies of width and structure of tree rings were used in Chaps. 6–8

cell number (15). Standardization of individual cell files was done to compare tree rings with different cell numbers. Cell size chronologies were obtained for the first cell in each tree ring (representing the beginning of the growth season), the fourth cell (earlywood formation), and the 12th cell (formation of latewood) (Fig. 6.3). Local chronologies were calculated by averaging measurements made for individual trees (12–19 trees for tree ring width, 15–20 for maximum latewood density, five for cell sizes) and chronology statistics were calculated (Table 6.1). Correlation coefficients of the chronologies with pentad temperatures (five-day blocks) were calculated. Long instrumental records (1936–1990) of daily temperature from the nearest meteorological stations (Berezovo, Khatanga, Olenek, Chokurdakh) were used. In addition, monthly temperature and precipitation data available for a longer period

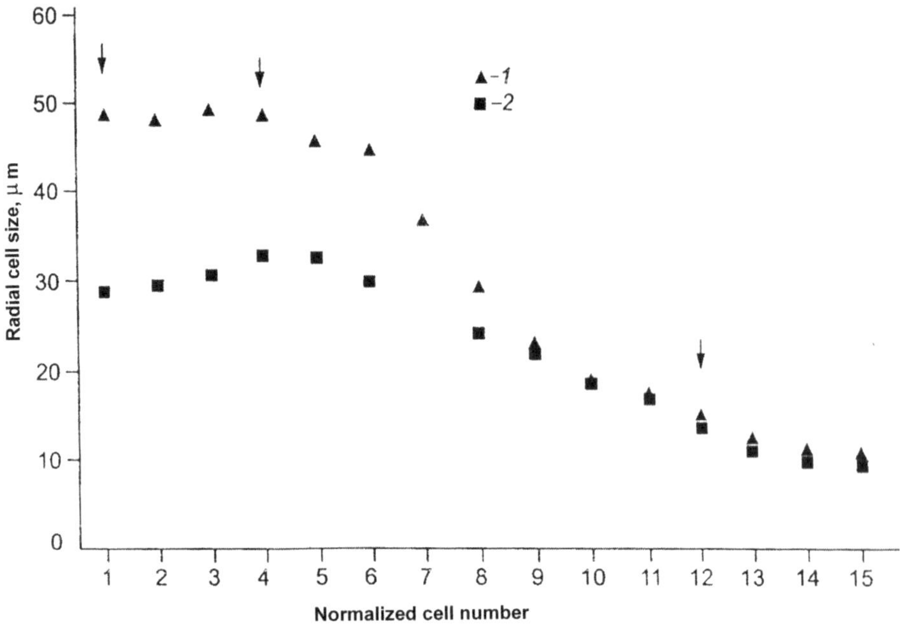

Fig. 6.3. Typical normalized tracheidogram of tree rings formed in relatively warm (*1*) and cold (*2*) years. The *arrows* show cell sizes corresponding to positions used for cell chronology construction

Table 6.1. Statistical characteristics of larch chronologies of tree-ring width (*TRW*) index, cell size (μm), cell wall thickness (μm), and maximum wood density (g/cm^3) at the northern timber line. Tracheidograms were standardized to 15 cells

Index	Average and standard deviation	Coefficient of variation	Sensitivity coefficient	First order auto-correlation
East Taimyr				
First cell size (cell 1)	38.7±3.29	0.085	0.089	0.20
Earlywood cell size (cell 4)	42.5±4.31	0.101	0.088	0.43
Latewood cell size (cell 12)	17.8±2.70	0.152	0.132	0.42
Maximum cell wall thickness	3.8±0.81	0.213	0.224	0.25
Maximum density	0.754±0.098	0.131	0.146	0.04
TRW index	1.0±0.34	0.340	0.410	0.11
Lowland of river Indigirka				
First cell size (cell 1)	43.9±2.64	0.060	0.060	0.19
Earlywood cell size (cell 4)	45.6±3.06	0.067	0.063	0.29
Latewood cell size (cell 12)	16.4±2.25	0.157	0.147	0.28
Maximum cell wall thickness	3.5±0.67	0.191	0.189	0.27
Maximum density	0.745±0.095	0.128	0.130	0.15
TRW index	1.00±0.34	0.340	0.39	0.24

(1887–1990) from other meteorological stations (Salekhard, Turukhansk, Verkhoyansk) were used.

Correlation of pentad temperature data with tree-ring width chronologies shows (Fig. 6.4) that the most important interval of the growth season when temperature influences cell production is quite short: four pentads (17 June to 6 July) with significant correlation for the western site, five pentads (17 June to

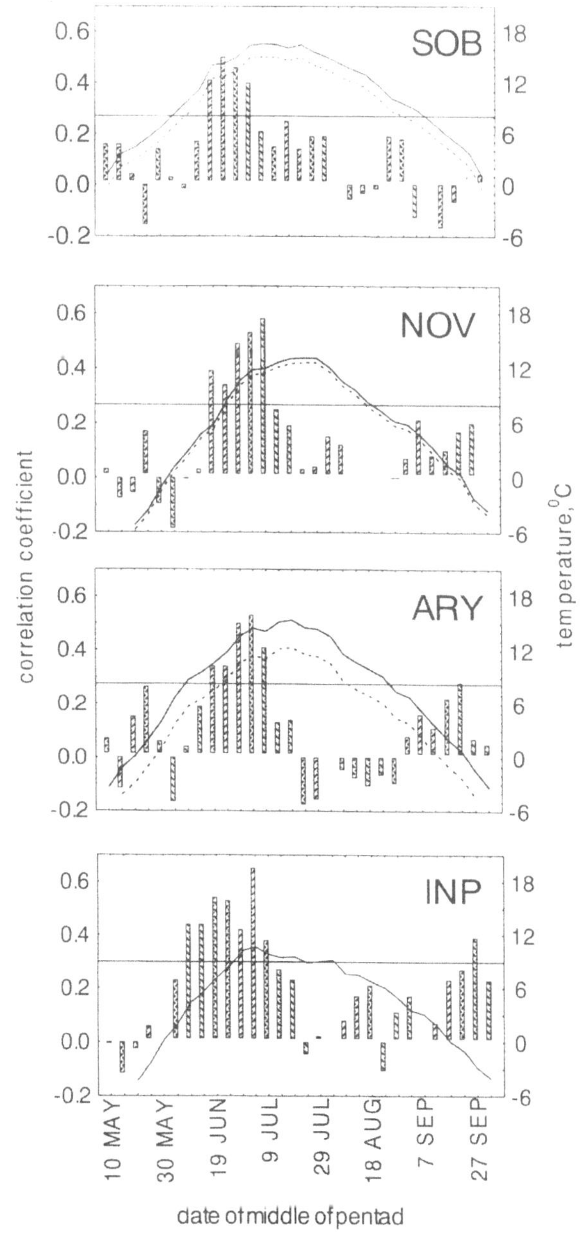

Fig. 6.4. Correlation coefficient between temperature of pentads and tree-ring width indices of four tree-ring chronologies (significant coefficients are *above* the solid horizontal line. Long-term seasonal changes in temperature is shown by the *solid lines* (for weather station sites) and *dashed lines* (corrected to studied sites)

11 July) for the sites from the Taymir and Anabar regions, seven pentads (7 June to 6 July) for the site from the Indigirka river. This interval is made up mainly of the period of early season temperature increase. The mean temperature (averaged for several years) of the first pentad, which shows a significant positive correlation with tree-ring width, decreases from west to east: 10 °C for the western site, 6–7 °C for the sites in the middle of the latitude transect, 4 °C for the eastern site. The fraction of the season with significant correlations between tree ring width and temperature is much shorter than the period with a temperature higher than 5 °C, increasing from 21 % at the western site through 35 % in the middle sites to 50 % at the eastern site.

The size of the first cell in the tree ring correlates significantly with temperature in one pentad (12–16 June) for the Taymir site and with four pentads (22 May to 11 June) for the Indigirka river site (Fig. 6.5). The most important intervals for earlywood cells shift to later dates (17-21 June, 2-21 June for the two sites, respectively); and the highest correlations with latewood cell sizes are later still (7–11 July, 7–26 June, respectively). Pentads when latewood cell wall thickness shows significant correlation with temperature are the same as for tree-ring width (cell production) at the Taymir site. For the Indigirka river site, latewood cell wall thickness correlates with almost all the period with a temperature >0 °C. The period when temperature influences maximum latewood density is longer, for example, 4 June to 2 October for the Indigirka river site.

Tree-ring width is defined by the number of cells in the tree ring and the cell sizes. Tree-ring width variability has a particularly strong influence on the variability of mass accumulation in the stem of trees from the Subarctic. Cell size characterizes the rate of seasonal cell production. These tree-ring parameters for trees from the northern timberline are defined by the temperature regime of a short interval at the beginning of the growth season (mainly June and the first ten days of July).

The seasonal dynamics of temperature varies to a great extent from year to year. For example, the date after which the temperature stays consistently above 0 °C can differ by up to 25–30 days. Data on the physiology of cambial activity indicate that, in conditions where temperature strongly limits tree radial growth, the temperature must be higher than some threshold and the thawing of the soil upper layer after snow melt must have begun so that radial growth can commence (Kandelaki 1979; Tranquillini 1979; Kandelaki and Dem'yanov 1982; Schweingruber 1996). Therefore, it is not strictly appropriate to correlate tree-ring parameters with temperatures for pentads whose dates are fixed relative to the calendar. Doing this will result, in some years, in tree-ring data being compared with the temperature of a period when there is no growth and, for other years, when growth is in an active, or even a very active, phase. It makes more sense to correlate tree-ring structural parameters with pentad temperatures where the dates of the pentad are fixed relative to the date of cambial initiation defined by tree physiology and climatic data.

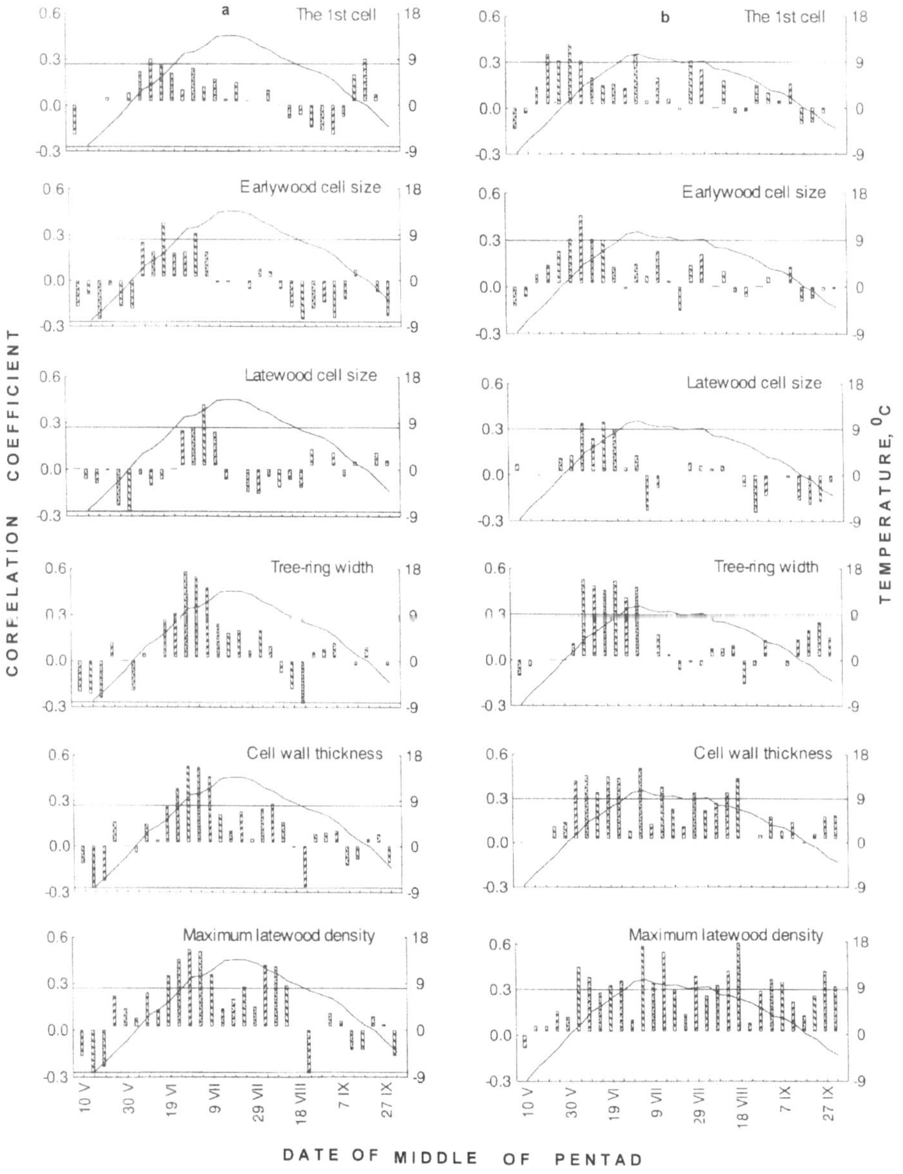

Fig. 6.5. Correlation between temperature of pentads and parameters of tree rings for the Taymir (**a**) and the Indigirka River (**b**) sites. Significant correlation coefficients are >0.3 ($P<0.05$). *Line* Mean temperature (for several years)

As soil temperature is a very important parameter for tree growth on permafrost soil (Pozdnyakov 1986), an attempt was made to connect the date of the beginning of growth and the date of snow melt. The date of snow melt was calculated from temperature and winter precipitation data according to simple methods (Kuzmin 1961). The dates calculated correlate very well with available observed snow-melt data (R=0.82; n=16, F=28.8; P<0.0001) and their standard deviation, maximum, and minimum values are very close. The size of the first cell is significantly correlated with the temperature of two pentads before the date of snow melt, the dates of the pentads being fixed relative to the date of snow melt rather than relative to the calendar (Fig. 6.6). This is consistent with the data on the stage of cambial initiation (swelling) before the production of new xylem cells, which is connected with the beginning of new needle development. One can often see trees near the northern timberline with new needle development when the soil is still under snow (Shiyatov 1969; Gorchakovsky and Shiyatov 1985). The size of earlywood cells correlates with the temperatures of several pentads just after snow melt; and the size of latewood cells with temperatures of three to five pentads after snow melt, obviously during the period when these cells are being produced. The correlation of temperature with cell wall thickness and maximum latewood density changes when pentad dates are fixed relative to snow melt rather than the calendar. These two tree ring structure parameters are dependent primarily on temperatures in the second half of the season. This is consistent with knowledge of the physiology of cell wall growth. Variability in tree-ring width is explained by the temperature of the pentads immediately after snow melt (the first part of the growth season). A high temperature during the first part of a season leads to the formation of a wider cambial zone and consequently to higher cell production throughout the season.

So the main controlling factors of seasonal growth and tree-ring structure formation in northern timberline trees are early summer temperature and the date of snow melt, which influences the date of cambial initiation. Multiple regression models of tree-ring width indices calculated with early summer temperature and snow-melt date as independent variables show strong agreement with instrumental data (Table 6.2). A comparison of the mean ring widths of years with very early and very late snow melt confirms the influence of snow-melt date on tree-ring width (Table 6.3). These data indicate that tree-ring width indices are higher in years with earlier snow melt.

Thus, in high latitudes, all growth processes are under the control of temperature: cell production, tracheid radial diameter, and cell wall thickness (density). Tracheid production and tracheid radial dimension are completely determined by early summer temperature. This means that the target of the temperature effect is the cambial zone. The data for maximum latewood density, however, disagree with this conclusion. They show a significant correlation with both the early summer temperature and the temperature during August–September. Is this a real contradiction? As we now know from previ-

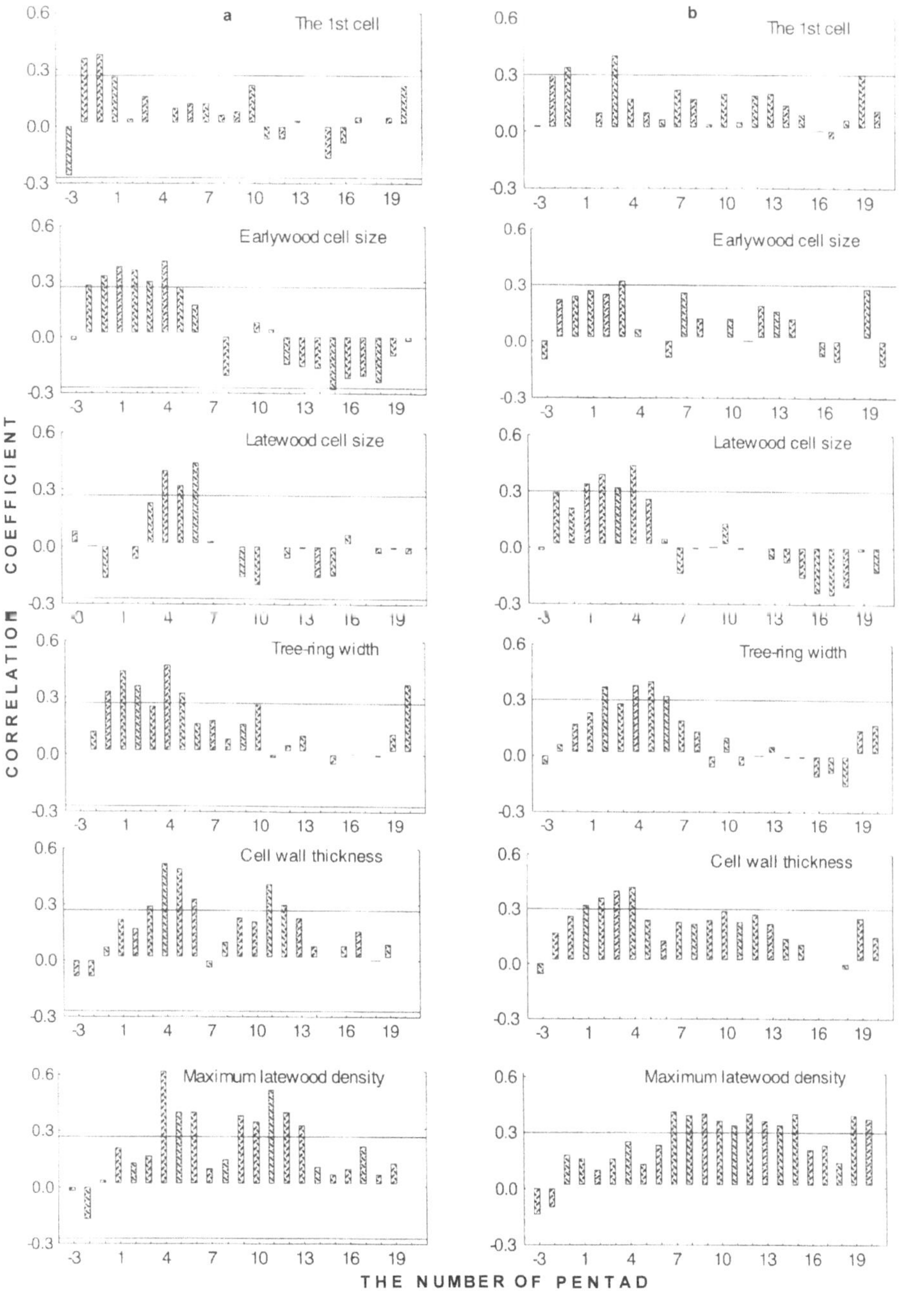

Fig. 6.6. Correlation between temperature of pentads calculated according to the date of snow melt and parameters of tree rings for the Taymir site (**a**) and the Indigirka river site (**b**). Significant correlation coefficients are >0.3 ($P<0.05$)

Table 6.2. Parameters of multiple regression model of tree-ring width indices with early summer temperature and date of snow melting as independent variables (Vaganov et al. 1999). *T* Taymir region (Khatanga), *I* Indigirka region (Chokurdakh)

Region	Coefficients of regression model		*R*	R^2	*F*	*P level*
	Temperature	Date of snow melting				
T	0.089	0.009	0.71	0.504	20.3	0.00001
I	0.121	0.014	0.73	0.539	22.8	0.00001

Table 6.3. Tree-ring width indices for years with early and late snow melting (Vaganov et al. 1999). *T* Taymir region (Khatanga), *I* Indigirka region (Chokurdakh), *SD* Standard deviation

Region	Early snow melting			Late snow melting		
	Number	Mean	SD of years	Number	Mean	SD of years
T	12	1.25	±0.19	10	0.88	±0.25
I	9	1.06	±0.28	9	0.92	±0.22

ous chapters, the enlargement and wall-thickening stages require several weeks to complete tracheid differentiation. If latewood tracheids are produced in the cambial zone in mid-July, they have to spend at least 2 weeks to expand and then 3–4 weeks to deposit wall material (see Chaps. 4, 5). So, they need a minimum of 6 weeks to complete the maturation processes. Taking into account that the rate of expansion and wall deposition may depend on current temperature too, in spite of potentially pre-determined tracheid diameter and cell wall thickness, the time for complete maturation may exceed 2.0–2.5 months. The main question is: can they realize this potential in current weather conditions in the second part of a season or not?

The relationship between the duration of the growing season (determined as the period with a temperature higher than some critical level) and maximum latewood density is not particularly strong (R=0.325; P<0.01; Fig. 6.7). But the long-term trends of growing season duration and maximum density are similar. The most important factor for maximum density is the seasonal course of temperature. We examine three different years when a low density was associated with a short season at high latitude in the Taimyr region of northern Siberia (1958), a high density was associated with a short season (1979) and a low density was associated with a long season (1952) in Figs. 6.7, 6.8. These three variants can be explained easily in terms of the seasonal course of temperature. In 1958, growth started later (in mid-July) and latewood tracheids produced by the cambial zone at the beginning of August

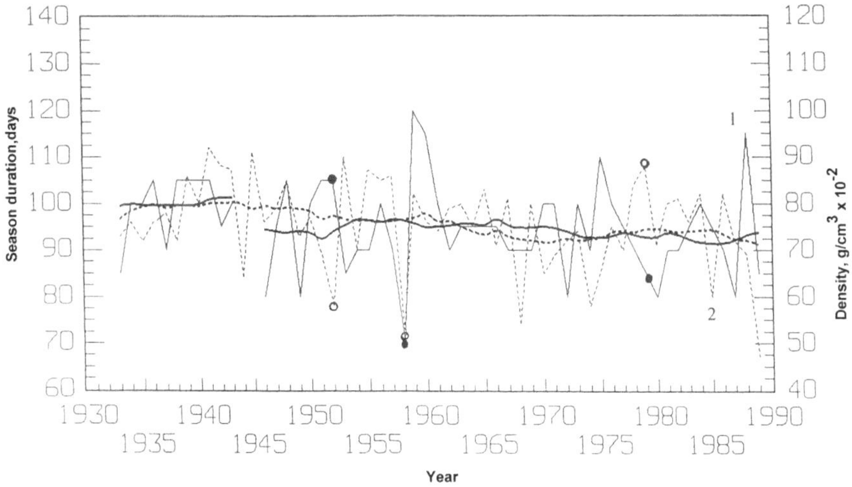

Fig. 6.7. Inter-annual variation of season duration (*1*) and maximum density in larch tree-rings (*2*)

could not complete wall thickening because the temperature fell below zero at the beginning of September. A typical "light ring" with a low maximal density was formed. In 1979, growth started in the second part of June and latewood tracheid formation was associated with high temperature in second part of July, which allowed cells to move rapidly through enlargement and the wall-thickening processes. As a result, a high-maximum-density latewood resulted because of this July acceleration, in spite of the short season. In 1952, growth started at the end of June, but the temperature was low throughout the long season. Latewood tracheids were produced (presumably in mid-July) at a time of low cambial activity, limiting their final radial dimension and cell wall thickness. The lower temperature during their maturation may also have served to reduce the final cell wall thickness. Latewood with a low maximum density resulted. Exactly the same conclusion was reached from an analysis of the features of "light" rings in black spruce (*Picea mariana)* at the northern treeline in northern Quebec, Canada (Wang et al. 2000). They divided "light" rings into three groups: weak, middle, and strong, according to the value of maximum density and considered different mechanisms of "light" ring formation: (1) delayed spring or late start of cambial activity, (2) cool summers, when all physiological processes may be slowed down and latewood tracheids may be immature at the end of the growth season, and (3) early-ending autumn because of the shortened length of the growing season. They assumed the first mechanism to be a major factor causing "light" rings. The same mechanism turned out to be the cause of the higher frequency of "light" rings in peatland (Lavoie and Peyette 1997). Most of the strong "light" rings were associated with cold springs after volcanic eruptions (Filion et al. 1986). The

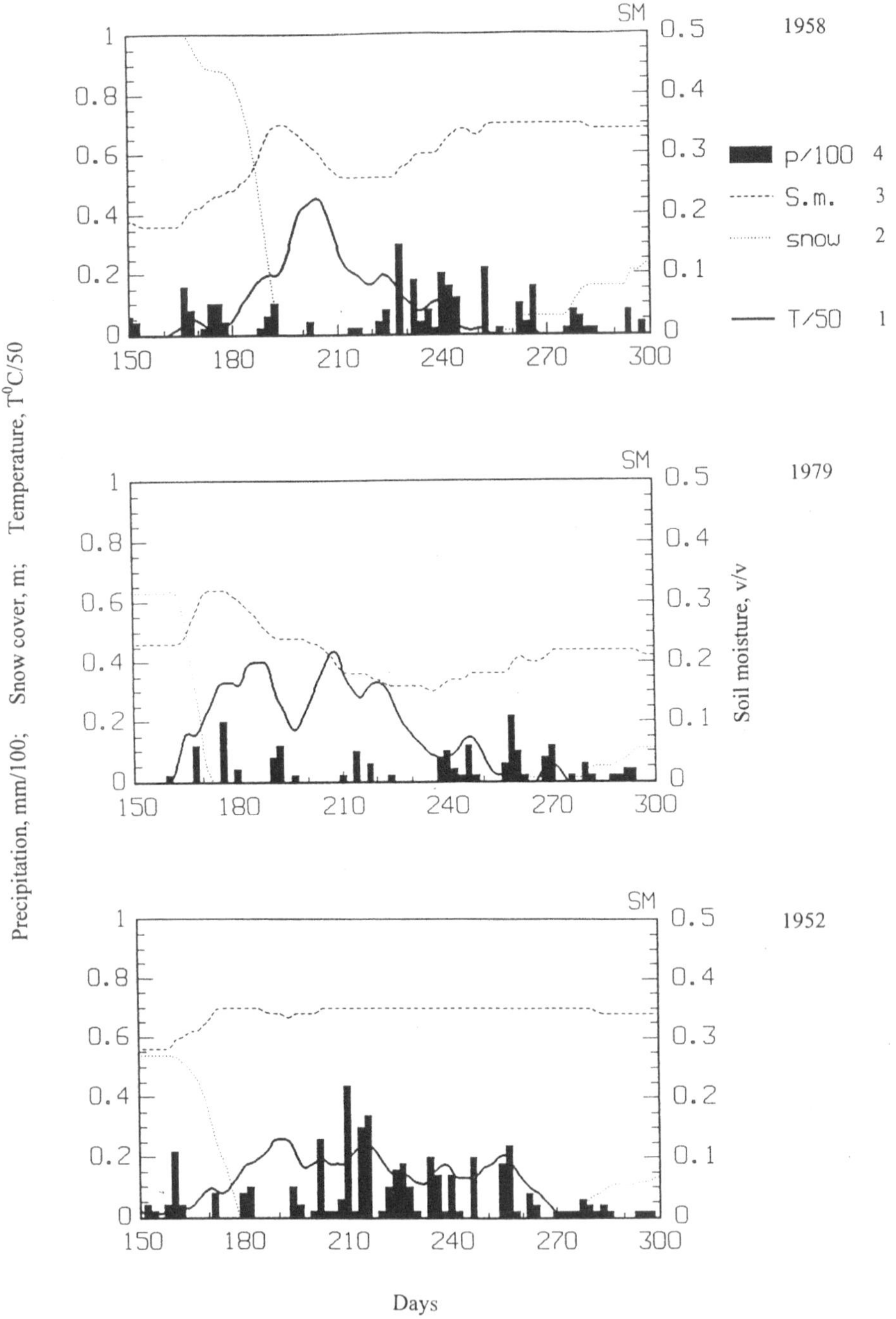

Fig. 6.8. Climatic conditions (*1* temperature, *2* snow cover, *3* soil moisture, *4* precipitation) for 3 years with different temperature regimes (see explanation in text)

early-ending autumn pattern was the main cause of spruce "light" rings at the timber line in the Alps (Gindl 1999).

Our analyses of larch in the Taimyr region suggest that a significant correlation of maximum latewood density with June–September temperature is limited, at high latitudes, to extremely cool or short growing seasons. In fact, if we exclude years with low temperature in August–September (about 20 % of total years), the correlation between pentad temperatures and maximal density is not significant in the second part of August and September (Fig. 6.9). If we move southward where the duration of the growing season is longer and

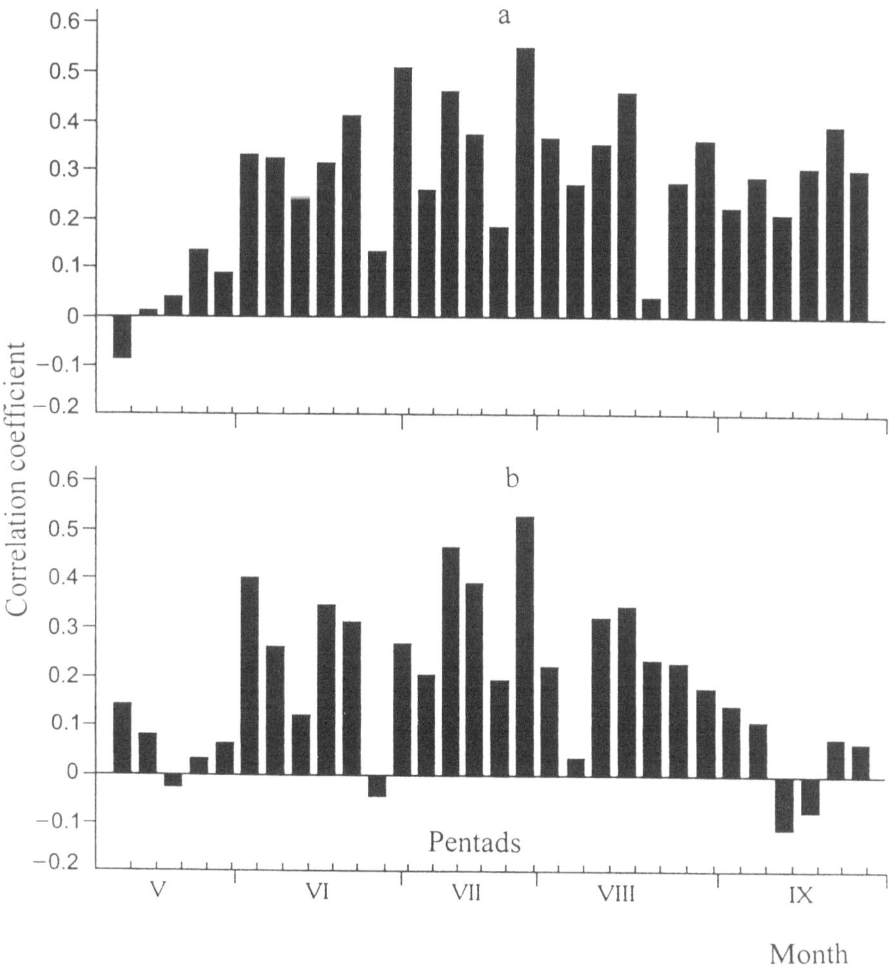

Fig. 6.9. Correlation between temperature of pentads and maximum wood density for all years (n=54; **a**) with measurements of light ring formation and for years (n=45; **b**) without measurements

sufficient for latewood tracheids to complete maturation and realize the potential obtained when they leave the cambial zone, there is no significant correlation between maximal density and temperature for a long interval within a season. This is especially the case between maximum density and late summer–early autumn temperatures. Even at a lower latitude (63° North), tree-ring width and maximum density of larch tree rings are correlated with pentad temperatures for much of the season (Fig. 6.10; Kirdyanov 1999). Tree-ring width and maximum density show a significant correlation with temperature for the same period at the beginning of the growing season. Yasue et al. (2000) analyzed radial diameter, cell wall thickness, and wood density (maximum density) of the last-formed tracheids in tree-rings of *Picea glehnii* grown in Teshio Experimental Forest in northern Hokkaido, in relatively mild climatic conditions (44° 57' North, 300 m above sea level; Fig. 6.11a). Maximum wood density showed a strong dependence on cell wall thickness, which was, in turn, only correlated with June–July temperatures (Fig. 6.11b). In this situation, cell division in spruce occurs from the beginning of May to the middle of August; and the thickening of the cell wall continues until the end of September (Yasue et al. 1994). If cell wall thickness were determined by weather conditions during the thickening process, both it and the maximum wood density of the last-formed tracheids should be correlated with the August–September temperature. In fact, the significant correlation is found with the temperature during June and July, the most active period of the season when most cells are produced. Therefore, in this case too, the degree of

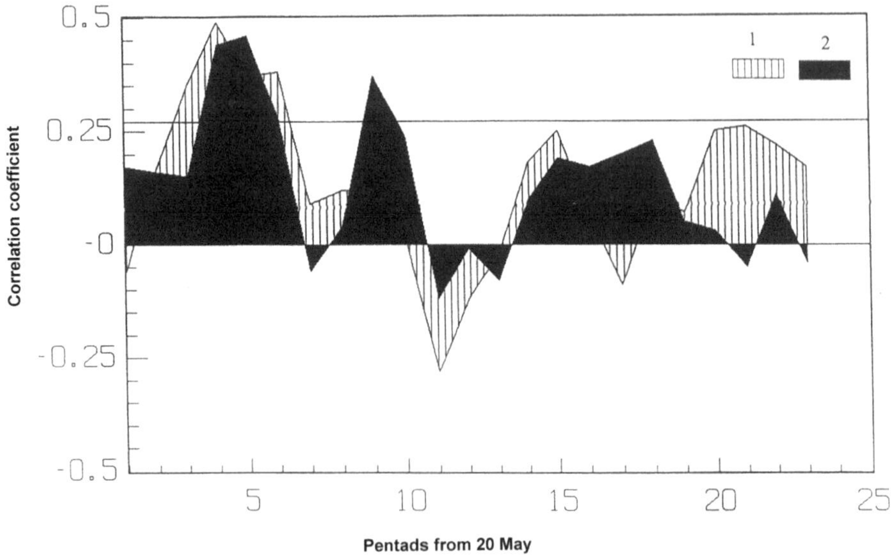

Fig. 6.10. Correlation of tree-ring width index (*1*) and maximal density (*2*) with pentad temperatures for larch tree rings from mid-taiga region

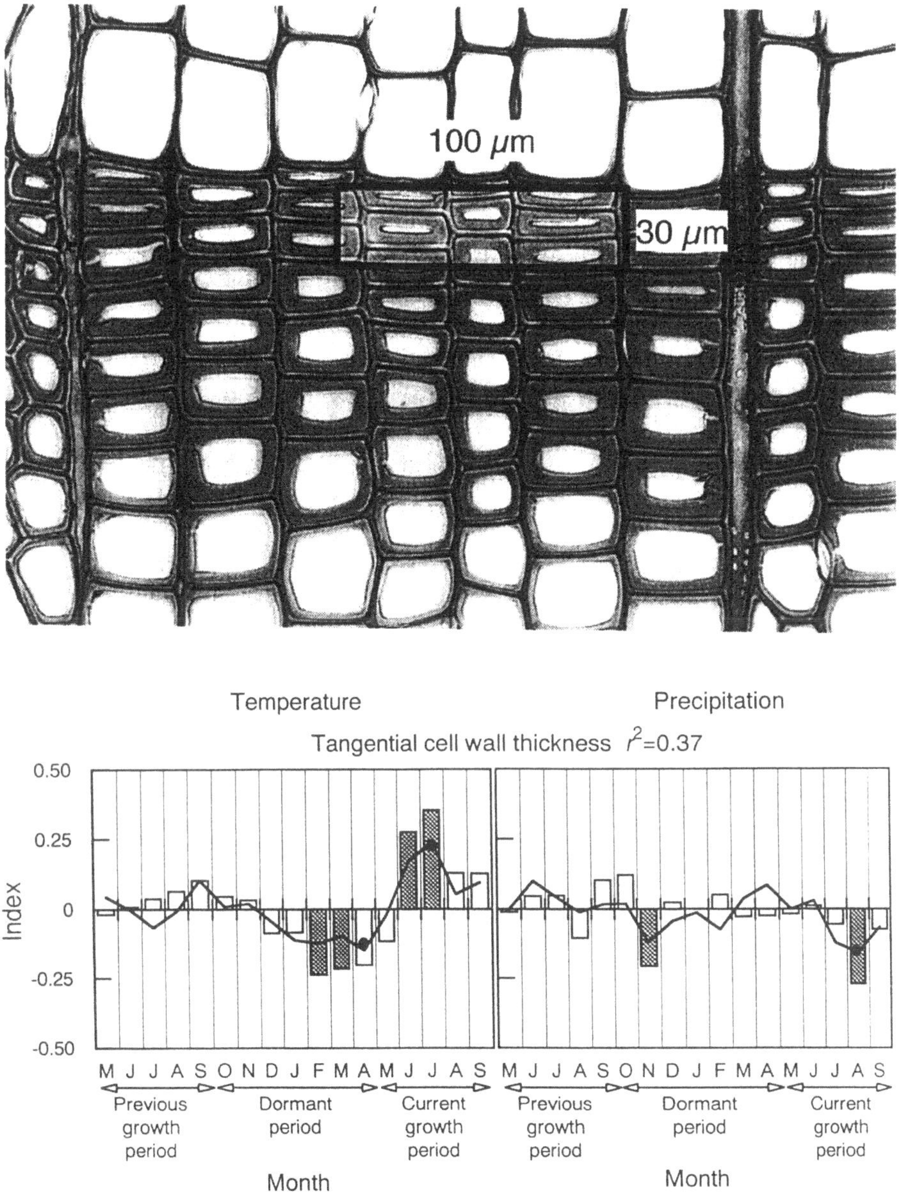

Fig. 6.11. **a** Light micrograph showing the outermost part of an annual ring of *Picea glehnii*. *Rectangle* indicates the size of the scanning slit for X-ray densotometry. **b** Climatic responses of the chronology of tangential cell wall thickness of the three last-formed tracheids in annual rings of *P. glehnii* in Teshio in northern Hokkaido, obtained from response function (*lines*) and simple correlations (*columns*), calculated for the period 1901–1990. The *shaded columns* and *black circles* indicate significant variables ($P<0.05$). r^2 Squared multiple-correlation coefficient of the regression equation of the response function (Yasue et al. 2000)

cell wall thickening is controlled by events in the cambial zone. Similar mechanisms may operate for many chronologies in western North America (Briffa et al. 1992b), eastern North America (Conkey 1986), western Europe (Briffa et al. 1988) and the Himalayan region (Hughes 1992, 2001). Typically the most consistent relationship is between maximum latewood density and temperature in the first weeks of the growing season, although links to late summer or early fall temperatures may also be strong (Fig. 6.12). Such late summer influences may represent cases where the potential tracheid dimension and cell wall thickness dictated in the cambial zone are not realized. This might occur, for example, when late-season conditions end the development processes early, or deprive the growing xylem of the resources needed to fulfill the "quota" determined in the cambial zone earlier in the year.

It is clear from these results that some combination of tree-ring structure parameters (tree-ring width, maximum latewood density, tracheid diameter in early- and latewood zones) describe the seasonal temperature regime at high latitudes and high elevations in mid-latitudes in more detail than with tree-ring width alone. These additional records cast light not only on the date of growth initiation and the temperature for shorter periods during tracheid

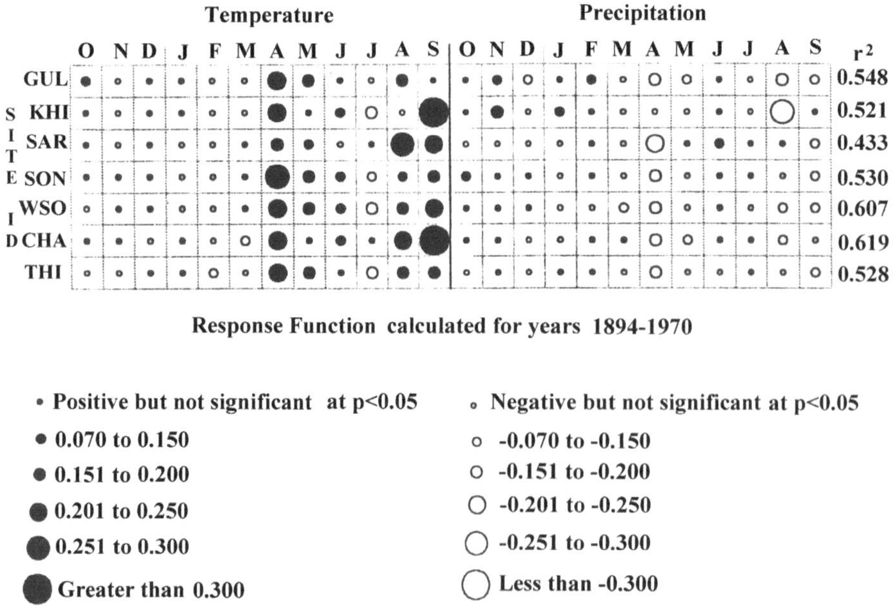

Fig. 6.12. Response function elements for seven *Abies pindrow* maximum latewood density chronologies in Kashmir, calculated for the period 1894–1970, using meteorological data from Srinagar. Response function elements indicate the proportional change in the tree-ring variable associated with unit change in the meteorological variable (Fritts 1976). The r^2 values indicate the proportion of tree-ring variance accounted for by the response function. After Hughes (2001)

production but also the mean temperature in the maturation period. For example, Panyushkina et al. (2003) used larch near the northern limit in Yakutia to derive temperature reconstructions back to the mid-seventeenth century for early summer (2 June through 11 July) and the rest of the summer (2 July through 30 September) from chronologies of N (total cell number) and latewood cell wall thickness, respectively. Interestingly, these two reconstructions vary in parallel from AD 1642 until 1978, after which early summers have been consistently cool (compared to the rest of the summer) to an extent that is unique in the past 350 years. This is seen not only in the internal structure of the larch tree rings, but also in the instrumental record. There is clearly considerable potential to derive such detailed descriptions of the seasonal march of temperature at such temperature-limited sites at high latitude or high elevation.

6.4 "Differential Tracheidograms" in the Analysis of Weather Conditions Within a Season

For each specific year, a tree-ring tracheidogram has its own features. By superimposing many normalized (for example to 30 cells) tracheidograms for a single tree but many years on one chart, a description of the seasonal course of growth is obtained, independent of the length or weather conditions of a specific growth season. Another variant would be to make such a composite tracheidogram, not using the individual values for all years at year-cell position, but only the maximum values (Fig. 6.13). This tracheidogram gives an estimate of the largest values of the cell dimensions that may be obtained by that tree under the optimal combination of light, temperature, and moisture availability. Any real tracheidogram will be located lower on the Y-axis than the maximal one. We will consider in general the dependence of cell size in various portions of the tree ring on external and internal factors:

$$D_R = f(W, T, t, \{z_i\}) \tag{6.1}$$

where W is humidity (available moisture, precipitation), T is temperature, t is tree age, $\{z_i\}$ is a set of ecological factors, characteristics of tree growth vigor, etc. Clearly, age and ecological factors have longer characteristic times of change than many climatic factors, so they may be considered constant for quite a long period of time (several years, several decades). Then Eq. 6.1 may have one of two forms:

$$D_R = f_1(W, T) + f_2(t, \{z_i\}),$$
$$D_R = f_1(W, T) * f_2(t, \{z_i\}) \tag{6.2}$$

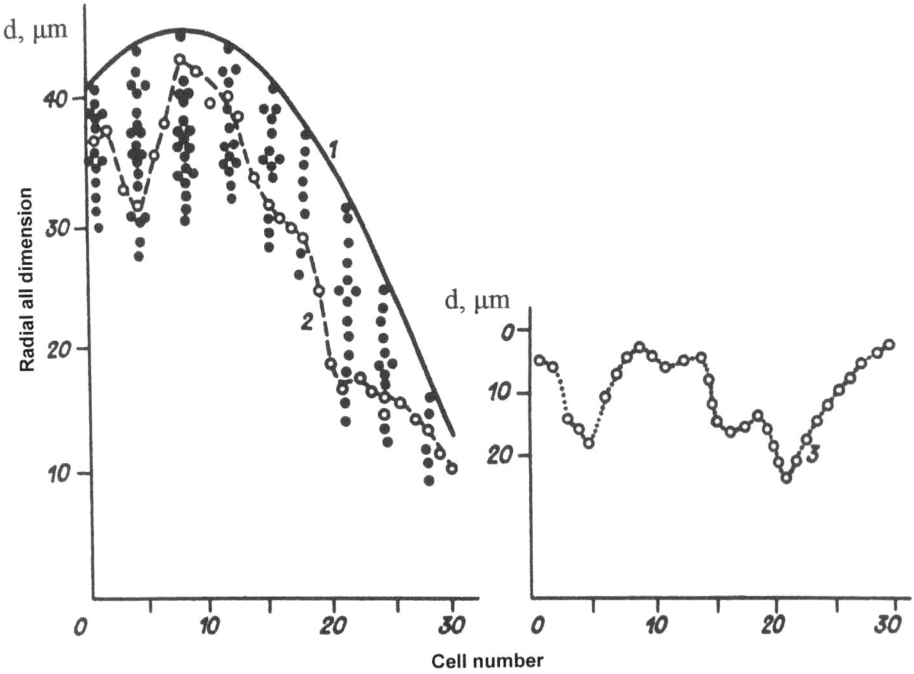

Fig. 6.13. Method of "differential" tracheidogram: *1* maximum values, *2* real tracheidogram, *3* "differential" tracheidogram, *points* cell size in tracheidogram of different tree-rings

Naturally, cell size in maximal and real tracheidograms will depend mostly on the same factors. That is why, if we measure the difference (or relation) between cell sizes in maximal and real tracheidograms for each portion of a tree ring, the resulting value will characterize the extent of the difference of growth conditions in a particular part of the season from those needed for maximal growth rate.

We will consider tracheidograms of pines growing in wet and dry sites (Fig. 6.14). In the first case, the water content of soil is always in the optimal zone, so the real size of the cells is mostly defined by fluctuations of temperature and this means that the "differential" tracheidogram shows the temperature deviations from the optimum. For pines in dry locations, such a tracheidogram reflects the influence of fluctuations of both temperature and water content. Hence, if we superpose the dynamics of single-year "differential" tracheidograms, it is possible to estimate (in cell sizes) the extent to which temperature and humidity growth conditions are sub-optimal in different parts of the season. The comparison of such "differential" tracheidograms for pines of two sites of the Beya Forestry District (in the steppe zone of southern Siberia) located only about 6 km apart shows that, in some years and in some intervals of the growth season, the "differential" tracheidograms from the two

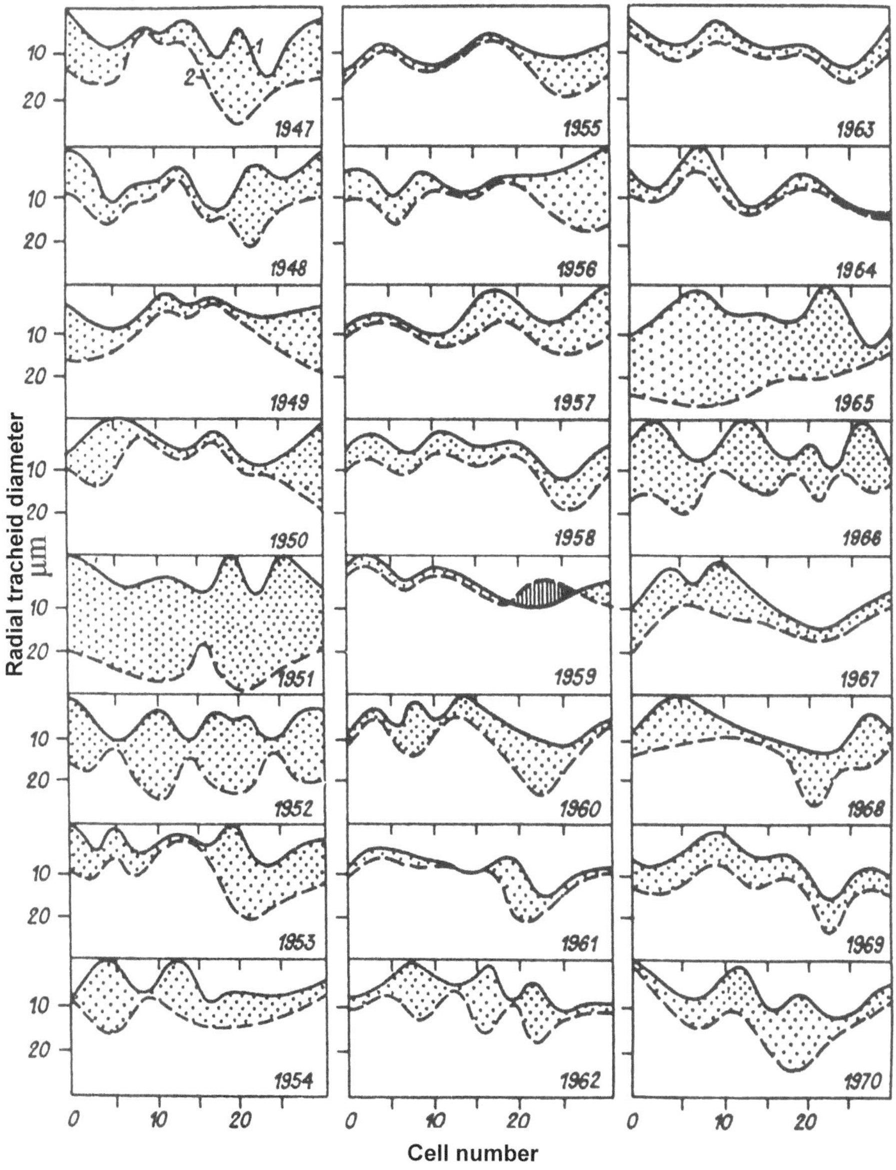

Fig. 6.14. "Differential" tracheidograms for pine tree-rings of calendar years from moist (*1*) and dry (*2*) sites

sites practically coincide. Deviations from optimum in these intervals of a season are defined by temperature fluctuations. In other periods, the "differential" tracheidograms are sufficiently different, but in this case the tracheidogram of dry-site pines is practically always lower than the tracheidogram of the wet-site pines. The lower the portion of the "differential" tracheidogram of a wet-site pine is in relation to the maximal tracheidogram, the colder were the conditions in the period of cell growth of the corresponding portion of the tree ring; and the greater the differences between the two tracheidograms are, the greater the moisture deficit in that part of the season.

In a careful analysis of "differential" tracheidograms, one can see that the growth seasons are characterized by a considerable variety of hydrothermal regimes. For example, the beginning of the growth season in 1947 was characterized by high temperature and a noticeable deficit of moisture, but by the middle of the season, temperature was close to optimal and the "differential" tracheidograms practically coincided (evidently, due to plentiful rain). In the very middle of the season, the temperature rose further, the moisture deficit increased and the divergence of the curves increased. The subsequent short-period rise of temperature further increased the difference between the curves.

Only at the beginning of the last third of the season did plentiful rain, combined with a temperature decrease, lower the moisture deficit. Four temperature rises up to the optimum (determined by the "differential" tracheidogram of the wet-site pines) characterizes 1966, as did also a decrease of moisture deficit in the dry-site pines resulting from lower temperature and an increase in precipitation.

Differential tracheidgrams permit a quantitative description of the specific weather conditions of a growth season and, in particular, the identification of the factor which is most responsible for the variation in cell size and growth rate change in a particular part of the season. Differential tracheidograms also reflect the reaction of the tree to changing conditions, depending not only on current, but also on antecedent, conditions. Moisture deficit might be different at the beginning of a season, might be compensated by precipitation over a few days, and might not occur at all during a season if the initial humidity is high and the precipitation is plentiful and evenly distributed. The decrease in cell size and growth rate might be caused by water deficit in some cases and by declining temperature in others; and differential tracheidograms allow one, in principle, to distinguish between these cases.

It is possible to divide a season into, for example, three periods and to give general estimates of their hydrothermal conditions from the character of the differential tracheidograms: warm and moist, cold and moist, hot and dry, cold and dry.

6.5 Tree-Ring Anatomy as an Indicator of Climate – Seasonal Growth Relations in a Monsoon Region – an Example of Growth Limitation by High Temperatures and Intra-Seasonal Drought

In contrast to growth conditions limited by one factor, tree growth in wet subtropical regions may be affected by temperature, water, and solar radiation because the growth season is long and there are favorable conditions during much of it. One growth-limiting factor may be replaced by another; and in the same intervals of a season in different years this limitation may result from different factors. Moreover, growth conditions may be optimal for most of any one season. This leads to weak correlations between tree-ring width and monthly climatic data (Park 1993; Park and Vaganov 1996; Park et al. 2000). We will consider the usefulness of tracheidograms in studying the growth–climate relationship on the within-season scale in *Pinus densiflora* growing in a monsoon region.

The study site was located in Wolak National Park, University Forest of Chungbuk National University Plantation on shallow (20–30 cm) sandy soil. The seasonal kinetics of tree-ring growth measured by dendrometer and by the sampling method differ considerably (Fig. 6.15). The dendrometer indicates high stem growth during May (about 60 % increase) and less intensity of growth rate during June and July. According to these data, the cessation of tree-ring growth occurs at the beginning of August, whereas on thin sections of tissue samples, the formation of new cells continues until mid-September

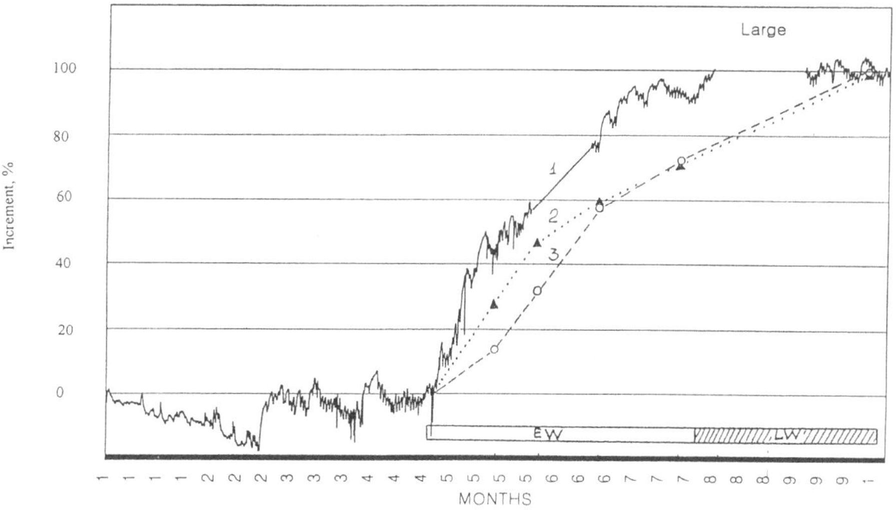

Fig. 6.15. Seasonal tree-ring growth measured by: dendrometer (*1*), number of cells in tree rings (*2*), and tree ring width dynamics (*3*)

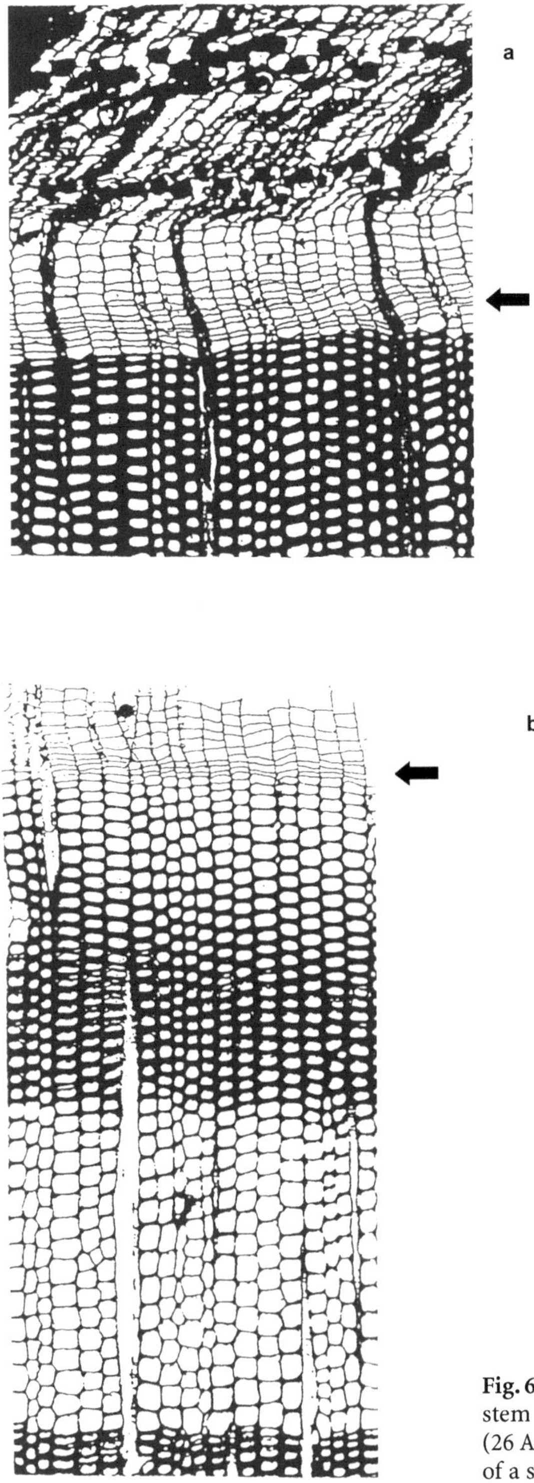

Fig. 6.16. Wood cross-sections taken from stem of *Pinus densiflora* at the beginning (26 April; **a**) and end (16 September; **b**) of a season. *Arrow* indicates cambial zone

(Fig. 6.16). Both dendrometer and thin-section data show that production of new xylem cells starts at the end of April when the average daily temperature reaches 10–11 °C. The production of new tracheids as measured on thin-sections was higher during May and the beginning of June, then decreased from the end of June, and remained low until mid September. So, the duration of the period of tracheid production was about 150 days. In the first part of the season, the growth curve of the number of tracheids outstrips the curve of tree-ring width because the earlywood tracheids need several weeks to enlarge. In the second part of the season, both curves draw together because the latewood tracheids enlarge faster.

The data in Table 6.4 show the portions of the tree ring formed during different months. The largest growth rate occurs in May (a little less than 40 % of total tracheid production). The growth rate decreases in June and reaches the lowest level in July and August (only a quarter of seasonal tracheid production occurs in these two months). We compare these data with the standardized curves of tracheid dimension along the files within complete tree-rings of *P. densiflora* (tracheidograms; Fig. 6.17). The radial tracheid dimension increases from the first tracheid produced to the tracheids that occupy the middle part of the tree-ring (tracheid dimension increases from 35 μm for the first cells to 45 μm in the middle of the tree ring). This increase in radial tracheid size corresponds to a high growth rate of cell production and tree-ring width during May. Then, tracheid dimension decreases very rapidly to the level of 25-28 μm and then very slowly decreases to the tree-ring boundary. This portion of the tree ring corresponds to the second part of the growth curve, which has a slower rate from mid-June to mid-September. So, qualitatively, the variations of radial tracheid dimension are in accordance with the measured changes in seasonal growth rate of tree-ring width and tracheid production.

Figure 6.18 shows typical tracheidograms of tree-rings formed in 1978 and 1979 in stems of five *P. densiflora*. These years are chosen as contrast years

Table 6.4. Percentage of total annual tree ring formed during different months in red pine, *Pinus densiflora*, from Korea

Month	Percentage of total formed
April	10
May	38
June	18
July	12
August	12
September	10

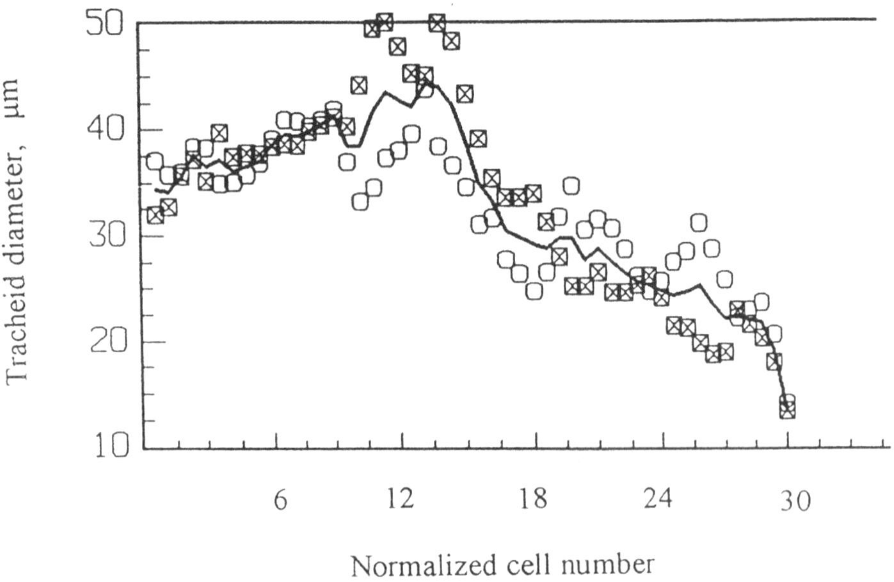

Fig. 6.17. Radial tracheid size variation along the two files of the 1997 annual ring. *Solid line* Average

because *P. densiflora* trees show wide tree-rings in 1979 and narrower in 1978. First of all, one can see similar tracheidograms in a particular year formed in different trees, in contrast to the varied tracheidograms formed in different years. This latter situation is especially visible in the average tracheidograms (Fig. 6.18, right). Some additional information can be extracted from average tracheidograms. In 1979, the earlywood tracheids have a wider radial diameter (about 48–50 µm) than in 1978 (40-43 µm). The transition to latewood in 1978 was very sharp and most of latewood tracheids have a radial diameter of about 22–24 µm. In comparison, in 1979, the transition to latewood was gentle; and the latewood portion occupied about one-third of the tree ring, with tracheids of radial diameter about 27–28 µm. *P. densiflora* trees formed their narrowest rings in 1988. Tracheidograms for this year indicate that the average earlywood tracheids have radial diameters of about 40 µm and show a fast transition to the latewood, which occupies 45 % of the tree ring with tracheids of 22–23 µm radial diameter.

Inter-annual variations of tree-ring width index show a high and significant correlation with the average tracheid diameter (Fig. 6.19). The coefficient of correlation is 0.73 (F=26.8, P<0.001). Small tree rings are associated with a low average tracheid diameter and vice versa. There is a similar long-term trend in tree-ring width index and average tracheid diameter which shows a growth rate decrease at the end of the 1980s and beginning of the 1990s (Fig. 6.19). The significant correlation between tree-ring width index and average tracheid diameter means that: (1) intra-seasonal variations in growth

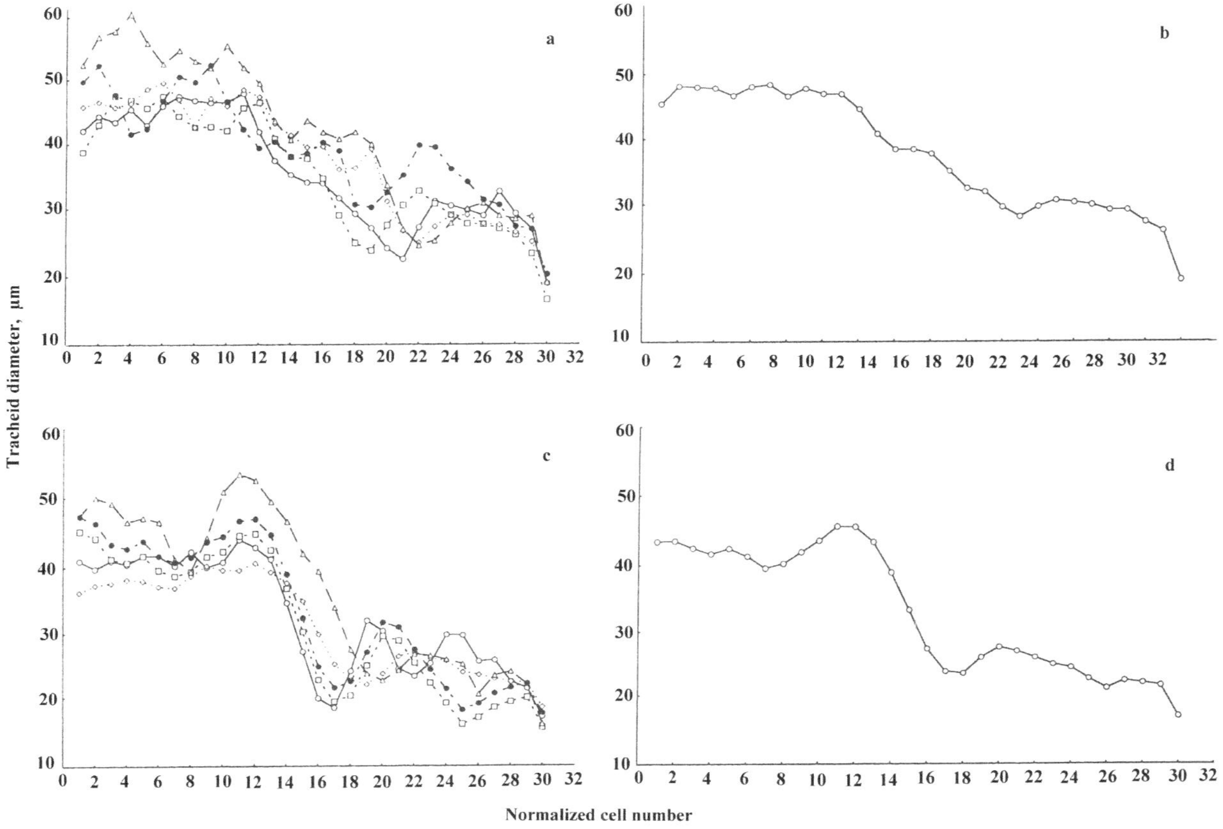

Fig. 6.18. Standardized tracheidograms formed in five different trees during 1978 (**a**) and 1979 (**c**) and averaged for all five trees (**b**, **d**) accordingly

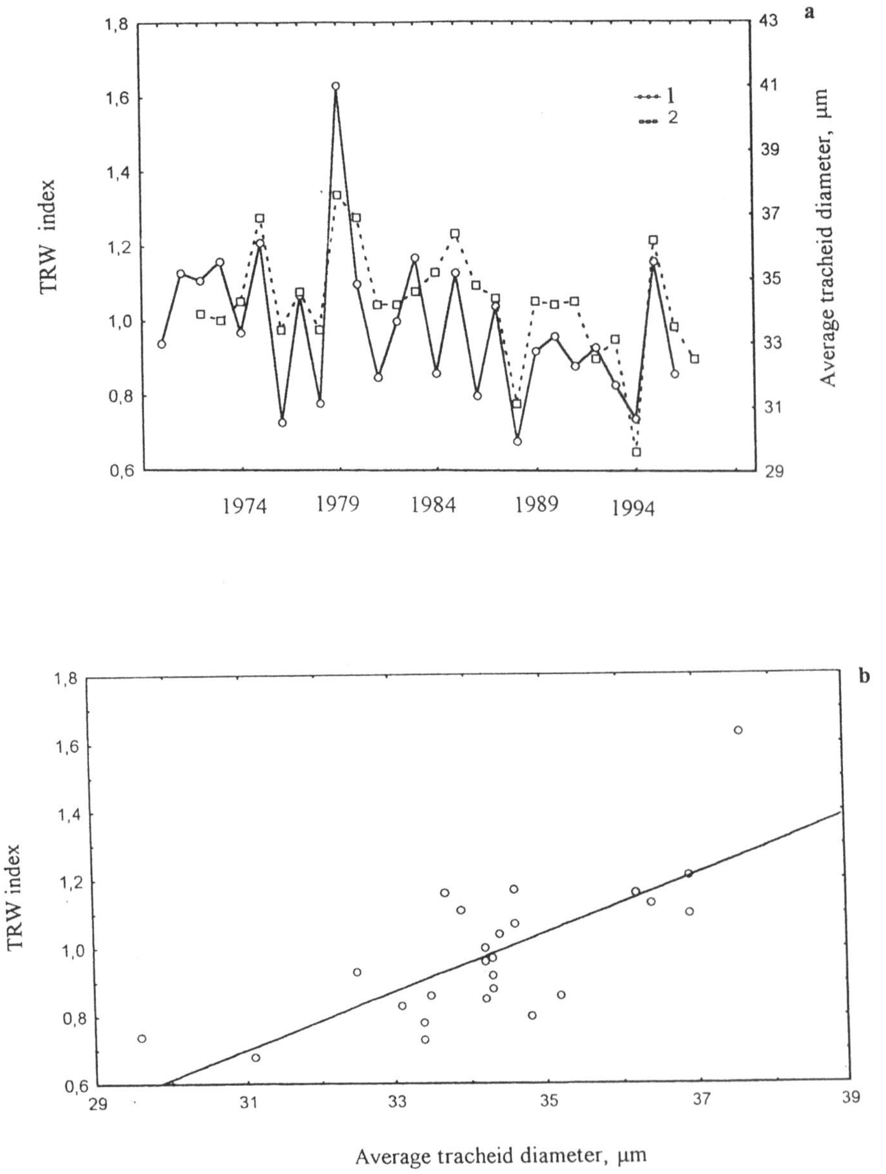

Fig. 6.19. **a** Inter-annual variation of tree-ring width (average for 15 trees; *1*) and average tracheid diameter (averaged for five trees; *2*) for red pine trees. **b** Plot of tree-ring width index vs average tracheid diameter

rate and xylem production are fixed according to variations in radial tracheid diameter, (2) these intra-seasonal variations are integrated in total seasonal xylem cell production and tree-ring width.

The variations in tree-ring width index represent the influence of climatic conditions integrated over the whole season. For such a long season as in the study case (about 150 days), this integration accumulates and mixes the influences of different factors that limit or stimulate growth rate in different intervals of the season.

We will focus on tracheid diameter variations as a possible tool to study the influence of climatic factors on growth rate with a resolution higher than annual. At the first stage, we correlate the radial tracheid diameter in standardized tracheidograms (using "cell" chronologies) with temperature and precipitation averaged for 10-day intervals (decades; Tables 6.5, 6.6). All correlations with temperature are negative; and practically all correlations with precipitation are positive. This means that the water availability in soil mostly strongly affects the seasonal growth rate in these conditions where *P. densiflora* is growing on a shallow layer of sandy soil. The negative correlation with temperature, even at the beginning of the season, indicates that higher temperature increases the water loss from soil due to increased transpiration and evaporation. Another interesting feature of the correlations is the significance of climatic conditions in the first part of a season in relation to the second part. This is not so surprising if we take into account the results of direct measurements of seasonal tree-ring development. More than 60 % of the tree ring is formed during the period from May through the beginning of July. The significant correlations shift in accordance with dates and relative position of tracheids within tree-rings (Tables 6.5, 6.6). The temperatures of April and May are the main influences on earlywood tracheid diameter (with relative position from the beginning of the tree ring to the mid-part). The temperatures in June and the beginning of July are the main influences on the transi-

Table 6.5. Significant coefficient of correlation between tracheid diameter and decadal temperature for *Pinus densiflora* in Korea

Decade	Relative position (%) of tracheid within tree ring								
	5–10	20	30	40	50	60	70	80	90
10–20 Apr	–0.40	–0.35	–0.43			–0.37			
20–30 Apr			–0.44						
1–10 May				–0.42	–0.47				
10–20 May					–0.36			–0.63	
20–30 May	–0.35								
1–10 Jun									
10–20 Jun						–0.38	–0.42		
20–30 Jun					–0.35				
1–10 Jul					–0.59	–0.62	–0.47		
10–20 Jul						–0.50			
20–30 Jul								–0.35	
1–10 Aug									
10–20 Aug									
20–30 Aug									

Table 6.6. Significant coefficient of correlation between tracheid diameter and decadal total precipitation for *Pinus densiflora* in Korea

Decade	Relative position (%) of tracheid within tree ring								
	5–10	20	30	40	50	60	70	80	90
10–20 Apr	0.53	–0.35							
20–30 Apr	0.53			0.40		0.47		0.35	
1–10 May	0.47	0.45							
10–20 May									
20–30 May			0.42						
1–10 Jun			0.35						
10–20 Jun			0.54						
20–30 Jun									
1–10 Jul									
10–20 Jul									–0.43
20–30 Jul									
1–10 Aug							0.47		
10–20 Aug							0.42		
20–30 Aug								0.42	

tion zone and latewood tracheids (relative position 50–80% across the tree ring). The highest correlation occurs between the temperature in the first decade of July and the diameter of tracheids located in the transition zone and the beginning of the latewood zone (relative position 50–70%). This is exactly the interval of the season just before the monsoon commences. We can also note that the April temperature and precipitation significantly affect the dimensions of tracheids formed in the transition zone and at the beginning of the latewood zone. This means that rapid water loss in the interval before the start of xylem growth negatively affects the diameter of tracheids formed later in the ring. Correlations between tracheid diameter and decadal climatic conditions support the results of standard correlation function analysis of *P. densiflora* tree-ring width index with monthly data (Fig. 6.20).

For a better visual demonstration of the influence of current climatic conditions on seasonal growth rate and tracheid dimension, we use the method described earlier (see previous paragraph). We derive maximal tracheidograms that correspond to optimal growth conditions in different intervals of the season (Fig. 6.21a). This curve reflects only the general seasonal trend of tracheid dimension changes. An average tracheidogram for a specific growth year will always be located lower than the maximal one. Deviations in tracheid dimension between the maximal and specific tracheidograms will reveal those portions of the annual tree ring (and relative seasonal intervals) for which conditions differ markedly from the optimal. Figure 6.21b shows the resulting difference tracheidograms for two years. In these years, the width of annual tree rings is the least. It is clearly seen that growth at the very

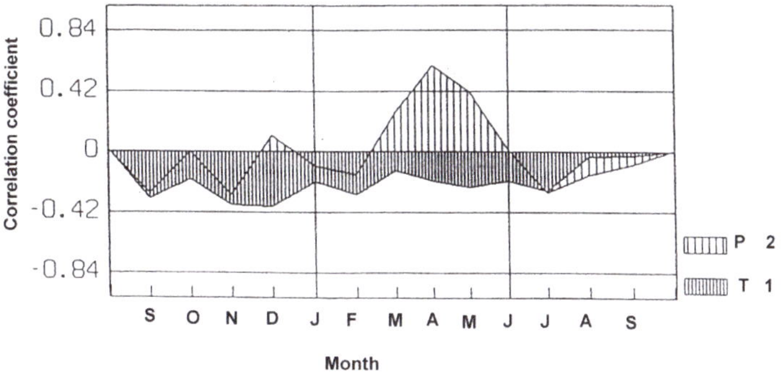

Fig. 6.20. Correlation of tree-ring width with monthly temperature (*1*, *T*) and precipitation (*2*, *P*) for *Pinus densiflora*

beginning of the season in these years is already far from optimal. In the middle of the annual ring deviations in tracheid dimension become largest, then in the transitional and latewood zones of the annual ring the difference between real and maximal tracheidograms decreases, and towards the outer boundary of annual ring the difference becomes insignificant. Such seasonal variations in differential tracheidograms accurately reflect the seasonal variations of soil moisture in the studied stand in the first half of the season, that is, before the monsoon. With the arrival of the monsoon rains, soil moisture achieves saturation faster. This does not, however, lead to an analogous increase of tracheid dimension up to the maximum.

Maximum air temperature values occur in Korea at the start of the monsoon in July and August. As the data from repeat sampling allow tracheid production to be assigned to a certain interval of the season (see Table 6.3), it is possible to compare tracheid dimension and temperature in those intervals (Fig. 6.22). The deviation of values is very large, but it is easy to see important patterns. The tracheid dimension of earlywood increases from temperatures of 6–8 °C to the optimum 16–18 °C. The earlywood tracheid dimension decreases remarkably with further increases in temperature. The maximal sizes of latewood tracheids increase with increasing temperature, with a higher optimum temperature range of 23–25 °C. The deviation of each point from the maximal curve on the graph is obviously conditioned by soil moisture influence, as this factor also influences tracheid dimension. The greatest suppression of tracheid dimension by high temperature occurs in the transition zone.

The maximal tracheidogram represented in Fig. 6.21 shows smooth changes of tracheid diameter during the season. At the very beginning of growth, the tracheid diameter has a slight increase, then it smoothly decreases towards the outer boundary of the tree ring, achieving its lowest value at the boundary. These smooth changes correspond to the seasonal change of solar radiation. However, if we take into consideration the data in Table 6.3, in which the relative proportions of annual rings formed in different months of a season are shown, tracheid diameter achieves maximal values in May. This

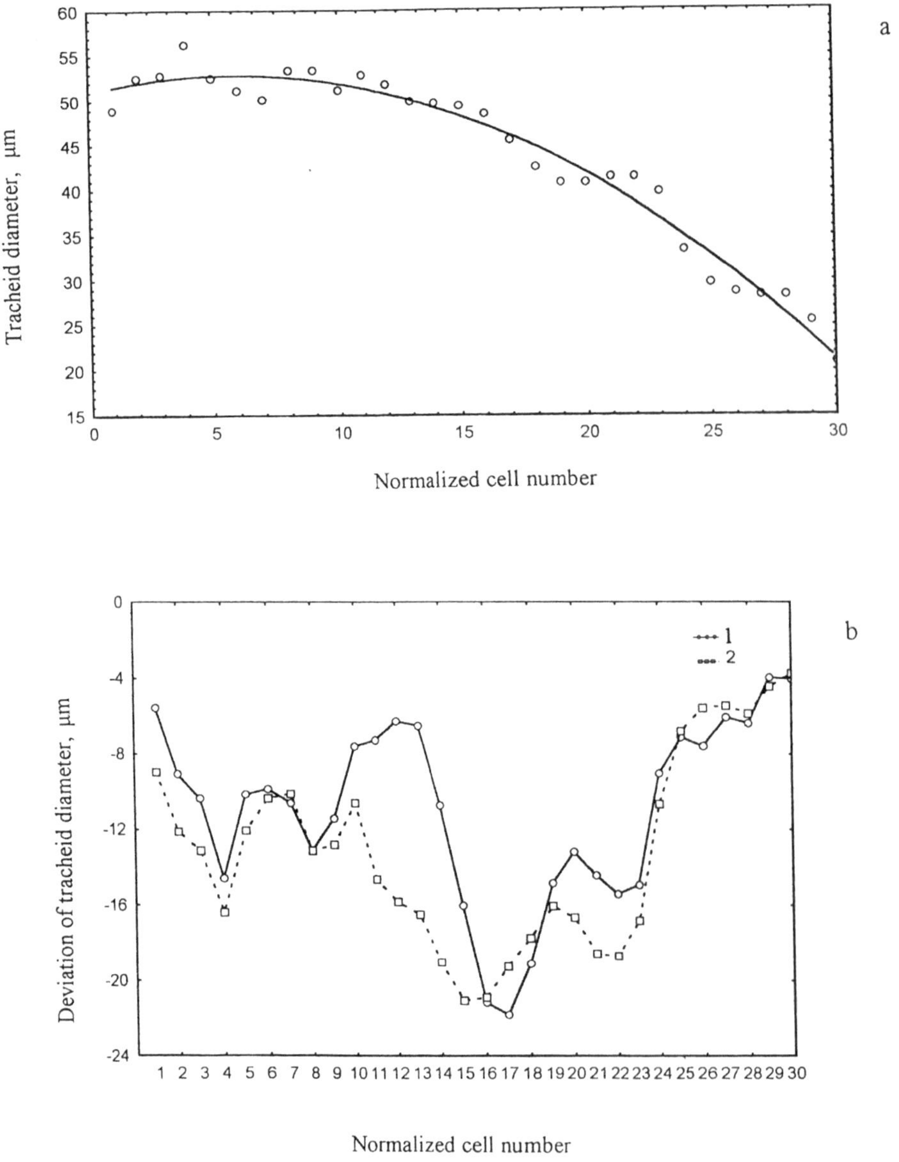

Fig. 6.21. Maximal tracheidogram (**a**) and differential tracheidograms (**b**) for 2 years (*1*, *2*; detailed explanations in text)

means that intra-seasonal changes of tracheid diameter in a maximal tracheidogram cannot be completely defined by day-length changes or by the quantity of solar radiation which reaches maximal values in the third decade of June. However, changes of tracheid diameter in a maximal tracheidogram fully correspond to a change of $E^{*}dE/dt$, where E is solar radiation intensity

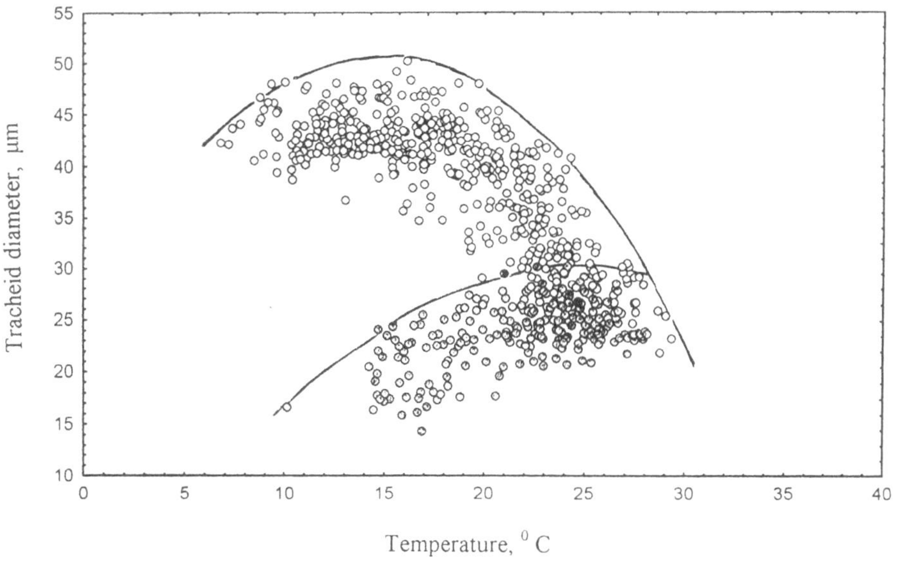

Fig. 6.22. Cumulative plot of tracheid diameter vs temperature for all tracheid measurements of *Pinus densiflora*. *Open circles* Earlywood, *circles with dots* latewood. *Upper line* Maximal earlywood tracheid diameter, *lower line* maximal latewood tracheid diameter

and dE/dt is its first derivative (Fig. 6.23). The correlation between the two values corresponds to the level of a functional fit (R=0.96; Fig. 6.23). We note that, in the given case, E and dE/dt are standardized in relation to maximal values and are completely positive. The positive linear relationship shows that tracheids reach their maximal diameter under the combination of high solar radiation and relatively small changes in day-length that occurs in Korea in May.

Thus, the analysis of anatomical changes of tracheids in annual tree rings of *P. densiflora* formed in years of different weather conditions shows the following: the seasonal trend in tracheid diameter reflects seasonal changes of growth rate and tracheid production. Tracheids with maximal diameter are formed during May at the highest rate of tree-ring growth. Seasonal changes in tracheid diameter are positively linked with solar radiation and day-length changes. Soil moisture is also positively correlated with tracheid diameter during the season. The relationship of tracheid diameter with temperature is more complicated. In the range up to 16–18 °C (for latewood cells up to 23–25 °C) it is positive, but with further temperature increase seasonal growth rate and tracheid diameter are suppressed.

We can consider these data as a source of additional information on the influence of climatic factors on the seasonal growth of *P. densiflora* in a monsoon climate. It is quite evident that the measurements and dendroclimatic analysis of annual tree ring width are insufficient to find out how the main

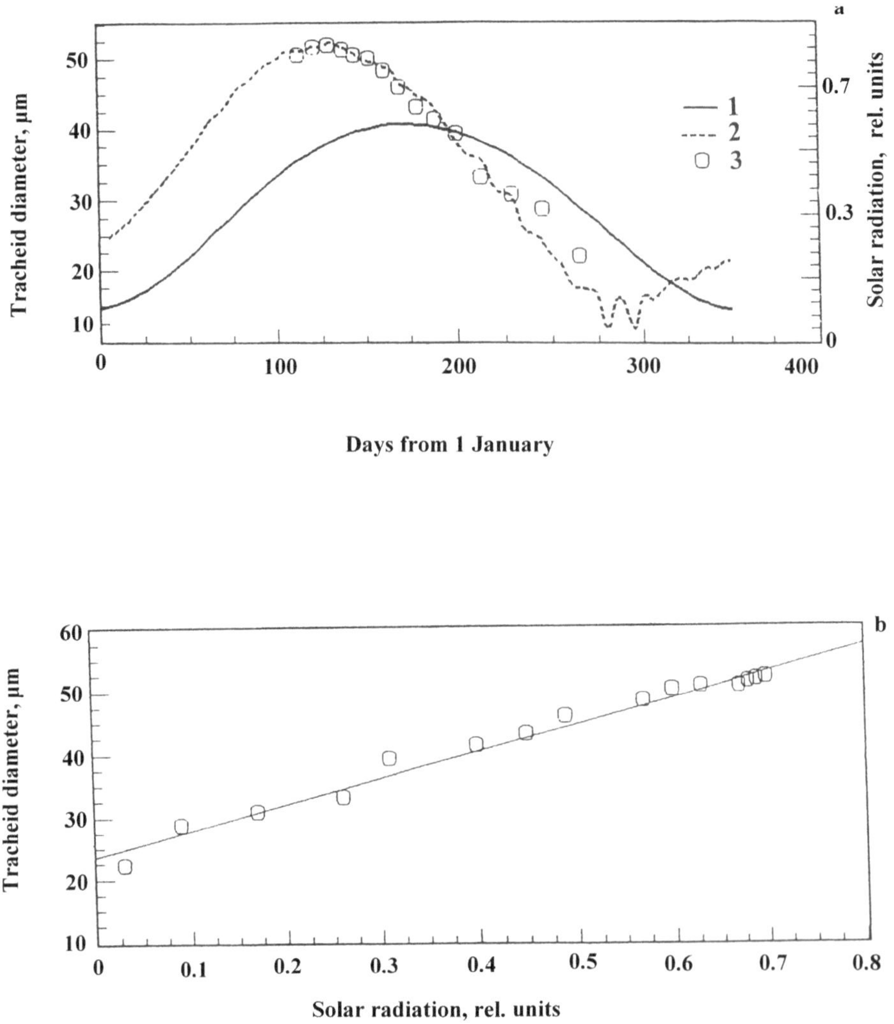

Fig. 6.23. **a** Seasonal variations of solar radiation (*1 E*) and solar radiation index (*2* $E^*\mathrm{d}E/\mathrm{d}t$) in comparison with maximal tracheidogram (*3*). **b** Solar radiation index ($E^*\mathrm{d}E/\mathrm{d}t$) vs tracheid diameter

climatic factors – temperature, moisture, and solar radiation – influence the dynamics of *P. densiflora* growth. As the tree-ring anatomy data show, seasonal variations of these factors are fixed in the kinetics of growth and tracheid production and are reflected in changes of tracheid diameter. The link between growth rate and tracheid diameter gives results identical to those obtained for the conifers of the forest-steppe and taiga zones and for *P. ponderosa* growing in conditions of high moisture deficit (Vaganov et al. 1985; Fritts et al. 1991; Park 1990; Shashkin and Vaganov 1993). The close correlation of the seasonal trend of tracheid diameter with solar radiation changes (E^*dE/dt) is interesting from the viewpoint that it confirms the results of a number of previous studies on the influence of solar radiation intensity, day-length, and photoperiod on radial tracheid dimension in conifer seedlings (Larson 1962, 1964; Richardson 1964; Creber and Chaloner 1984; Kozlowski and Pallardy 1997). In this research, it is shown that not only intensity but also the day/night duration ratio influence the anatomical characteristics of the tracheids formed: length, radial diameter, and cell wall thickness. That is why our data show that seasonal growth rate and tracheid diameter are closely correlated, not only with solar radiation intensity (E) but with its changes (dE/dt), which can be considered as a close analogue of photoperiod (Kozlowski and Pallardy 1997).

The positive correlation of tracheid diameter with precipitation, as well as the analysis of "differential" tracheidograms revealing the seasonal kinetics of soil moisture decrease before the rainy season is completely consistent with other work. There is a long dry period before the onset of monsoon rains in the southwest United States. An analysis of seasonal kinetics of *P. ponderosa* growth showed a gradual decrease of radial tracheid diameter through this period and typically the formation of false rings with small tracheids and thin cell walls (Fritts 1976). Wilpert (1990, 1991), analyzing variability of European spruce (*Picea abies*) tree-ring anatomy in conditions of moisture deficit, linked the decrease of tracheid diameter to the increase of soil moisture deficit and even obtained a quantitative fit of soil water potential with the "differential" tracheid diameter. No doubt soil moisture deficit plays an important role in the suppression of tracheid diameter and transition to latewood formation in annual rings of *Pinus densiflora*.

The nature of the statistical relationship of tracheid diameter variation and temperature shows that there is an optimum. The influence of temperature on the rates of seasonal processes (photosynthesis, production of xylem and phloem, height, growth, etc.) is widely represented in publications concerning plant growth (Gates 1980; Kramer and Kozlowski 1983; Kozlowski and Pallardy 1997; Barnes et al. 1998). Such results, however, have mostly been obtained for young trees growing under controlled conditions. In our research they are obtained indirectly – by a comparison of temperature during short intervals with the diameter of tracheids formed during those intervals. These results also confirm that tracheid diameter fixes (or "records") the changes in the xylem growth rate during the corresponding intervals of the season.

We will stress some features of the link between tracheid diameter and temperature. For the part of Korea studied, the average daily temperature begins to exceed the optimum of 16–18 °C in early June. So, in June the growth rate begins to be slowed down by high temperatures, even if the soil moisture is sufficient for optimal growth. In July and August the limiting influence of high temperatures is still greater, though monsoon rains provide abundant water in soil that is practically saturated. The tracheids of transition and latewood are formed in these months. There may be a higher optimum for tracheid production and hence radial dimension in latewood as compared with earlywood (Fig. 6.22).

Thus, the changes of tracheid dimension inside the annual tree rings of *P. densiflora* record the changes of seasonal growth rate and the influence of current climatic factors. In these conditions, the seasonal changes of tracheid diameter depend on the course of solar radiation, soil moisture, and temperature. Whichever one of these factors limits growth varies at different times in the season. Several main results may be noted. At the beginning of the season, growth acceleration is conditioned by a temperature increase, but the period of its influence is short, as optimal values are 16–18 °C and these are usually achieved by the end of the first decade of May. In the following period up to the end of June (start of the rainy season), tree-ring growth rate is limited by low soil moisture. With the beginning of the rainy season, some retardation of growth rate is produced by the increase of effective solar radiation, and slowed down additionally by the high seasonal temperatures. Precipitation in April–May, i.e. in the period prior to cambial activation and in the initial period of cambial activity, accelerates production and the increasing diameter of earlywood tracheids.

The observed intra-seasonal changes in rate and tracheid dimension are integrated by inter-annual variations of tree-ring width. Variations of tracheid diameter inside tree rings of the same calendar years of growth in different trees are of similar character; and different years differ in the tracheidograms formed. A clear view of such differences is given by "differential" tracheidograms. The observed correlation link between tree-ring width index and tracheid diameter on an inter-annual basis is relevant to the kinetic reconstruction of tree-ring seasonal development by tracheidograms, which is important not only for a better understanding of the physiological mechanisms of wood formation but for the prediction of those changes in seasonal growth kinetics which can be expected in the case of significant climate changes.

6.6 Conclusions and Discussion

A conceptual scheme of the environmental control of xylem formation has been outlined. This has been based on the findings reported in Chaps. 3–5, which point to the the cambial zone as the target of environmental control, and the strong relationships between cell production, expansion, and cell wall thickening.

The basic conceptual scheme of the environmental control of xylem differentiation should have certain characteristics. These include an ability to treat the activities occurring in the cambial zone as the targets of environmental control, albeit within a hierarchical structure. In this structure, signals from a high level, for example the whole tree or the canopy, control the operation of key processes at lower levels, for example the tissues. In this case, we propose that the key process is cell production in the cambial zone. It, in turn, controls the outcome of other processes taking place at the tissue level, for example the extent of cell expansion and cell maturation.

Several examples from very different regions, from the forest-tundra in the North, through the steppe zone of southern Siberia and the semi-arid American southwest to monsoon Asia have been examined in the light of this scheme. In each case, as well as for a number of statistical studies of much wider geographical extent, it seems useful to think in the terms we have described. In particular, it is possible to explain observed patterns of cell dimensions within and between growing seasons rather simply, if one assumes the primacy of the environmental control of cell production over the final products of expansion and maturation.

7 Modeling External Influence on Xylem Development

7.1 Introduction

One of the most effective approaches to understanding nature is by the creation of models of processes or events, if we use the term "model" in a broad sense. Each well known equation in physics is a quantitative model of the corresponding process. In biology and ecology, many of the models used seem more complicated because they include many processes and their inter-relations. Many of them also have a hierarchical structure. Statistical models which only evaluate the probable relations between the parameters of processes and forcing mechanisms are still useful. There are several classifications of models used in ecology (Odum 1971; Fedorov and Gilmanov 1989). Modeling has several advantages: (1) complete quantitative description which allows one to repeat the procedure with the same results, (2) possible comparison of final or intermediate output with experimental measurements in order to improve the model status and hence our knowledge, (3) the possibility to plan experiments with exact goals based on model structure, (4) prediction (Fedorov and Gilmanov 1989).

Considering the conceptual scheme of Denne and Dodd (1981), Ford (1981) wrote: "If we are to build a model of environmental influence on xylem production then we must be prepared to specify this influence through the intermediate pathways of the growth processes. The problem of modeling then becomes one of producing mathematical formulations of the relationships between environment and growth which can be tested experimentally". He further proposed that such a model could not be developed because there were two large problems:

1. Are cell division, expansion and wall-thickening completely independent processes? Or can some of the apparent differences in their control be resolved simply by considering other quantities, as Benthel (1964) has suggested? Certainly if we wish to consider the diffusion of material to the developing xylem, then total density of wall formation is likely to be an important parameter to measure.

2. Does the tissue change in character as its rate of growth alters? Again if we are to consider a growth model incorporating diffusion processes, then we should investigate possible changes in ray density and size and possible differences in the probability of death of new radial files or in the cell growth rates of established ones in relation to the proximity of cambial cells to radial files of tracheids.

He concluded that, at the time of writing (the beginning of the 1980s), the theory and measurement systems to produce a testable, function-based model of xylem production and differentiation were absent. Savidge's review (1996) arrived at a similar conclusion. However, there were several publications on modeling of xylem production in relation to environmental factors between these dates (Vaganov et al. 1990; Fritts et al. 1991; Shashkin and Vaganov 1993). The rapid development of dendroclimatology and dendroecology in the 1970s and 1980s resulted in a large body of observational material on the environmental influences on tree growth in different regions and climates and many sophisticated statistical analyses of these effects. In spite of this, those working on xylem formation at the tree and tissue levels have not commonly used the results and data of dendroclimatic analysis whereas dendrochronologists have mainly focused on statistical models that are easily testable by reference to widely available meteorological data. So, to solve the first problem declared by Ford, it will be necessary to combine the experimental data from tree-ring width, cell, and density analysis to produce a simple scheme and model for calculating tree-ring formation in a changing environment. The second aspect of the discussion concerning the simulation modeling approach is the following. It is a widely accepted point of view that a model has to include all knowledge available at the time of its formulation. A model, however, is a tool that allows us to summarize our knowledge about a process, taking into account only the most important interactions. Then, at the stage of model usage, we can compare the model results and observations to reveal the agreements and divergence between them. If agreements explain the studied processes well enough for our specific purposes, we conclude that the model is adequate and mathematically describes the main patterns of the processes of interest. If not, we have to improve the model, and so on. However, if the model explains some basic processes but cannot explain others, we may incorporate some improvements (some additional relationships) into the model that could be expected to be sufficient for explanation. Coming back to Ford's (1981) conclusions, we may say that we have enough results and the power to build a model of the climate–tree-ring relationship, because the first problem is solved and it is not necessary to solve the second problem in the context of tree-ring–climate modeling, except perhaps in the context of age-dependent relationships. In order to set the context for the description of the Vaganov–Shashkin (VS) model, we give a brief summary of the use of statistical models in dendroclimatology and their advantages and limitations

(Sect. 7.2). Earlier mechanistic simulation models of tree-ring formation are introduced in Sect. 7.3. Then the VS model, which is the primary focus of this book, is introduced (Sect. 7.4). It is designed to simulate seasonal growth and tree-ring formation, using daily meteorological data. Particular attention is given to the modeling of cell growth within the cambial zone and the production of new xylem cells, as this is the key process through which, in the model, the later stages of tree-ring formation are controlled. Examples of the application of the model in a range of very different environments are discussed (Sect. 7.6), namely, near the northern timber line, in the taiga and steppe zones, in a semi-arid climate, and in a monsoonal climate. Conclusions are presented and discussed in Sect. 7.7.

7.2 Statistical Models in Dendroclimatology and Dendroecology: Their Advantages and Limitations

A special period in the development of dendrochronology began in the 1960s with the application of the methods of multivariate statistics to the study of the relationships between climate and the radial increment of woody plants (Fritts 1962, 1976, 1991; Fritts et al. 1971; Hughes et al. 1982; Cook and Kairiukstis 1990). Monthly mean air temperature and monthly total precipitation were most often used as the candidate climate-predictors because they are, by far, the most widely available climate data for the largest range of locations on the Earth's land surface over the longest period of years. The length of these records is important because, for any derived statistical model to be valid, it must have the largest possible ratio between the number of cases (years) and the number of variables (months or seasons per year per climate parameter). Usually, climate data were used not only for the growing season, but also for the months of the preceding fall, winter, and spring. Within the general scheme modified by Fritts et al. (1971) for application to dendroclimatology, regression models are used in which the climate variables are the predictors and the tree-ring variable, for example ring-width index, is the predictand. These regression models are known as "response functions'. Although matrices of correlations or partial correlations between radial increment and these climate variables may be used to test the linearity of relationships, they do not provide a tool that should be used for prediction.

The development of a predictive model requires the calculation and testing of some kind of regression model. Several problems arose during the development of these models. Two are common to all regression models – the risk that a likely important predictor has not been considered and the perils of extrapolation, that is applying the regression model to a combination of predictor values outside the range found in the data used in its development (Mitropolsky 1971). The second of these may be avoided by the exercise of self-discipline by the users of the model – never extrapolate, or if it is unavoidable, mark the extrapolations clearly. The effects of the first, the omission of a vital predictor, may be assessed by testing the model's performance under a range of conditions. Two further problems of special relevance to the development of climate-tree ring response functions were: (a) lack of independence between succeeding years' ring measurements, and (b) frequent and sometimes strong multi-collinearity between the predictors. Problem (a) is important because it violates a basic assumption of all regression procedures, namely that the residuals are independent from case to case. This was initially dealt with by forcing the prior year's ring-width indices into the regression model to provide lagging predictors (e.g. Fritts 1976). Later, improved understanding of the time-series properties of tree-ring data (Meko 1981; Guiot et al. 1982; Cook 1985) led to techniques for the removal of the persistence from tree-ring data prior to the derivation of response functions. This was done by calculating the residuals from some time-series model of the data and using them as the predictand in the response function. Problem (b) was important because multicollinearity of the predictors violates an important assumption made when interpreting multiple regression analyses in terms of the original predictor variables. This assumption is that the predictors are independent of one another, and so may be treated as being additive. When inter-correlated predictors are offered to a stepwise multiple regression, the type often used in such situations, a variable important to tree-ring growth may be excluded from the model simply because one with which it is correlated has already entered the equation. If the equation is being developed solely for prediction, this is not a major problem, but it may invalidate interpretations. Similar limitations apply to the simple correlation function. The problem of multi-collinearity has other consequences, for example, the sampling distributions of the regression coefficients can become so wide as to lead to poor predictions (Wilks 1995).The problem of multicollinearity between predictors was dealt with very elegantly by Fritts et al. (1971) when they used a modification of orthogonalized regression analysis. Before entry as candidate predictors, the original set of predictors is converted to a set of new, *uncorrelated* variables so that they satisfy this assumption of multiple regression analysis. Because this is done using principal components analysis (also known as empirical orthogonal function analysis), no variance is lost in the transformation and indeed it may be completely recovered by applying the appropriate matrix algebra. This means that, after deriving a regression equation to calculate the ring-width index from objectively selected principal components of the temperature and precipitation data, it is possible to derive a set of response function elements analogous to partial regression coefficients, expressed relative to the original units of temperature and precipitation. Thus, a response function element of +0.2 for, say, April temperature means that an increase (decrease) of 1 °C in April temperature will produce a proportional increase (decrease) of 0.2 (20 %) in ring-width index, relative to the mean ring-width index in the calibration period, all other variables being held constant (Figs. 6.11b, 6.12). This therefore retained both the capacity to develop a regression model that conformed to the main assumptions of this analysis and a method of displaying the form of this model in an understandable manner in terms of familiar climate variables. The number of statistically significant response function elements is determined by the conditions of growth, the species of a tree, the duration of the climate series used for comparison (calibration), the quality (homogeneity) of climate data, and so on.

Consider, for example, response functions obtained for tree-ring chronologies of larch and pine trees located along the Yenisey meridian from the northern to the southern limit of forest vegetation (Fig. 7.1) At the northern limit of the forest (Fig. 7.1, FT, forest tundra) only June–July temperatures have significant positive relationships with the tree-ring width index, with the relative influence of July temperature being slightly greater than that of June. In the northern taiga (Fig. 7.1, NT) the relative contribution of June temperature is much higher than that of July, perhaps because of earlier activation of cambium and an earlier start of the growth season (Vaganov and Park 1995; Vaganov et al. 1996a–c). There is a significantly negative response function element for the April prior to growth, for which we have no explanation, and a negative response function element for the February prior to growth. Snowy winters appear to decrease the value of the annual increment. To explain this influence, it is necessary to assume that the larger the snowpack at the end of the winter, the more slowly soil temperatures increase at the time of the spring thaw, delaying the start of the growth season and finally influencing the formation of the cambial zone and the total production of wood (Vaganov et al. 1994, 1999). In the middle taiga, June temperature has a significantly positive influence, as in the northern taiga, and a similar negative element for April temperature. Southern taiga (Fig. 7.1, ST) has a significantly positive response function element for July, whereas the forest steppe (Fig. 7.1, FS) response function shows positive influence of both temperature (January prior to growth) and precipitation (May in the year of growth). There is also a significant negative response function element for the temperature of the prior year. Finally, in the growth conditions of the steppe zone May–July and September–October, the temperatures of the previous year have clearly significant negative coefficients and significantly positive precipitation coefficients for April–June. We do not have an explanation for the positive January temperature element but can see that, in this drier environment, high temperatures at the end of the prior growing season, just before the autumn freeze, may accelerate soil drying and lead to an earlier moisture deficit in the next growing season. This in turn would affect the production of tracheids (Vaganov 1989; Shashkin and Vaganov 1993). By the same token, the positive element for May precipitation is not surprising. This small set of response functions serves to illustrate that it is possible to use response function analysis to identify general trends in tree-ring response to climate.

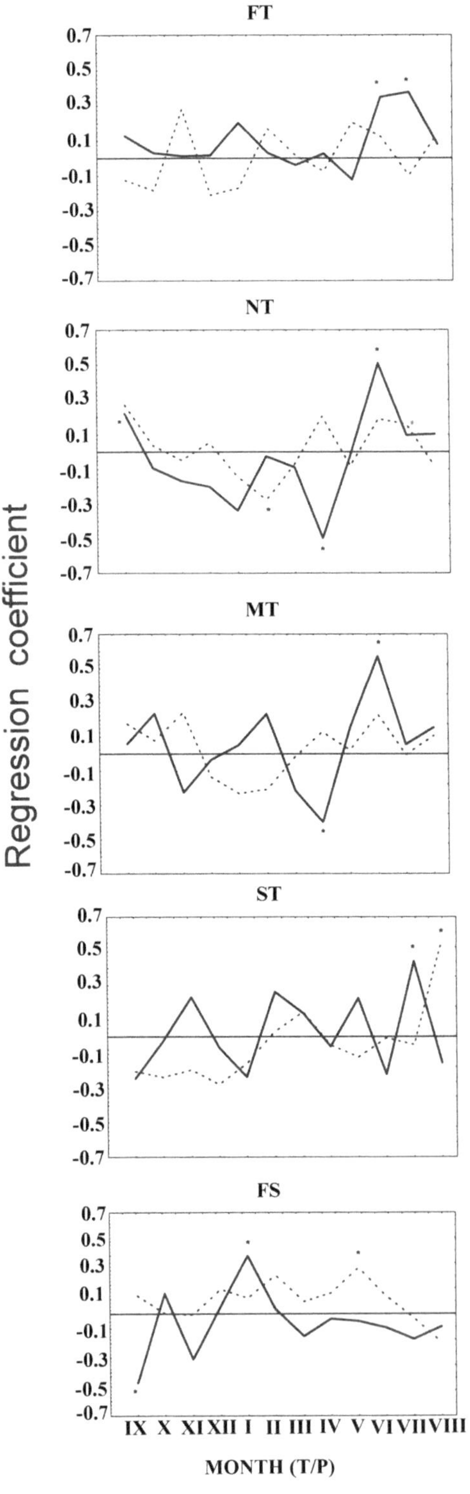

Fig. 7.1. Response functions of tree-ring width with monthly temperature (*solid line*) and precipitation (*dotted line*) derived for regional chronologies in different taiga regions along the Enisey transect. *FT* Polar timberline, *NT*, *MT* and *ST* northern, middle and southern taiga accordingly, *FS* forest-steppe zone. *Stars* indicate elements significant at 95 %

A much fuller treatment of this is given by Fritts (1974), who calculated response functions for 127 tree-ring chronologies of conifers from a wide range of conditions in western North America. He reports that "The median percent of tree growth variance accounted for by climate is approximately 60 % to 65 %." He used an objective clustering approach to group the response functions and showed that the differences in response functions may be largely attributed to site factors, such as aspect, elevation, and latitude, rather than to species differences.

It is important to remember the difference between an empirical response function and an ideal response function that might be derived from experiments on physiology of woody plant growth (Kramer and Kozlowski 1983; Kimmings 1997; Kozlowski and Pallardy 1997; etc.). "Physiological" response functions for all factors of environment without exception (temperature, light, humidity, CO_2 content, concentration of nutrients in soil, etc.) would either have the form of sigmoid curves with saturation or of bell-shaped curves with an optimum of the factor's influence. The latter, for example, is characteristic of the curves of growth rate response to temperature (Gates 1980). As temperature reaches some optimal value, the growth rate continuously increases, then in the zone of the optimum it reaches maximal values, and only after a further temperature increase (often to more than 23–25 °C) does the growth rate decrease. If we compare such a response with the statistical response function to the temperature for trees from a steppe zone, it is easy to see that even far from optimal temperatures of the prior September appear to influence the growth rate (tree-ring width) negatively. As discussed above, this is not a direct effect, but rather an indirect effect through soil moisture. The physiological response to moisture in soil usually has a shape of a sigmoid curve and, once the optimum is reached, a further rise of moisture content does not yield increased growth rate. Indeed, it may produce a reduction in growth if water-logging and anaerobiosis in the soil ensue.

The extensive dendroclimatological literature contains many cases where a naive reading of statistical response functions would appear to contradict physiological understanding. In a number of these cases, the biological and ecological interpretations of the statistical response functions are complicated. One important reason for this is that they represent a combination of physiological responses to climatic variables which are transformed by the operational environment of the tree – the surrounding stand conditions, soil, aspect, slope, elevation, regional climate, and so on (see Fritts 1976). Therefore, in a sense, the statistical response functions are the ecological response functions of trees to climatic variables. So they must be interpreted in their ecological context. This is also one of the reasons why it is perilous to interpret response functions calculated for individual sites. By comparing response functions for several or many sites in a region, it is possible to infer which patterns of response are general to a species in that region, and which may be specific to individual sites, as in the work of Fritts (1974) referred to above. In addition to its value in understanding the response of tree rings to climate, the examination of regional sets of response functions has another important purpose. It is an important stage in identifying the common climate signal in the trees of a region and disentangling it from site-specific factors.

The selection of situations where tree-ring growth responds simply and directly to a single strongly limiting factor is an essential procedure in dendroclimatology. Thus, in search of a moisture-related signal, trees growing in semi-arid regions on steep slopes with freely draining substrate will be used. If summer temperature is the target, trees from regions with cool, moist summers will be chosen, and, most probably, maximum latewood density used instead of ring width. As a result, a high proportion of all published dendroclimatological results show remarkably simple linear correlations between the tree-ring variable and some climate variable. Such situations are selected to maximize climate signal, not to represent the response of the forest. In such simple cases, physiological response functions can explain the statistical features of tree-ring chronologies and their associations with climatic variables. This is particularly the case in situations where the tree response to climate is consistently dominated by a single climatic factor or closely linked set of factors.

In a dry region, this could be the case for trees growing in situations such as that indicated in the upper right panel of Fig. 7.2. Because the tree is growing on freely draining material in a dry region, there is a rapid response of available soil moisture to precipitation, whose effect is not "damped" by intrinsically more slowly changing water sources, such as ground water. As a result, ring width is strongly influenced by inter-annual variations in precipitation and tends to show large and sharp inter-annual fluctuations. Such a "sensitive" tree-ring series may be contrasted with that produced from trees such as that shown in the upper left panel of Fig. 7.2. Here, as a result of the tree's location on level ground, or even in a hollow on poorly draining material, moisture is more consistently available from year to year and so water is not usually limiting to the formation of the tree ring. As a result, there is very little variation in ring width from year to year and the ring-width series is described as "complacent". Moreover, there is at most a weak relationship between moisture availability and ring width in this range of conditions. As can be seen from the lower panel of Fig. 7.2, both these series are produced by the same physiological response to differing environmental situations. Thus, under conditions of higher water availability, the proportionate increase in ring growth for each additional unit of water is very small or even zero, whereas when water is scarcer, the gradient of the relationship is much steeper, producing the kinds of difference seen also in Fig. 2.14.

Note that, although the overall form of the underlying physiological relationship is asymptotic, over the range of moisture availabilities to which the "sensitive" tree is almost always subject, the relationship is often effectively linear, as witnessed by the many highly significant correlation coefficients

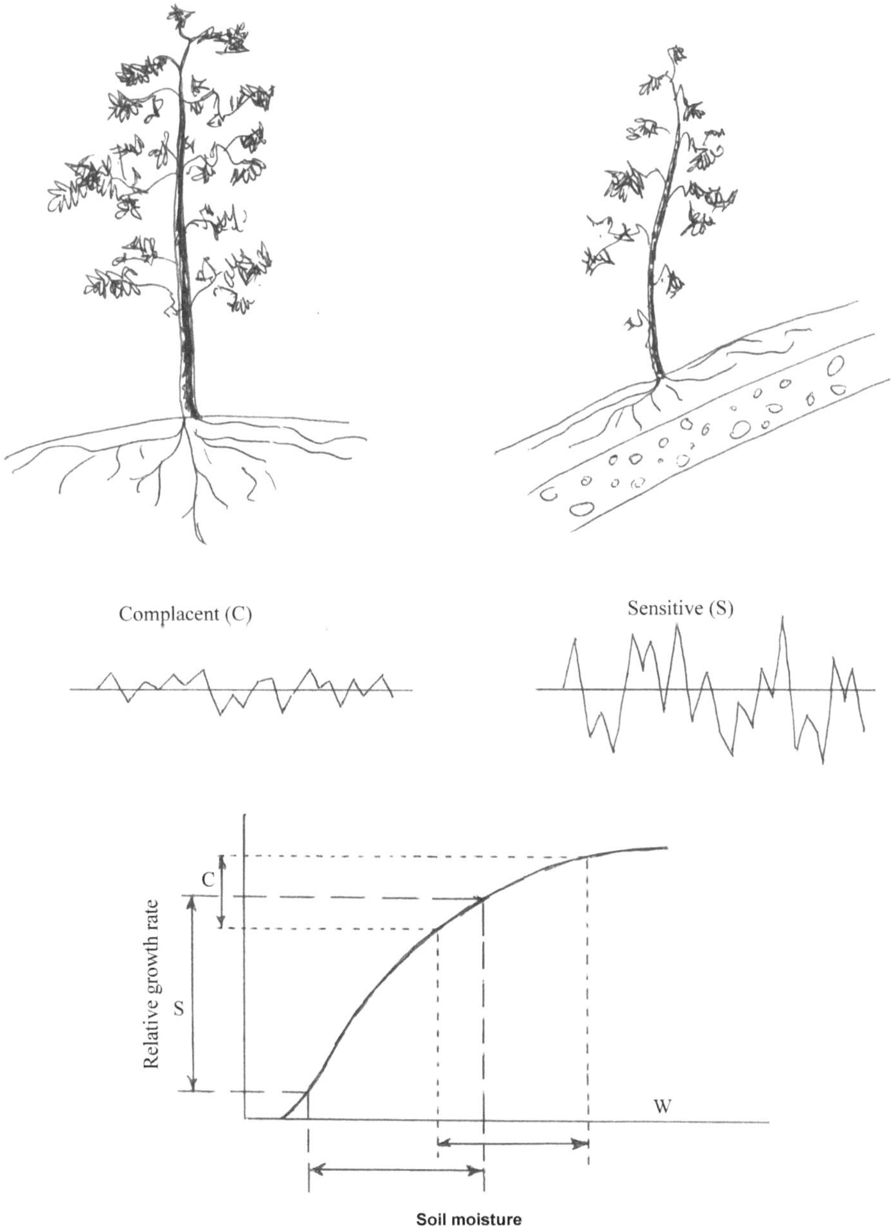

Fig. 7.2. Diagram illustrating the different sensitivity of tree growth to climatic variability (water regime) due to the non-linear response of growth rate to different values of an environmental variable

recorded between chronologies of trees from such sensitive situations and dominant climatic factors. See, for example, Fritts (1991) for evidence of the prevalence of such strong linear relationships in tree-ring chronologies specifically collected and prepared for dendroclimatic use. In fact, the selection of regions, sites, species, and individual trees likely to produce such "sensitive" tree-ring series are all vital stages in the development of dendrochronological data and are the main explanation for the large number of strongly linear relationships found in that literature.

Even without this system of selection, however, linear relationships are surprisingly common. For example, LeBlanc and Terrell (2001) calculated correlations between the radial growth of *Quercus alba* L. at 128 sites in the eastern United States and several climate variables. To their surprise they found that "correlations with early growing season *P*, *P*/*T* and *P*/*PE*, were almost as strong and spatially consistent as correlations with the more complex *AE*/*PE* ratio", where *P* is precipitation, *T* is temperature, *PE* is potential evapotranspiration, and *AE* is actual evapotranspiration.

Of course, it is perfectly possible to develop statistical response functions that do not assume linear relationships between tree-ring growth and climatic variables. Techniques available for this include neural networks, fuzzy regression, and intersecting response surfaces (Graumlich 1993; Guiot et al. 1995; Woodhouse 1999; D'Odorico et al. 2000; Zhang et al. 2000; Ni et al. 2002). Each of these techniques has its own strengths and weaknesses, including the most general drawback of the statistical "response functions" which is that, as mentioned above, they are only valid for the range of combinations of predictor values found in the period of calibration. This limits the applicability of statistical response functions for problems of forecasting the variability of radial tree growth because their use is only valid within the range of conditions seen in the calibration period.

Clearly, statistical response functions that use monthly, or even for example weekly, blocks of climate data cannot deal adequately with situations where a major seasonal transition occurs sometimes in one month or block, and sometimes in another. So for example, thawing of snow at the end of winter may have important consequences for the onset of the growth season, but in a particular northern hemisphere region it might occur in April one year and in May the next. Similarly, the date of onset of monsoon rains in tropical and subtropical environments may vary by many days and so pass from one element of a response function to another. It is important to remember that, until recently, almost the only climate data freely available for most of the Earth's landmass for a long enough period (several decades) were monthly mean temperature and total precipitation. This explains the emphasis on monthly data, in particular temperature and precipitation, in the dendroclimatological literature. Daily data for a wider range of variables has only been available for a more limited set of meteorological stations. As it is very likely that any process-based model will need daily meteorological input, the statistical response function approach is much more widely practicable. With the advent of more extensive datasets of daily and even more frequent meteorological observations, particularly those derived from the reanalysis of the full range of observed variables, including those from balloons and satellites, this limitation has lessened.

The limitations of statistical response functions are well known, indeed sometimes exaggerated. They have been applied usefully in many situations and provide a valuable first analysis of the climate response of tree-ring variables when used appropriately and with caution. One of their major uses has been in the design of statistical transfer functions used to develop quantitative, testable, reconstructions of climate variables in the past from tree-ring data. Whereas a response function is an equation for calculating a tree-ring variable such as ring-width index from climate data, a transfer function is an equation or algorithm for calculating climate data from tree-ring data. A response function is typically calculated for a single tree-ring chronology using some local meteorological data. A transfer function, in contrast , may well use many tree-ring chronologies, perhaps on a continental or even hemispheric scale, to calculate either a single climatic series, or a time-series of the spatial patterns of the chosen climate variable. The details of their use and their strengths and weaknesses are largely outside the scope of this book, but are reviewed elsewhere (Fritts et al. 1971; Hughes et al. 1982; Cook and Kairiukstis 1990; Fritts 1991; Cook et al. 1994, 1999; Mann et al. 1998).

It would be a mistake to think that dendroclimatologists base their work on statistical methods to the exclusion of biological and ecological considerations. For example, Fritts (1976) gives great emphasis to examining the biological reasonableness of response functions before using them in the design of transfer functions. The VS model, and no doubt its successors, provides a valuable additional tool for this stage of climate reconstruction, the evaluation of biological reasonableness. Similarly, just as climate reconstruction from tree rings must have a sound biological basis, so tree biology must be able to assimilate and benefit from dendroclimatology and the unique body of data and results on tree growth of many species in regions and habitats worldwide that it provides. In this case too, models such as the VS model provide a bridge.

7.3 Mechanistic (Simulation) Models of Xylem Development

7.3.1 Wilson and Howard's Computer Model for Cambial Activity

Wilson and Howard (1968) published the description and application of a model of cambial activity in which they fully developed the idea of independent influences of environment at all stages of tracheid differentiation. The model simulates the daily processes of cambial activity in conifers during a growing season, and uses a "combination of 'rules' that specify the behavior of cells in a radial file as they pass through the various phases of differentiation" (p. 77). The construction of the model repeats the processes of cell division within the cambial zone, cell enlargement, and cell wall-thickening. The basic inputs of the model are the values for 14 variables; and each of these variables depends on time on a daily basis. So, in reality, to calculate the kinetics of cell production and differentiation by the cambial zone, it is necessary to input the following table of units (Table 7.1). What variables are chosen as most important in the model? To describe the process of cell division, there are rates of radial enlargement (dividing cells; V1) and the maximum radial diameter (also for dividing cells; V2). According to the terminology used in Chap. 3, V1 is equal to the rate of production or inverse to the cell cycle. The next variables determined are the maximum number of xylem mother cells (V3) and the maximum number of phloem mother cells (V3). The initial cell is also specified in the model by the rate of radial enlargement (V4), maximum radial diameter (V2), same as mother cells, the rate of elongation (V5), and the maximum length (V6) to determine when pseudotransverse division occurs and the probability for a daughter cell from periclinal division of the initial becoming a phloem mother cell (V7). There are other parameters to describe the processes of radial enlargement and cell wall-thickening: daily rate of enlargement (V8), maximum radial diameter (enlarging cells; V9), primary wall thickness (V10), rate of cell wall-thickening (V11), and maximum wall thickness (V12). Each of the variables changes during the growing season and these changes form the daily input of the model. Strictly speaking, the real input is the number of variables multiplied by the number of days (or short intervals, as in Table 7.1) in a growing season. Most of them have to be determined from seasonal measurements of wood thin sections. So Wilson and Howard's (1968) model gives a "mimic" image of tree-ring formation and cannot be used without seasonal measurements. The results of the calculations are shown in Figs. 7.3, 7.4. Figure 7.3 shows the time-dependent changes in the number of cells at different stages of differentiation and especially shows (Fig. 7.3, arrows) the mid-summer drought effect that decreases cell production and tracheid enlargement. The effects of mid-summer drought on tracheid diameter are seen better in Fig. 7.4, where the simulated and measured within-tree-ring variations of radial tracheid diameter are shown.

Table 7.1. Values for input variables used to develop the red pine ring (*Pinus resinosa*). The length of the cell cycle is half the maximum diameter of the mother cells (5 μm in this example) divided by the rate of enlargement for dividing cells (not given in this table). *XMC* Xylem mother cells, *CWT* cell wall thickness (from Wilson and Howard 1968)

Day Number	Dividing cells		Enlarging cells		Maturing cells	
	Maximum number of XMC	Average length of cell cycle (days)	Rate of enlargement (μm/day)	Maximum radial diameter (μm)	Rate of CWT (μm/day)	Maximum CWT (μm)
0	3	–	–	–	–	–
1–5	7	6.2	5	40	0.2	2
6–10	7	5.6	5	43	0.3	2
11–15	9	5.0	6	39	0.3	2.1
16–20	11	4.6	7	38	0.4	2.1
21–25	10	6.6	6	36	0.4	2.2
26–30	9	10.6	5	36	0.35	2.1
31–35	8	14.3	4	36	0.3	2.4
36–40	7	18.4	3	37	0.3	2.2
41–45	6	20.0	2.5	31	0.3	2.1
46–50	5	20.6	2.5	30	0.3	2.0
51–55	4	24.0	2	28	0.25	2.0
56–60	4	24.0	2	22	0.2	2.4
61–65	4	24.4	2	20	0.2	2.5
66–70	3	31.3	1.75	19	0.2	2.6
71–75	3	31.3	1.75	20	0.2	2.5
76–80	7	6.0	4	22	0.2	2.5
81–85	7	5.9	4	23	0.3	2.6
86–90	7	5.8	4	22	0.3	2.8
91–95	6	5.4	2	19	0.3	3.0
96–100	5	7.4	3	25	0.25	3.2
101–105	4	10.3	2.5	20	0.2	3.4
106–110	4	12.5	2	19	0.2	3.6
111–115	4	14.6	2	20	0.2	3.8
116–120	4	14.6	1.5	20	0.2	4.0
121–125	3	13.2	1	22	0.2	4.5
126–130	3	15.6	1	16	0.25	4.0
131–135	3	17.8	1	12	0.25	3.5
136–140	3	19.2	0.5	10	0.2	3.0
141–145	3	20.8	0.5	8	0.3	2.5
146–150	3	–	–	4	0.2	2.5

According to Table 7.1, in mid-summer drought the average duration of the cell cycle in the cambial zone increases up to 3–4 times, the number of xylem mother cells decreases and the rate of radial expansion also decreases. These changes are completely associated with the decrease in tracheid radial size produced during the mid-summer drought.

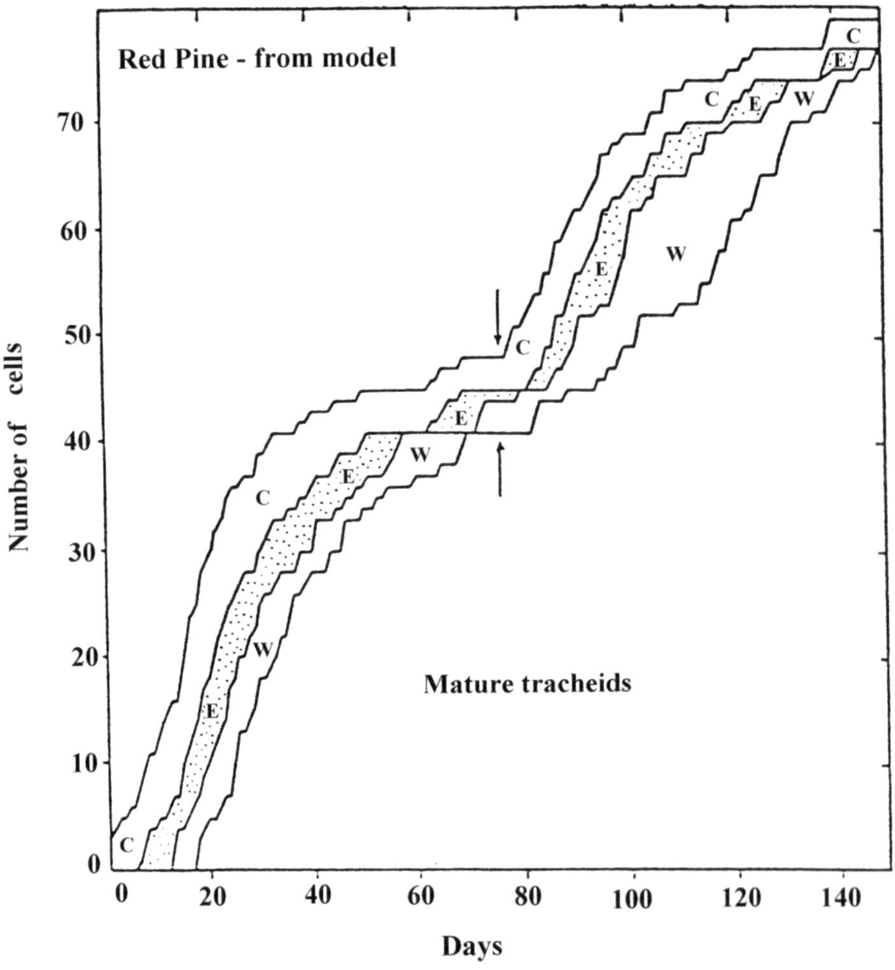

Fig. 7.3. Simulated development of a model red pine annual ring. *C* Cambial zone, *E* (*stippled*) enlarging zone, *W* zone of cell wall-thickening. *Arrows* End of simulated mid-summer drought. Note that the zones of enlargement and wall-thickening disappear when no new cells enter these respective phases. Read vertically, the graph gives the number of cells in each zone of differentiation for each day in the season. For instance, at day 40 there are two cells in the zone of wall-thickening, six cells in the zone of enlargement, and seven cells in the cambial zone. Read horizontally, the graph shows how long each cells was in the phases of enlargement and cell wall-thickening. For instance, cell number 50 spent 2 days in the phase of enlargement and 6 days in the phase of wall-thickening (Wilson and Howard 1968)

The authors assert the value of the model in the following ways: (1) it can generate new information, (2) it can test the "rules" used, and (3) it can include other sets of variables at other levels, such as environmental conditions and the concentration of growth regulators that may directly determine the values for the model's variables. It seems questionable that, from such a

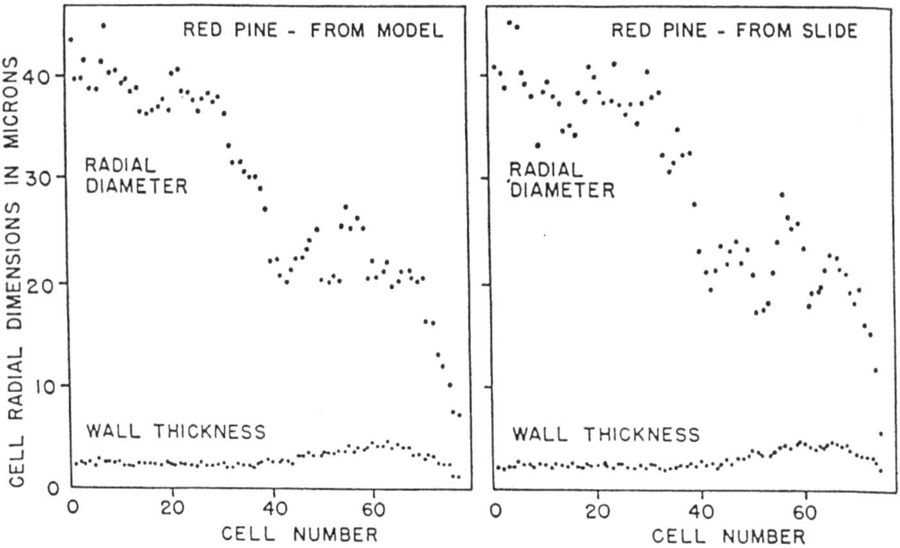

Fig. 7.4. Radial diameter and cell wall thickness of tracheids from real and model rings of red pine that passed through a mid-summer drought. Measurements for the model ring were taken directly from the output of the last day of season, after using the input from Table 7.1. Measurements from a microslide for the real ring are the average of five adjacent radial files in transverse section (Wilson and Howard 1968)

description, the model can produce new information outside that already input to the model. All processes in the model are almost fixed ("rules", in fact), so there is very little variability in the output. Point 2 can be applied to several variants of the model because the rules can be changed from the governing maximum number of xylem cells, maximum cell diameter and cell wall thickness (as used by Wilson and Howard 1968) to determine the stage of differentiation of the cell by its distance from phloem (according to Whitmore and Zahner 1966) or to specify the length of time that the cell can remain at each stage of differentiation (according to Wodzicki 1962). The most questionable is point 3, because to incorporate environmental or hormonal control of xylem development it is necessary to describe the influence of each factor(s) on the full set of variables (12) and on a daily basis. With this structure, the real size of the input is the number of variables multiplied by the number of days of a growing season and multiplied by the number of equations describing the environmental effects. The calibration and verification of such a model becomes impossible. This is the reason why Wilson and Howard's model is not widely used further and is not applied to the description and understanding of the processes of tree-ring formation in conifers growing in different conditions as well as in different climatic zones.

Later, Howard and Wilson (1972) slightly modified their model by including a stochastic element. They calculated tracheid production, radial expansion, and wall-thickening for different radial files using the following innovations: (1) each value in the stochastic model is selected from a statistical population of values (selected by Monte Carlo sampling) where the standard deviation is about 10 % of the mean value, (2) the maximum number of xylem mother cells allowed in the phase of division for each day is described by a multinomial probability distribution. The results of the model processing are shown in Figs. 7.5, 7.6. The mean tracheid diameter and cell wall thickness measured and computed are very similar (because they are also in the input), but it is more interesting to see the two-dimensional picture of different files

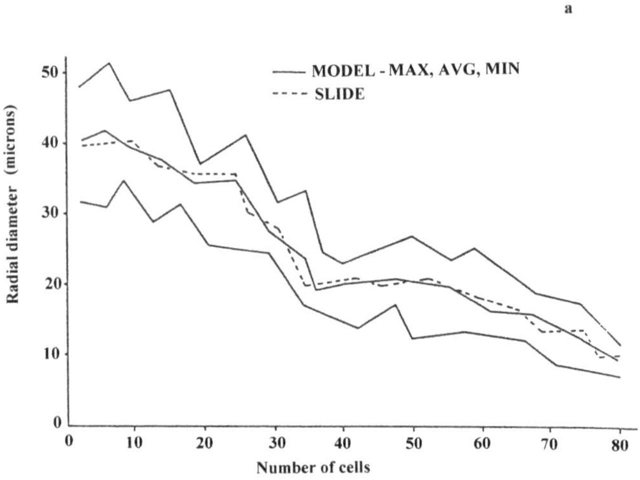

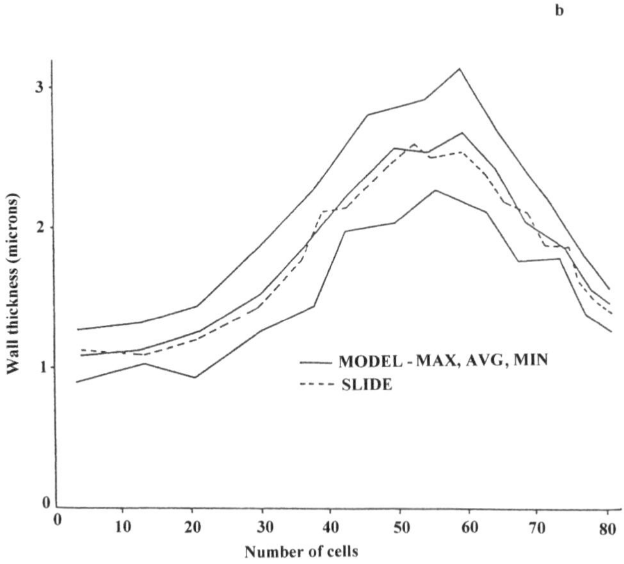

Fig. 7.5. Within annual ring variations: **a** cell radial diameters from a red pine radial file from a slide and from 16 simulations of the same file by the model (cell number is counted from the previous latewood), **b** cell wall thickness from a slide and from the model for the same file as in **a** (Howard and Wilson 1972)

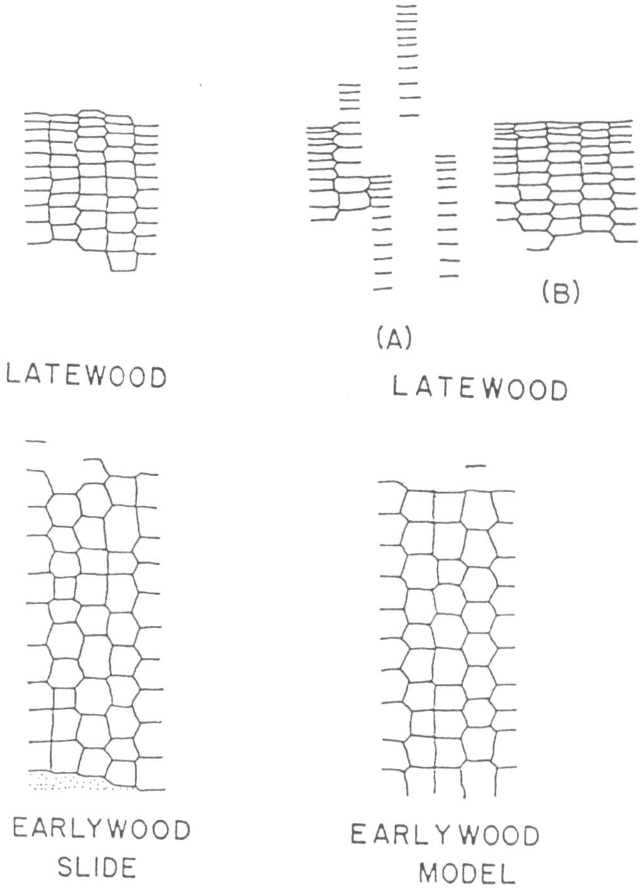

Fig. 7.6. Comparison of adjacent files in real wood from a slide with 16 simulated files. See text for rules for converting radial dimensions from the model to show cell geometry. Cell wall thickness has been omitted. In latewood from the model, files are: **a** aligned on the earlywood and **b** aligned on the latewood (Howard and Wilson 1972)

produced by the model (Fig. 7.6). In the earlywood, the shapes of tracheids in different radial files are similar to those in the slide, but in latewood the adjacent files have different cell numbers as well as different radial diameters of tracheids, so the different files do not match each other. If the last latewood tracheids of adjacent files are aligned, then the picture becomes realistic and very similar to the slide. This stochastic modification of the model shows clearly that the "mimic" model reflects the seasonal growth of the tree ring, but does not overcome the model limitations mentioned above.

At approximately the same time, Wilson (1973) published a paper with a description of a "diffusion model" for tracheid production and enlargement in conifers, where he tried to reduce the input parameters of the model and incorporate some "integral" parameters describing the dynamic diffusion gradient. In this modification, all growth processes were related to a single dynamic diffusion gradient arising from the source on the phloem side of the file. He assumed the concentration of the source (CS) and the growth: concentration factor (SL) as the main integral parameters, but for their input he had

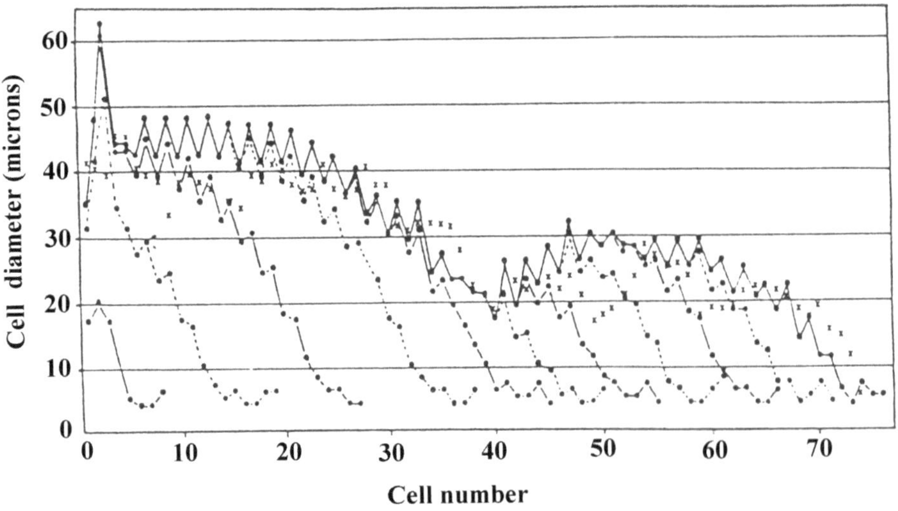

Fig. 7.7. Simulation of the production of a red pine annual ring. Cells are numbered in the order of production during the season. Diameters of each cell are shown at 10-day intervals with each day marked by a curve. At day 10 there were seven cells in the ring, all enlarging. The first completely enlarged cells appeared at day 30 (when there were 27 cells in the ring). Enlarging cells are marked by *dotted* or *dashed lines*. Cells that have stopped growing are marked by *solid lines*. Enlargement of each cell can be followed vertically. Each *X* marks the final cell diameter, measured from a real red pine annual ring formed during a midsummer drought (the cause of the dip in cell diameter for cell no. 40; Wilson 1973)

to describe it on a daily (5-day interval) basis. The model output gave, for example, the tracheid diameter of each cell at different stages of differentiation (Fig. 7.7) through the growing season. The tracheidograms in this figure are very similar to those called "instantaneous tracheidograms" (see Chap. 4). Note that the main limitation of the model was not overcome in this version because, although the number of parameters was reduced using CS and SL, it was necessary to describe these new integral parameters as changing during a growing season to calculate, for example, the decrease in tracheid diameter during a midsummer drought. Moreover, this version of the model does not use biologically distinct characteristics but uses those that cannot be explained or related to measurements. The assumption of the hormonal evidence for "diffusion" parameters have only theoretical interest and are in contradiction to some experimental data (Sheldrake 1971; Wodzicki 1971; Savidge 1996).

7.3.2 Stevens's Model With Slight Modifications

To overcome these limitations partially, Stevens (1975) proposed a computer program for simulating cambial activity and tree-ring growth, based on the input of 22 parameters (partly constant, partly time-dependent). He assumed that the half of the constant parameters which do not change markedly in Wilson and Howard's model, or which change with little loss of accuracy in output, can be considered as constant. The other half are described as time-dependent parameters. There is a daily rate of radial growth of a cambial initial, a daily rate of radial growth of xylem mother cells, a daily rate of enlargement and cell wall-thickening, etc. These parameters are similar to those in Wilson and Howard's model that have daily variations (see Table 7.1). According to the measurements, Stevens (1975) used the single-maximum or double-maximum form of curves to describe time-dependent variables quantitatively (Fig. 7.8). The model outputs look like those in the previous model and largely capture the variations in tracheid diameter and cell wall thickness (Fig. 7.9). It should be noted, however, that these parameters are used as input parameters in the model (daily maximum radial diameter, daily maximum cell wall thickness). Thus, Stevens's modifications of the model and computer program change the numerical daily presentation of time-dependent parameters of Wilson and Howard's model to continuous curves. No real improvement has been made that would incorporate environmental or hormonal influences; and the main limitations of the model so far as it may be applied to dendrochronology remain.

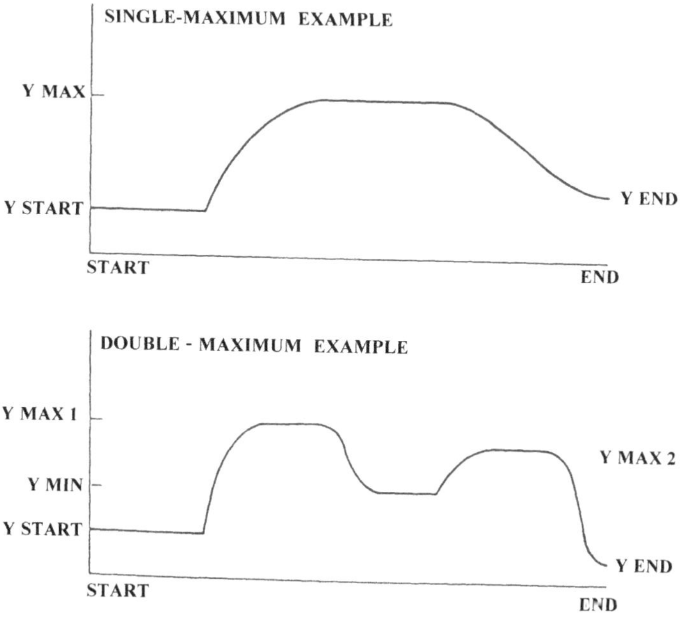

Fig. 7.8. General forms of time variation (redrawn from Stevens 1975)

Fig. 7.9. Sample plot of cell diameter vs cell position (redrawn from Stevens 1975)

The failure of these early models to achieve wide application to the study of tree-ring dynamics in dendrochronology may be attributed to the limitation that Ford (1981) perceived – the impossibility of modeling cambial activity at that time. One can agree on this statement completely if a model must deal with the environmental influence on each stage of tracheid development. If, however, the reasoning given in the previous chapters of this book is accepted, it is possible to reduce most of the uncertainties as well as the number of targets for environmental influence on tracheid differentiation. This can be done in the context of the simple simulation model described below.

7.4 The Vaganov–Shashkin Simulation Model of Seasonal Growth and Tree-Ring Formation

The model is based on these hypotheses:

1. The main target of external influence is the cambial zone, the zone of actively dividing cells. The external influence affects the linear growth rate of cambial cells (and the cell cycle).
2. The main external factors affecting the growth rate of cambial cells are temperature, light, and soil moisture. The model input is widely available daily data on temperature and precipitation. No other climatic data need be incorporated into the model. If necessary, other data can be used.

3. In calculations of growth rate, the principle of limiting factors is used, i.e. growth rate at a certain interval (time) of a season cannot be higher than allowed by the factor that is most limiting.
4. The variations of growth rate in the cambial zone mainly pre-determine the anatomical characteristics of the tracheids being formed (radial diameter, cell wall thickness).
5. The model deals with the simulation of variations in growth rate and tree-ring structure due to current climatic changes (within the season). However, the model simulates only climatically induced tree-ring width and structure variations.

There are two main blocks in the model: (1) computations of relative growth rate in relation of current changes in temperature, light, and soil moisture, and (2) a sub-block which uses the computed relative growth rate as input to cambial activity and computes the production of tracheids and their anatomical features throughout a season (Fig. 7.10). There are two limitations on model output:

1. The model in fact outputs the variations of the relative values of tree-ring width and absolute values of tracheid radial diameter and cell wall thickness as determined by within-seasonal variations of climatic conditions. The absolute value of tree-ring width may also depend on other factors. They are: the position of the tree in the stand, competition with neighbors, tree age and size, features and mineral composition of the soil, and other factors which are considered constant during the modeled time-scale. This assumption is permissible because in dendroclimatology each tree-ring series is normalized, detrended, and combined in a strongly replicated sample to emphasize the common climatic "signal" and de-emphasize individual and stand effects (Fritts 1976; Shiyatov 1986; Cook and Kairiukstis 1990). Therefore, one of the main model outputs is the standardized tree-ring width series. This may be considered the series of common, climatically-produced variations shared by the trees at the site.
2. The calculated anatomical structure of a tree-ring does not usually depend on the number of cells produced during the growing season. This means that tracheidograms of tree-rings formed in different trees with different

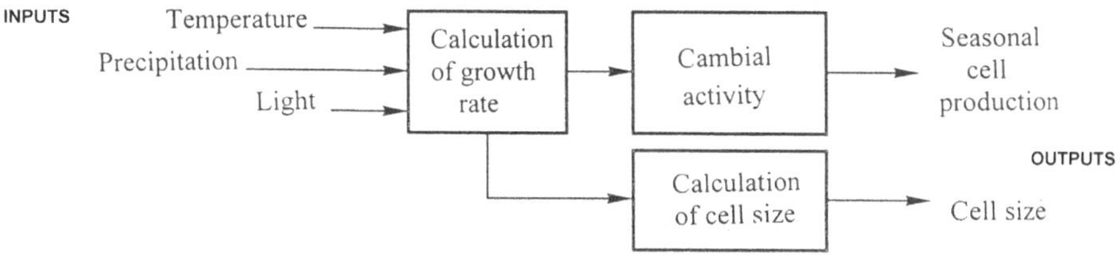

Fig. 7.10. Calculation scheme of the Vaganov–Shashkin simulation model

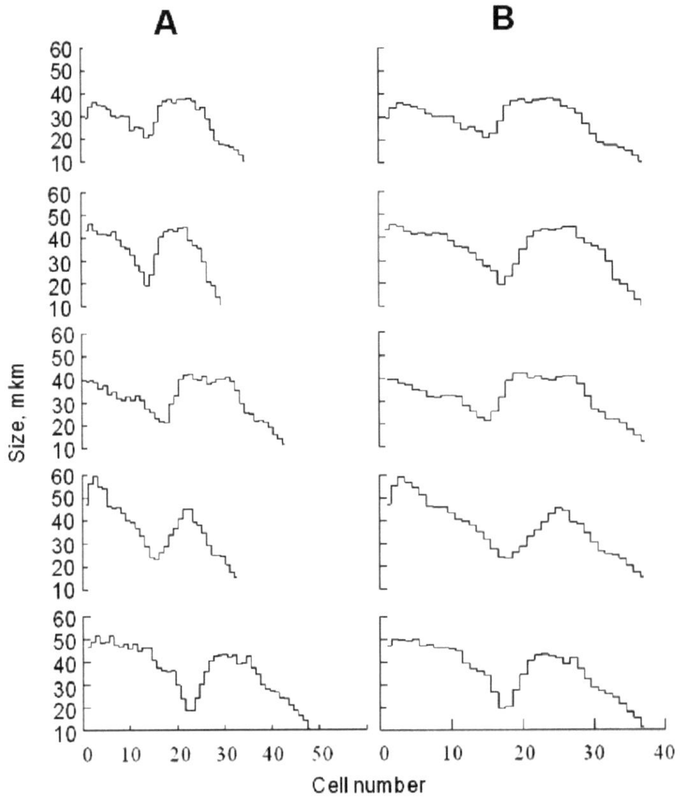

Fig. 7.11. Tracheidograms of five tree rings of *Pinus ponderosa* with different tracheid numbers in the same year before (**a**) and after (**b**) standardization (semi-arid conditions)

absolute ring width values are the same (Vaganov et al. 1985; see also Chap. 2). Tree-ring structure (the series of tracheid diameter, cell wall thickness, and wood density from the first to last cells of the ring) depends on variations of relative growth rate during a growing season. This can break down in cases where only very narrow rings of a handful of cells are formed, as near the latitudinal timber line, or near the lower forest border in dry conditions. Figure 7.11 shows several tracheidograms of rings formed in different trees in the same year with an intra-annual severe drought. As one can see, in different trees the number of tracheids formed before the drought season is different, possibly because of the late initiation of cambium or a slow growth rate in the first part of a season. The model, in effect, calculates the mean anatomical structure.

7.4.1 Growth Rate Dependence on Current Climatic Conditions

In the model, growth is considered primarily as cambial activity that is related to the structure of the cambial zone. To compute it, a special block was designed whose output is the growth rate at different times in the growing season; and this growth rate is input to the other special block of the model, the cambial activity block. The calculated growth rate manages the variation of the rate of division (or rate of cell radial growth) of initial and xylem mother cells, and finally, the rate of production of cells by the cambial zone. We will examine how this block was designed.

It was assumed in the model that the seasonal growth of a tree depended on three main factors: solar irradiation, which mainly determines the rate of photosynthesis, temperature, and soil moisture. These three factors determined the daily production of cells by the cambial zone in the model. The calculated value of growth rate $G(t)$ is the function of three variables: daily temperature, daily soil water content, and daily solar irradiation:

$$G(t) = g_E(t) * \min\{g_T(t), g_W(t)\} \tag{7.1}$$

where $g_E(t)$, $g_T(t)$, and $g_W(t)$ are the partial growth rates, calculated independently from solar irradiation, temperature, and water content in soil. All these functions are normalized and change in the range from 0 to 1. According to biological sense and experimental data, the functions of $g_T(t)$ and $g_W(t)$ must be unimodal, i.e. must have the optimal values as well as rising and decreasing branches (Fritts 1976; Kramer and Kozlowski 1979; Gates 1980; Lyr et al. 1992). The experimental data indicate that, for this function, the best approximate fit is a polynomial function (Vaganov et al. 1990; Fritts et al. 1991). However, in practical use, the most appropriate form of equation-fitting was a piece-wise linear approximation (Fig. 7.12). A typical example for calculation of $g_T(t)$ is shown in Fig. 7.12. There are two main reasons for using a piece-wise linear approximation: (1) the results of calculations with linear form and polynomial form do not differ significantly, and (2) there are simple explanations for the main parameters of the curve in piece-wise linear form as, for example, minimum temperature for growth, range of limitation of growth

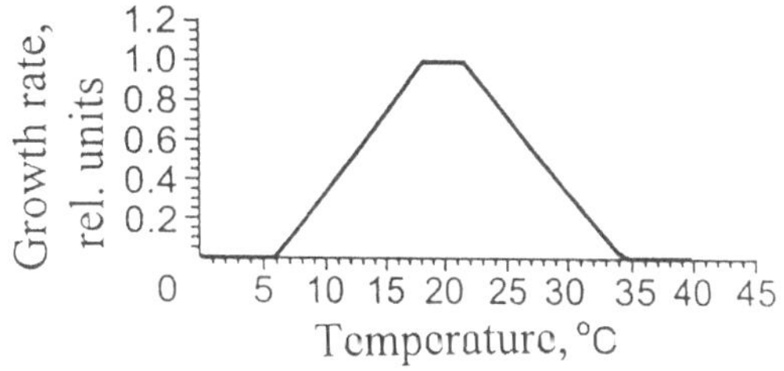

Fig. 7.12. Typical curve of growth rate response to the temperature used in the model

rate by low temperatures, range of optimal temperature of growth, range and maximal temperature when growth rate decreases and no growth occurs over the limit of high temperature.

The main complexity in the determination of $G(t)$ is related to the correct calculation of the seasonal dynamics of water content in soil from the daily data of temperature and precipitation. We apply a water-balance equation to calculate the dynamics of water in soil in a simplified form (Thornthwaite and Mather 1955; Alisov 1961):

$$\Delta W = f(P) - Ev - Q \tag{7.2}$$

where ΔW is the change in water content in soil of defined thickness per time unit, $f(P)$ is precipitation per time unit (day), Ev is the water loss in soil by transpiration, Q is runoff. Not all precipitation measured by weather stations gets into forest soil. Part is caught by the crown and part forms the surface runoff (Protopopov 1975). To evaluate the part of precipitation caught by crowns a linear dependence is often used (Protopopov 1975; Gash 1979). In this model, we describe the precipitation input to soil as a portion of the total daily precipitation ($k_1{*}P$). The coefficient k_1 is used in the model because its value may change with the site slope, mechanical properties of soil, tree species (crown characteristics), etc. Surface runoff depends on the amount of daily precipitation. It is assumed that the daily input of water to the soil cannot be higher than some critical level P_{MAX} (associated with completely saturated soil). Surface runoff also occurs when soil is completely saturated; and this is incorporated into the model of water balance. Thus, $f(P)$ is determined as:

$$\begin{aligned} f(P) &= k_1 P, \text{ if } k_1 P < P_{MAX} \\ f(P) &= P_{MAX}, \text{ if } k_1 P \geq P_{MAX} \end{aligned} \tag{7.3}$$

The value of Q is proportional to the water content in soil, W, so $Q = k_R{*}W$.

Transpiration of water by the tree crown is mainly a result of differences in water content between needles and air; and it is usually described by a diffusion equation. However, there are some uncertainties in the evaluation of the rate of crown transpiration related to: (1) water concentration inside needles depends on needle temperature which is not equal to the air temperature, and (2) the resistance to water diffusion from needles is not a constant value (Gates 1980). The calculation of transpiration rate is mainly based on Monteith's equation (Monteith and Unsworth 1990) or its modifications. These calculations require the exact evaluation of energy balance and description of stomatal conductance. It is not possible to do this in the model we propose, because it is based only on widely available meteorological data (daily temperature and precipitation). Therefore, we use a simplified equation to calculate the water balance in the model, where the transpiration is described as:

$$Ev = k_2 * G(t) * \exp(k_3 * T) \tag{7.4}$$

where k_2 and k_3 are the constant coefficients and T is daily temperature. This relationship is represented by an exponent, in accordance with the non-linear increase of water deficit in air with increasing temperature (psychrometric scale). The proportionality of transpiration to growth rate indicates that, in conditions with temperature or water limitations, the resistance to the diffusion of water from needles to air increases. This relationship also has clear physiological sense, because it is well known that transpiration is closely related to the growth rate of trees (large crown, higher growth rate, higher accumulation of organic material, etc.; Larcher 1980). The combination of coefficients may describe a large variety of natural tree species and growth conditions. For example, succulent plants may have a lower value of both coefficients because of some special tissue adaptation to water loss from leaves, whereas other plants may show higher values of the coefficients.

Trees growing in high latitudes in permafrost conditions may use water only from the thawing of a thin soil layer, the active layer. The process of thawing and water content in the active layer, some of which arrives in the current year and some of which is retained from the previous year, is taken into account only for high latitude regions. For other regions, the active soil layer $S_r = S_{THAW} =$ constant and is equal to the depth of the main root system. The rate of thawing is described as proportional to temperature (Kuzmin 1961) and exponentially decreases with the increasing thickness of the thawing layer:

$$S_{THAW}(t+1) = S_{THAW}(t) + a_1 * T(t) * \exp(a_2 * S_{THAW}(t)) \tag{7.5}$$

The thawing layer thickness decreases the partial growth rate $g_w(t)$ as

$$g_w(t) * S_{THAW}(t)/S_r \tag{7.6}$$

The calculated dynamics of soil thawing corresponds well to multi-year observations in permafrost regions (Pozdnyakov 1986) and to scattered data on soil thawing which are available from observations at northern latitude meteorological stations.

The third important factor affecting the growth rate is solar irradiation. Direct actinometric measurements are not provided for most meteorological stations. Therefore, the incoming radiation calculated for the model, which depends only on latitude, solar angle, and day-length (Gates 1980):

$$E = E_0 * (h_s * \sin\phi * \sin\theta + \cos\phi * \cos\theta * \sinh_s) \tag{7.7}$$

$$g_E(t) \equiv E / E_0$$

where E is incoming irradiation, E_0 is a constant calculated according to Eq. 6.13, p. 104 in Gates 1980), ϕ is latitude, θ is sun slope angle, and h_s is day length. The function $g_E(t)$ is equal to the ratio E/E_0. Following several earlier studies, growth starts when the temperature sum reaches some critical level (Lindsay and Newman 1956; Landsberg 1974; Valentine 1983; Cannell and Smith 1986; Hanninen 1990). This assumption is included in the model too.

Calculated $G(t)$ is the basis for day-by-day computation of seasonal tracheid production by the cambial zone, so these data are used as input to the cambial activity block.

7.4.2 Modeling of Cell Growth Within the Cambial Zone and Production of New Xylem Cells

In the model block that describes the changes of growth rate in the cambial zone, the process of differentiation is simulated for a single radial file from the kinetic point of view. This single file is taken to be average and hence representative for all tracheid radial files within the tree ring (Fig. 7.13). Growth in the number of cells in the cambial zone is caused by the division of initial and xylem mother cells. The rate of new cell production depends on the number of cells in the cambial zone and the rate of their division. Each cell within the cambial zone is characterized by two parameters: (1) position in the radial file relative to the initial j and (2) radial diameter $D(j,t)$. The process of division is

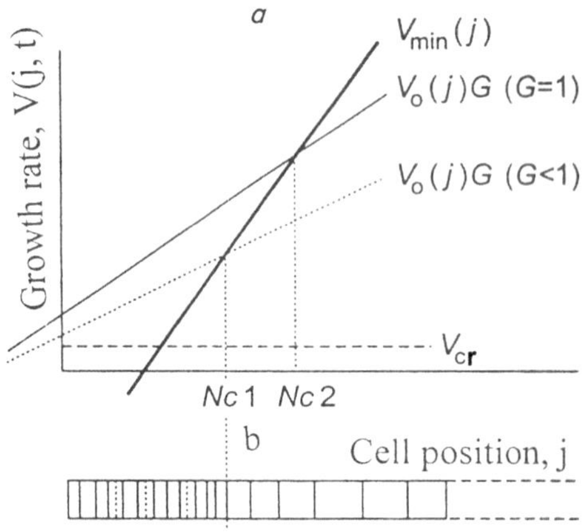

Fig. 7.13. Dependence of growth rate [$V(j,t)$; vertical axis] on cell position (j; horizontal axis) in the cambial zone (**a**); and simplified scheme of cell size changes in the cambial and enlarging zones (**b**). When the externally determined growth rate is at its maximum, the absolute growth rate varies with cell position j as in the line $V_0(j)G$ (G=1). When the externally determined growth rate is less than the maximum, a line such as $V_0(j)G$ (G<1) applies. If the growth rate $V(j,t)$ is less than the value of the line $V_{min}(j)$, the cell can no longer divide and leaves the cambial zone. $Nc1$ is the number of cells in the cambial zone when this applies to the case of G=1, $Nc2$ when G<1. If the growth rate of a cell, $V(j,t)$, falls below the horizontal line V_{cr}, dormancy follows. See text for further explanation

described in the model so that cell diameter increases continuously to a critical level at which division occurs. Both daughter cells have an equal radial diameter after division, which is half of the critical diameter, and they occupy adjacent positions: j and $j+1$. The duration between two subsequent divisions is determined by the rate of radial cell growth. The absolute growth rate $V(j,t)$ of each xylem mother cell depends on its position, as seen in the two lines labeled $V_0(j)G$ in the diagram. As j increases (towards the right of the graph), the growth rate increases. If the externally determined growth rate is at its maximum (i.e. when $G = 1$), the upper of the two lines applies. If growth is in some way limited externally (i.e. when $G < 1$), the lower, dotted, line applies. The growth rate of cambial cells determined only by position is $V_0(j)$, and $G(t)$ is the integral external growth rate, which is calculated according to the description in the previous paragraph. So:

$$V(j,t) = V_0(j) * G(t) \tag{7.8}$$

The cell loses the ability to divide and moves into the enlargement zone if its growth rate falls below the critical level $V_{min}(j)$ $[V(j,t)<V_{min}(j)]$, which is determined for each cell position. If $V(j,t)>= V_{min}(j)$, then the cell in this position is able to divide and is incorporated in the cambial zone. In the model, each of the cells goes through all phases of the cell cycle: G_1 cell growth, S (doubling of the genome) G_2 (preparation) M (mitosis, halving of the genome). Phases G_1, S, and G_2 together are called "interphase", but only in interphase may the cell lose the ability to divide and move to expansion or dormancy. If a cell moves to the mitotic cycle it will definitely complete it. At the end of the growing season, the remaining cells in the cambial zone move to the dormant state and these cells form the starting number of cells in the cambial zone at the beginning of the next year's growing season. This effect is handled in the model as the effect of the growth rate falling below V_{CR} (Fig. 7.13). In favorable growth conditions, the growth rate of all cambial cells increases and the cell cycle duration decreases. It also leads to an increasing number of cells in the cambial zone, because it includes more available positions for division and a shift of the boundary between the cambial zone and the zone of enlargement. This shift is determined by an increasing minimal growth rate $V_{min}(j)$ with the increasing position of a cell within the cambial zone (the results of Chap. 3, Sect. 3.6).

The right choice of the $V_0(j)$ and $V_{min}(j)$ functions is very important for the adequate description of the functioning of the cambium,. We assume that both are linearly increasing functions. The point at which they intersect is determined by the average number of cells in the cambial zone at a certain time. This number is shown as Nc in Fig. 7.13. The biological sense of $V(j,t)$ is obvious because it accumulates the changes in the growth rate due to external factors: temperature, light, and water content in soil. In spite of the formal description, the $V_{min}(j)$ function reflects the gradients of substances important to growth through the cambial zone, as in Wilson's (1973) model. These factors may affect the balance between the proliferation and differentiation of cells. This function, $V_{min}(j)$, is similar to the growth rate distribution within the cambial zone shown in Fig. 3.19 and discussed in Chap. 3. The angle of this linear function can be different for trees in different climatic zones. The calculated number of cells in the cambial zone (*Nc*) corresponds well to observed data of seasonal kinetics of cambial activity, the relationship between the average cell cycle, and the size of the cambial zone (*Nc*; (Bannan 1955; Wilson 1964; Gregory 1971; Vaganov et al. 1985). It follows from the model calculations that, if the duration between subsequent divisions decreases and the number of divisions increases, this leads simultaneously to a more or less constant average "life duration" of cells within the cambial zone during the growing season. This is also seen in the observed data (Wodzicki 1971; Table 3.3).

Note one more consequence that follows from such a quantitative description of cambial activity. Following the destiny of each cell originated from the first xylem mother cell reveals that each daughter cell has a different "life time" in the cambial zone, number of divisions, and size when it moves into the enlargement zone. If these differences, especially those from the last division, are maintained in radial tracheid size, we may find the explanation for cyclic variations of different radial tracheid diameters in single radial files (Ford and Robards 1976; Vysotskaya and Vaganov 1989; Djanseitov et al. 2000).

7.4.3 Calculation of Radial Tracheid Dimension and Cell Wall Thickness

In accordance with the hypothesis that radial tracheid diameter is mainly predetermined by the rate of cell production within the cambial zone and then modified slightly during the cell enlargement, the radial tracheid diameter in the model was calculated through the rate of the cell division. As follows from the analysis in Chap. 3, the radial tracheid diameter closely relates to the average rate of cell production in the cambial zone and controls the transition of the cell to the stage of enlargement. Thus the rate of cell production can be calculated from $G(t)$, which is determined by the average rate of cell production for the whole cambial zone. The flux of cells from the cambial zone to the enlargement zone is averaged and it is assumed that tracheids are produced in sequence of their position in a radial file. In that case, the rate at which each tracheid is produced is determined by the interval of time τ_i when two consequent cells move to the enlargement zone (see also Sects. 3.5, 4.4). The higher value of $G(t)$ corresponds to the lower value τ_i and vice versa. The number of tracheids in the tree-ring (*N*) and the duration of the season of production are determined by the cambial block of the model. The start date of a growing season is $t_{(S)}$, which is the time when *G* becomes higher than zero. The end of the season of cell production is the date $t_{(S)} + t_{(GS)}$, when the last cell moves from the cambial zone to the zone of enlargement. The time interval τ_i

between the transfer of two successive cells $(i-1, i)$ into the enlargement zone is calculated by:

$$dt = \int_0^{\tau_i} G(t_S + t_i + \alpha)d\alpha \qquad (7.9)$$

where:

$$dt = (1/N) * \int_{t_S}^{t_S + t_{GS}} G(t)dt$$

$$t_i = \sum_{k=1}^{i-1} \tau_k$$

The radial tracheid diameter of the i-st cell is calculated according to:

$$D_R^i = D_R^0 + K * N * (D_R^{MEAN} - D_R^0)/\tau_i \qquad (7.10)$$

where D_R^0 and D_R^{MEAN} are the minimal and average tracheid diameters correspondingly and K is a normalized multiplier calculated as:

$$K = \sum_{i=1}^{N} 1/\tau_i \qquad (7.11)$$

The model allows the calculation of cell wall thickness and then density variations within a tree ring, using two methods: (1) the empirical relationships between radial tracheid diameter and cell wall thickness (as shown and described in Chap. 5) and (2) the kinetics of cell wall-thickening related to the current weather conditions (mainly in regions with a short growing season, for example, near the polar timber line). The first method may be characterized as a combination of a simulation with an empirical approach, when a growth rate of tracheid production is determined from seasonal climatic data and is then used to calculate the tracheid diameter which determines the wall thickness and density profile. The second method is completely simulative, where the cell enlargement and cell wall deposition have been described as processes with rates dependent on current climatic conditions. The average rates of enlargement and wall deposition were defined from the data presented in Chaps. 4, 5 and were assumed constant for certain species. Maximal rate was defined as twice greater than average; and the real rate for each day was calculated as maximal rate*$G(t)$. The final radial cell dimension is restricted by the cell growth rate before the cell is passed to the zone of enlargement. Cell wall thickness is restricted by the total mass of material that may be deposited for the ultimate radial diameter (see Fig. 5.7). These assumptions led to similar results for the two methods of calculation of radial cell dimension and cell wall thickness. This is why we used the first method in further calculations.

7.5 Description of Model Parameters

The model is written as a computer program; and Table 7.2 shows the main parameters used.

Many of parameters used in the model have a simple biological meaning. For example, critical levels of temperature described above are: T_{min}, T_{opt1}, T_{opt2}, T_{max}, which describe the growth rate dependence on temperature and the minimal and maximal soil moisture content in the upper 500 mm of the soil (W_{min}, W_{max} correspond to wilt and fully saturated soil moisture content). Two coefficients, k_2 and k_3, determine the rate of transpiration. The first one depends on several factors, such as the resistance of leaves to water vapor diffusion from inner leaf (needle) tissue to outer air, the ratio of crown area per land unit, etc. The second is related to air water deficit, which depends on temperature. If we take into account the coefficients shown in Table 7.2, it means that maximal transpiration in the temperature range 10–30 °C is about 2.7–52.0 mm water per day, which is in good agreement with the observed data for different conifers (Whitehead and Jarvis 1981). The portion of precipitation that enters the soil (coefficient k_1) depends on the geographical area, slope, climate, soil type, etc. Its range is about 0.4–0.94 (Doley 1981; McNaughton and Jarvis 1983).

In the block of the model that describes the cambial activity, the main parameters are chosen based on known average kinetic characteristics (see also Chap. 3). Radial cell diameter in the cambial zone ranges over 5–10 μm (Larson 1994). The rate of a cell's passage through the G_1 phase of the cell cycle is critical. The minimal duration time of the cell cycle depends on the parameter V_0, the duration of G_1 and the duration of mitosis and equals 2 days. The sensitivity of cambial production to $G(t)$ alterations is determined by functions of $V_0(j)$ and $V_{min}(j)$. The angle of intersection of these two curves becomes smaller as the number of cells in the cambial zone increases; and both the total production and the rate of cycling within the cambial zone also increases (Fig. 7.13). The position of j, where both lines cross, is equal to the maximal size of the cambial zone at a certain moment in the growing season. In the case of Fig. 7.13 the line crosses at an average value 7.5 cells, so N_C is equal to 7.5 cells. The heterogeneity of cambial cell duration in the cell cycle increases with the increase of the slope of the function of $V_0(j)$. The critical growth rate V_{cr} determines the average number of cells in the dormant cambium, the minimal size of which depends on the crossing point of V_{min} and V_{cr}.

To evaluate the parameters shown in Table 7.2 in model calculations for concrete examples, two approaches may be used. The first one is possible if we have enough experimental information about the site conditions, tree growth response, and limit values of the main factors (temperature, soil moisture content), etc. In this unique case, all parameters used can be defined before

Table 7.2. Numerical value of model parameters

Model parameter	Description	Value
T_{min}	Minimum temperature (°C)	1–4
T_{opt1}	The first temperature optimum (°C)	18
T_{opt2}	The second temperature optimum (°C)	22
T_{max}	Limiting temperature (°C)	33
w_{min}	Minimum soil moistures (wilting moisture; v/v)	0.02
$w_{Ápt1}$	The first optimum of soil moisture (v/v)	0.2
w_{opt2}	The second optimum of soil moisture (v/v)	0.8
w_{max}	Limiting soil moisture (v/v)	0.9
S_r	Depth of root system (mm)	500
T_{melt}	Sum of temperatures for beginning of soil thawing (°C)	60
P_{melt}	Period of sum temperature calculation (days)	10.0
a_1	The first coefficient in equation of soil defrosting (mm/°C)	10.0
a_2	The second coefficient in equation of soil defrosting (1/day)	0.006
P_{max}	Maximum daily precipitation falling into soil (mm)	15.0
k_1	Part of precipitation falling into soil (dimensionless)	0.62
k_2	The first coefficient for calculation of transpiration (mm/day)	0.2–0.3
k_3	The second coefficient for calculation of transpiration (1/degrees)	0.18–0.22
k_R	Coefficient of water infiltration from soil (dimensionless)	0.001
T_{beg}	Temperature sum determining growth start (°C)	30
P_{beg}	Period of temperature sum calculation (days)	10
V_{cr}	Critical rate of cell transition to resting (μm/days)	0.04
b_1	Coefficients of growth rate function from position (μm/days)	0.42
b_2	$V_o(j) = b_1 + b_2 j$ (relative units; μm/days)	0.25
b_3	Coefficients of minimum growth rate function (μm/days)	−1.62
b_4	$V_{min} = b_3 + b_4 j$ (μm/days)	0.54
v_0	Growth rate in phase S, G_2 and M (μm/days)	1.0
Y_{G1}	Maximum cell size in phase G_1 (μm)	8.0
Y_S	Maximum cell size in phase S (μm)	9.0
Y_{G2}	Maximum cell size in phase G_2 (μm)	9.5
Y_M	Maximum cell size in phase M (μm)	10.0
d_t	Tangential tracheid size (μm)	40.0

making calculations. If the experimental material to define all or at least most parameters is absent in most instances, we apply the second approach, which may be described as iterative. Even a visual description of a site is sometimes enough to make qualitative evaluations of certain model parameters. For example, sandy soil has a lower water-holding capacity than clay and loam soils and less water is necessary to saturate it. This, in its turn, leads to a low value of wilt moisture content (see also Table 1.2) and to the transpiration rate in relation to temperature. One can assume that the amount of precipitation input to soil will be more in a flat area than on a slope. Some information could be obtained from a comparison of different tree species. So, ponderosa pine (*Pinus ponderosa*), which can grow in more arid conditions, must have a less steep dependence of transpiration on temperature than Siberian pine (*P. sibirica*), which prefers wetter conditions. From our data on larch tree-ring width, density, and cell size analysis, it follows that larch at the western Siberian polar timber line has a higher minimal temperature for growth than larch growing in the eastern part of the subarctic zone. In these conditions, we may assume a significant effect of temperature and snow melting on seasonal growth and an insignificant influence of soil moisture in most sites.

Accumulated qualitative data allow us to choose the approximate starting values of parameters, to make the first-step calculations, and to compare with the observed tree-ring chronologies. If possible, it is better to compare the tree-ring width and tracheidogram data for 5–6 years of growth. Also, if it is possible, we choose the 5–6 year-long segment which contains contrasting annual tree-rings (wide and narrow, with small and broad latewood zone, etc.). Then we gradually change one parameter and test the agreement between the observed and calculated values of tree-ring width and cell size. If the significance and correlation values indicate sufficient convergence between actual and model, then the calculations will be made with these fixed parameters for the whole period with available daily climatic data (for many places about 50–60 years). The correlation between the calculated and observed data for tree-ring width, cell size, and density indicates the quality of modeling and success in the evaluation of model parameters.

7.6 An Example of Model Application

7.6.1 Tree Growth and Formation of Annual Rings Near the Polar Timberline

We defined some parameters from prior statistical analysis and others were taken from the relevant literature. For example, the starting temperature of growth was defined as 6 °C for the Taymir region and 4 °C for the northern Yakutia region. The optimal temperature range was from 15 °C up to 19 °C; and the coefficient of soil melting was chosen from several years' observations at one northern meteorological station. This latter estimate was shown to be accurate because the seasonal calculation of thaw depth gave observed values from 40 cm to 60 cm, corresponding to reality. For most years, soil moisture did not limit growth rate because of low transpiration and sufficient precipitation. First, we discuss the results of calculations for a larch site (INP) in northern Yakutia, not far from Chokurdakh.

In high latitudes, the starting dates of cambial initiation vary greatly from year to year. The model calculation shows the key importance of dates of cambial initiation and early summer (mid-June to mid-July) temperatures for tracheid production, tracheid diameter, and cell wall thickness. A higher temperature at the beginning of the growing season caused an extension of the cambial zone, which in turn produced increased tracheid production. We will examine the calculations of growth rate and tracheid diameter within two tree rings formed in years of warm (1960) and cold (1962) summers (Fig. 7.14). In the warm year (1960), the temperature reached 0 °C at the beginning of June and cambium initiated in the second decade of June (after snow melt). A relatively high temperature corresponding to the highest solar irradiation in June through to mid-July determined the high growth rate of cells in the cambial zone. This combination of temperature and irradiation conditions provided the highest growth rate, increasing the radial tracheid diameter in earlywood (up to 56–58 μm) and producing a wide zone of latewood, which started to form at the end of July and matured during August under relatively high temperature until mid-September. In the cold year (1962), the starting date of growth shifted to the beginning of July because of a lower temperature and later snow melt. The growth rate during the rest of the season reached only 50 % of the maximum in spite of increasing temperature, because of decreasing solar irradiation. Only a few earlywood cells (tracheids) were produced under the slow growth rate, with a decreased radial diameter (40–42 μm); and 2–3 latewood cells were produced, with a smaller wall thickness because of the shorter duration of the season.

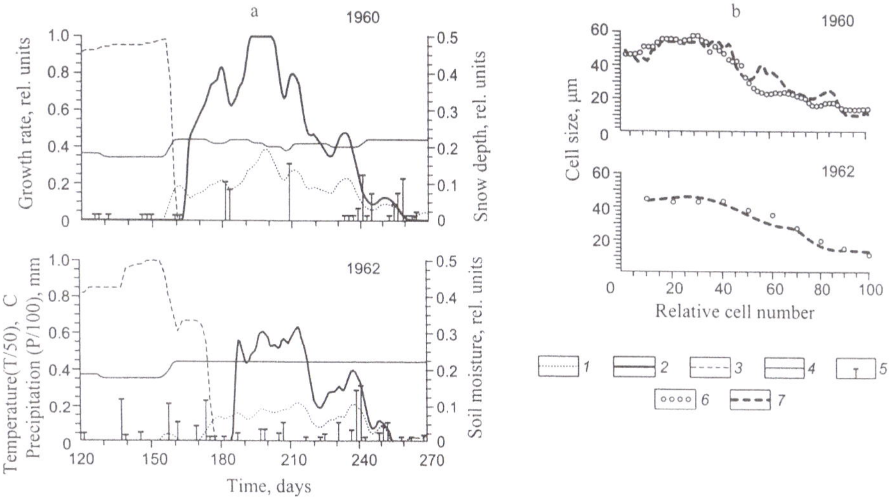

Fig. 7.14. Seasonal dynamics of climatic factors and calculated cell growth rate (**a**) and corresponding tree-ring tracheidograms (**b**) of larch trees (INP) in warm (1960) and cold (1962) years. *1* Temperature (*T*/50), *2* growth rate, *3* snow depth (relative units), *4* soil moisture, *5* precipitation (*P*/100 mm), *6* measured tracheidogram, *7* calculated tracheidogram

The calculated tree-ring width index values coincide quite well with the measured (R=0.67; P<0.0001; Fig. 7.15). This is somewhat less than the correlation obtained from a statistical model (R=0.79, P<0.0001; Hughes et al. 1999). It should, however, be borne in mind that the correlation for the statistical model reflects the training period of that model, whereas all but five years in the simulated case do not. The observed and calculated indices show a very strong common high-frequency variability, so that the coefficient of parallel run (the percentage of years in which the sign of first difference is the same in both series) is 86 %. The disagreements may have several causes: (1) a partially inadequate model (structure and parameters), (2) errors in daily data which can not be checked, (3) differences between actual weather conditions at the site where the trees grow and at the meteorological station.

The model highlighted why the starting date of cambial initiation is most important for the formation of wider rings at high latitudes. This is because, in June, temperature increase coincides with the highest solar irradiation. The two parameters of solar irradiation are relatively high at that time: flux and its differential (as an analogue of photoperiodicity). From the beginning of July, both parameters decline (the differential more steeply than the flux), which induces the cambial growth rate (obviously indirectly through canopy hormones) to transfer to the formation of latewood tracheids. Even the relatively high temperatures of July do not compensate for the decline of these parameters.

The effect of soil water content in the thawing soil layer on tree-ring formation in subarctic latitudes is an issue of special concern. The thawing layer is not deep. The development of a moss layer increases the thermoisolation of the soil and decreases the depth of the thawing layer (Abaimov et al. 1997). Water input from thawing soil and from precipitation may not compensate for the water loss due to evapotranspiration in the first part of the season if temperatures are high and there is thick moss cover. Growth rate will be subject to limitation by both temperature and drought in such cases. Our model calculations for one more site located further south, near Verkhoyansk, confirmed this assumption. This effect occurs infrequently, but it is enough to

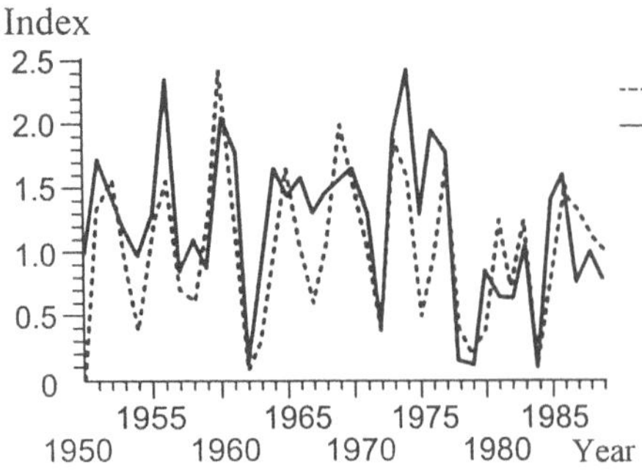

Fig. 7.15. Calculated (*1*) and measured (*2*) tree-ring width indices of larch trees from Chokurdakh region (INP)

explain the formation of narrow tree-rings, even though the temperature was close to optimal during most of the season (Vaganov et al. 1996b, c).

7.6.2 Examples of Modeling of Seasonal Growth and Formation of Tree Rings in the Middle Taiga Zone

In the taiga zone, seasonal dynamics of climatic conditions are combined with other factors more than at higher latitudes. This determines the changes of limiting factors within one season, as well as between subsequent seasons. The climatic response function may be complicated and difficult to interpret. In this particular case, we consider the application of the model to reconstruct the seasonal kinetics of growth and to calculate the main environmental factors for a year with observed data.

We chose a site (BOR) in the middle Enisey region of central Siberia with one year's seasonal climate, soil water content, transpiration, seasonal growth, and formation of annual rings in old Scots pine (*P. sylvestris*) trees (Fig. 7.16). This site was located in typical lichen-pine forest on sandy soil. Statistical analysis of a regional chronology with climatic factors revealed the influence of both temperature and moisture on inter-annual variability of tree growth, with the June temperature being particularly important (see Fig. 7.1; Arbatskaya and Vaganov 1997). The observations are used to define some important parameters in the model.

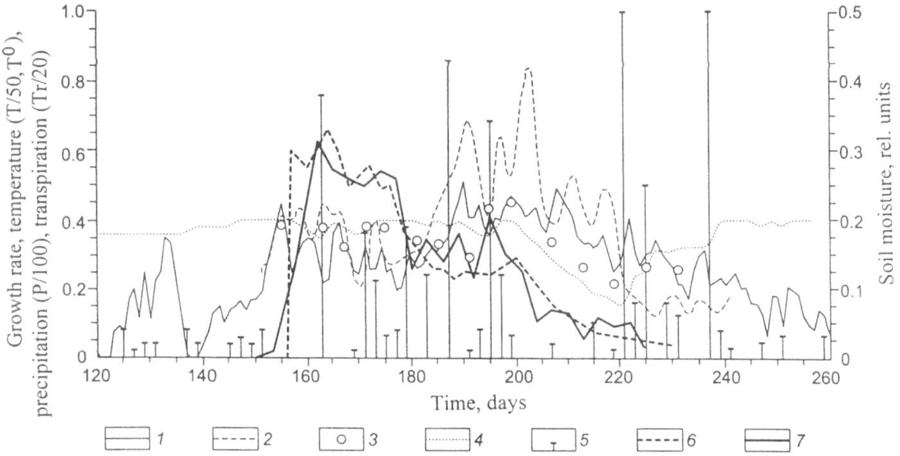

Fig. 7.16. Seasonal dynamics of environment, experimental and calculated growth rate of pine trees from middle taiga (BOR). *1* Air temperature (*T*/50), *2* transpiration (*Tr*/20), *3* measured soil moisture, *4* calculated soil moisture, *5* precipitation (*P*/100), *6* calculated growth rate, *7* measured growth rate

The temperature exceeded 5 °C in the third ten-day period of May and then it was relatively low during all of June, increased in July and the first half of August, and after that gradually declined (Fig. 7.16). A short interval in the season can be seen (end of May to first half of June) without precipitation. Nevertheless, winter precipitation provided the available water content in the soil in the first part of the growing season. The water content in the soil then declined from the end of June. The highest tracheid production occurred in June, then growth rate decreased and remained at a lower level during July. Growth ceased in mid-August, according to the observations. The measured tracheidograms reflect the features of the seasonal course of growth-rate changes (Fig. 7.17). So far, the observations allow us to define the dates and duration of the season of tracheid production; and we can use the tracheidograms to reconstruct the seasonal growth rate from the model and compare it with the observations. The reconstructed growth curve shows high similarity with the observed (R=0.78, n=35, P<0.001). The earlywood tracheids were formed during June, whereas tracheids of the transitional zone were pro-

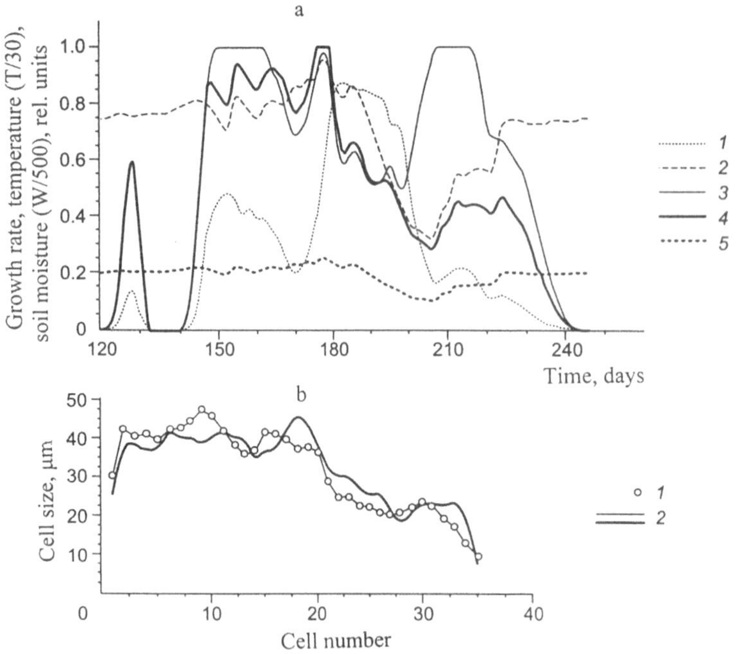

Fig. 7.17. Seasonal dynamics of climatic factors and calculated specific and general growth rates (**a**) and tracheidograms of tree rings (**b**) of pine trees (BOR). **a:** *1* Air temperature, *2* growth rate determined by soil moisture, *3* growth rate determined by temperature, *4* general growth rate, *5* soil moisture. **b:** *1* Measured, *2* calculated tracheidograms

duced by the cambial zone from the end of June. This transition was related to a decrease in both temperature and solar irradiation. More than half of the total tracheid production occurred during June. This was about 70–75 % of total ring width. In July, when solar irradiation (flux and its differential) declined, the latewood tracheids formed with a radial diameter of 20–25 μm, even when the temperature increased.

Figure 7.17 shows the results of the calculation of seasonal growth rate variations and variations in radial tracheid diameter. The initiation of cambial activity is closely connected with temperature and the calculated data correspond well with the observations – the beginning of active division in the cambial zone occurred in the third ten-day period of May. In June, the calculated growth rate was practically unlimited by soil water and temperature, and its value was close to the maximum (0.8–1.0). At the beginning of July, the growth rate began to decrease, due to a short interval of temperature decrease and especially a decrease in soil water content. From the end of July, the growth rate declined faster due to decreasing solar irradiation, even when the temperature and soil moisture became more favorable for growth. The calculated tracheidogram was similar to the measured one (Fig. 7.17b). The parameters defined by the one year's seasonal observations were then used for a calculation of the inter-annual variation of pine growth in this site (and region). The calculated and the measured tree-ring width series were rather similar (R=0.68, n=58, P<0.001). The model analysis indicated that, in the environmental conditions of the middle taiga, as in higher latitudes, the weather conditions of the first part of the season played the key role in determining the inter-annual variation in tree-ring width growth. To produce wide rings with a larger number of cells, the dates of cambial initiation as well as the temperature in June are important. The growth rate declines markedly in July and August, because solar irradiation declines in spite of temperature and soil moisture conditions that are optimal for growth. The statistical response function for this chronology (middle taiga) shows persistent but not significant negative effects of winter precipitation on tree-ring width and maximum density, as in the northern taiga (Fig. 7.1; Arbatskaya 1998; Kirdyanov 1999). There is much more winter precipitation in the middle taiga than in sites near the polar timberline (see Fig. 1.8), and snow-melting processes can influence the date of soil-thawing and the beginning of cambial initiation as well. Inter-annual variation in the date when the temperature reaches the critical level for growth initiation is less than at high latitudes (12-15 days), but the timing of snow melt may add 5–10 days to these variations, which finally may lead to an increase or decrease in growth of 30–40 % (Kirdyanov 1999).

Comparing the results of simulation in the middle taiga with those at higher latitudes, we note that the critical conditions for formation of wide tree-rings occurred during the first part of the season, as according to the date of initiation of cambial activity. If we apply these results to forecast the response of trees in the north and middle taiga to global climate change, we may conclude that this response will take place if global change on the regional level changes the early summer conditions. Direct and indirect evidence of this has now appeared in some publications (Groisman et al. 1994; Keeling et al. 1995; Myneni et al. 1997), where analysis of the annual cycle of atmospheric carbon dioxide concentration and remotely-sensed vegetation indices show the onset of plant growth several days earlier in high latitudes in response to rising spring temperatures in the past two decades.

7.6.3 Simulation of Annual Tree Growth and Tree-Ring Formation in Trees Growing in the Steppe Zone

For this example, the simulation modeling of inter-annual ring width and tracheidogram variations in trees growing in the dry conditions of the steppe zone in the south of the Krasnoyarsk region were chosen. The sampled site is characterized by a steep slope (up to 20 %), a thin layer of gravel and sand soil, and Scots pine trees (*P. sylvestris*) more than 150 years old. Standard dendroclimatic analysis revealed the importance of seasonal changes in temperature (mainly negative) and precipitation (positive) for inter-annual growth variations (Vaganov et al. 1985; Vaganov 1989, 1990). A response function from such a site is shown in the bottom panel of Fig. 7.1. The effect of precipitation was significantly greater than the effect of temperature. The main complexity of model application for such sites was related to the accurate calculation of the water content in soil from daily climatic data. To ascertain the correct determination of model parameters, we made some simulations with different parameter values.

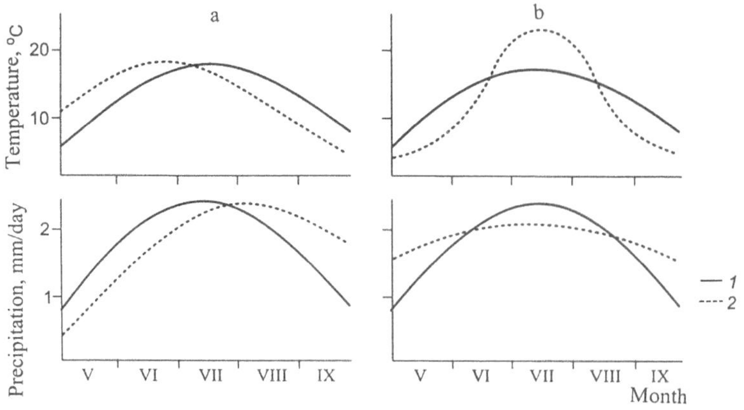

Fig. 7.18. Different variants of seasonal dynamics of climatic variables: *1* average long-term data, *2* temperature and precipitation data of dry spring, wet autumn (**a**) and hot July (**b**) variants

To test the model, calculations of seasonal kinetics of growth rate and the tracheid radial sizes produced were made for different scenarios of seasonal climate in this region (Fig. 7.18). The basic variant (curves labeled 1) corresponds to the long-term average seasonal course of temperature and precipitation. The others were modifications of this: dry and warm early summer and cold late summer, cold early summer with dry and warm late summer, dry and warm July, the whole summer cold and moist, drought in early summer, warm and moist summer. The calculations were made with constant parameters which were evaluated partially from publications and partially from our own experience. The three different values of initial water content in the soil layer were: low (80 mm in 50 cm layer), moderate (120 mm) and high (160 mm). Figure 7.19 shows representative examples of calculations of seasonal growth rate and tracheidograms for the variant with dry warm early summer and cold late summer under two initial soil water contents (low,

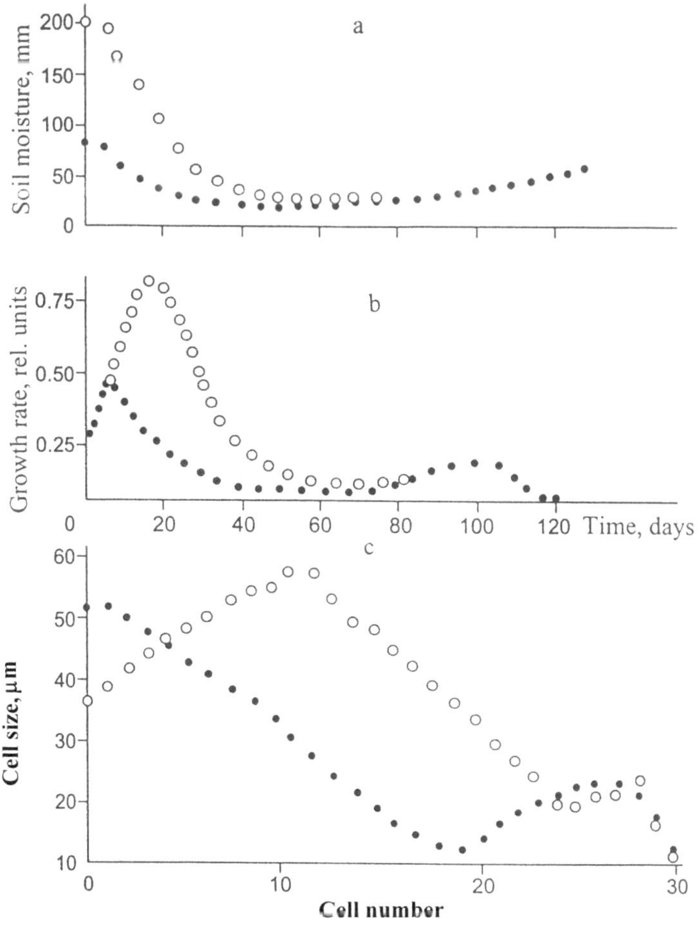

Fig. 7.19. Calculated seasonal dynamics of soil moisture (**a**), relative growth rate (**b**), and corresponding tree-ring tracheidograms (**c**) for dry warm spring, wet autumn seasons at low (*closed circles*) and high (*open circles*) initial soil moisture

high). In spite of the qualitative similarity of the calculated growth rate in both cases (fast increase at the beginning of a season, then fast decrease due to water content limitation), the resulting tracheidograms differ significantly. In a season with low initial water content, the growth rate was relatively high during a short period at the beginning of the season and then was strongly suppressed, leading to an increased fraction of small tracheids in the tree ring. Otherwise optimum conditions in the late summer only increase growth a little, because of the strong decline in solar irradiation. The higher temperatures and water supply did, however, produce a visible increase in the diameter of the last formed tracheids, giving the appearance of a "false ring". In the case with high initial soil moisture, the growth rate reached the highest level (up to 0.8–0.9) and a wide earlywood zone of many tracheids with large diameter was produced. The drought effect appeared late in the season.

Further qualitative analysis showed that similar growth rate kinetics and similar tracheidograms could be produced by different combinations of the seasonal climatic condition with initial soil moisture. For example, the variant with average climatic data and moderate initial soil moisture leads to similar results as the variant of cold wet summer with low initial soil moisture (Fig. 7.20).

These qualitative examples show the predominant effect of initial soil moisture on seasonal growth rate calculations in steppe zone sites. So, in multi-year calculations most errors may come from the incorrect calculation of soil water supply, based on the water balance equation (Eq. 7.2). The calibration of the model was made on available data on seasonal variation of water content in soil measured at the nearest agroclimatic station for several consecutive years. The result of calculations for four years is in very good agreement with the measured water supply in the productive soil layer

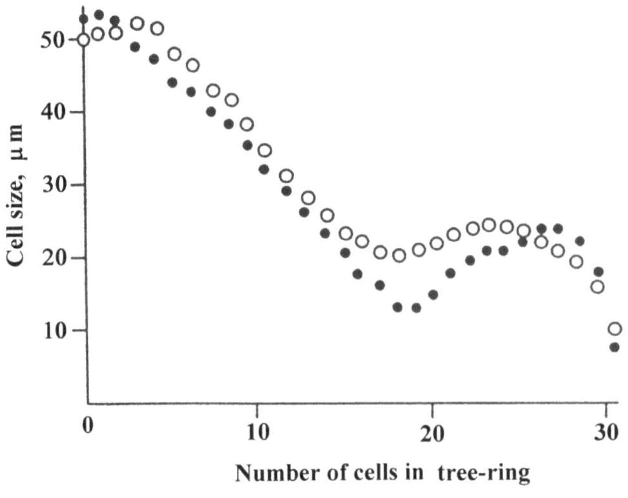

Fig. 7.20. Tracheidogram calculations of seasonal dynamics of climatic variables and initial soil moisture for dry warm spring and second wet part of season at mean level of initial soil moisture (*closed circles*) and for cold wet summer at low initial soil moisture (*open circles*)

(R=0.79, n=91, P<0.0001; Fig. 7.21). Note that, during winter, the soil moisture level is nearly stable (the soil is frozen) and there is a fast drying of the soil layer from the end of June to the beginning of July, due to the highest summer temperatures and hence highest water losses from soil by transpiration and evaporation.

The common scheme of calculations for an individual year is shown in Fig. 7.22. Analysis of the operation of the model indicates that the growth rate is determined by temperature at the beginning of the season. The response function for the forest steppe site in the bottom panel of Fig. 7.1 shows a slightly negative, statistically insignificant effect of temperature at this time, soil moisture begins to decline faster from the end of June and, until mid-September, the rates of growth and production of tracheids are limited by water availability. This is reflected in the statistical response function shown in Fig. 7.1. The date of growth cessation is mainly determined by decreasing solar irradiation and soil moisture rather than by temperature. Total annual xylem increment is closely related to the area under the calculated curve of growth rate. In Fig. 7.23, one can see a good agreement between the calculated and measured inter-annual variation in tree-ring width (R=0.69, n=43, P<0.005) and the calculated and measured tracheidograms of single years (R from 0.75 to 0.91). The narrowest rings are formed in years with a dry early summer (low initial soil moisture) and a hot summer (highest rate of water loss from soil, due to high temperature in June and July). The model captured these years (1945, 1946, 1965, 1974) quite well. In these conditions, growing pine trees are periodically subject to intra-seasonal droughts of differing severity. This leads to the formation of "false" rings with different anatomical features: abrupt and deep intra-seasonal reduction in tracheid diameter; smoothed with small deviation in diameter of tracheids formed after typical earlywood tracheids and latewood tracheids with slightly increased diameter as in "false ring" tracheids; scarcely visible "false rings" in earlywood due to a short, rapid reduction in temperature or soil moisture at the beginning of the

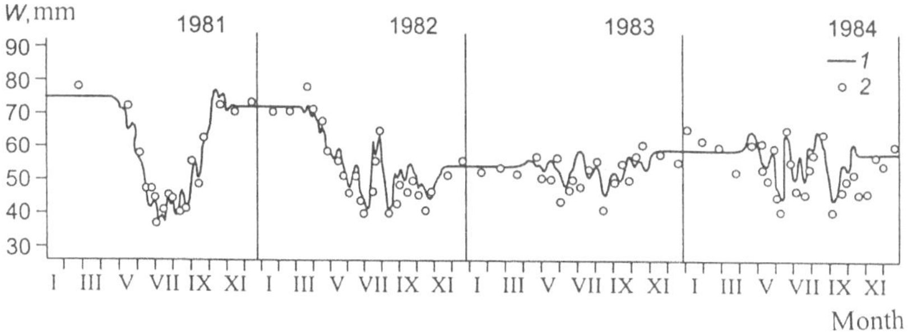

Fig. 7.21. Calculated (*1*) and experimentally measured (*2*) dynamics of soil moisture for four consecutive years (site in steppe zone of Khakasia)

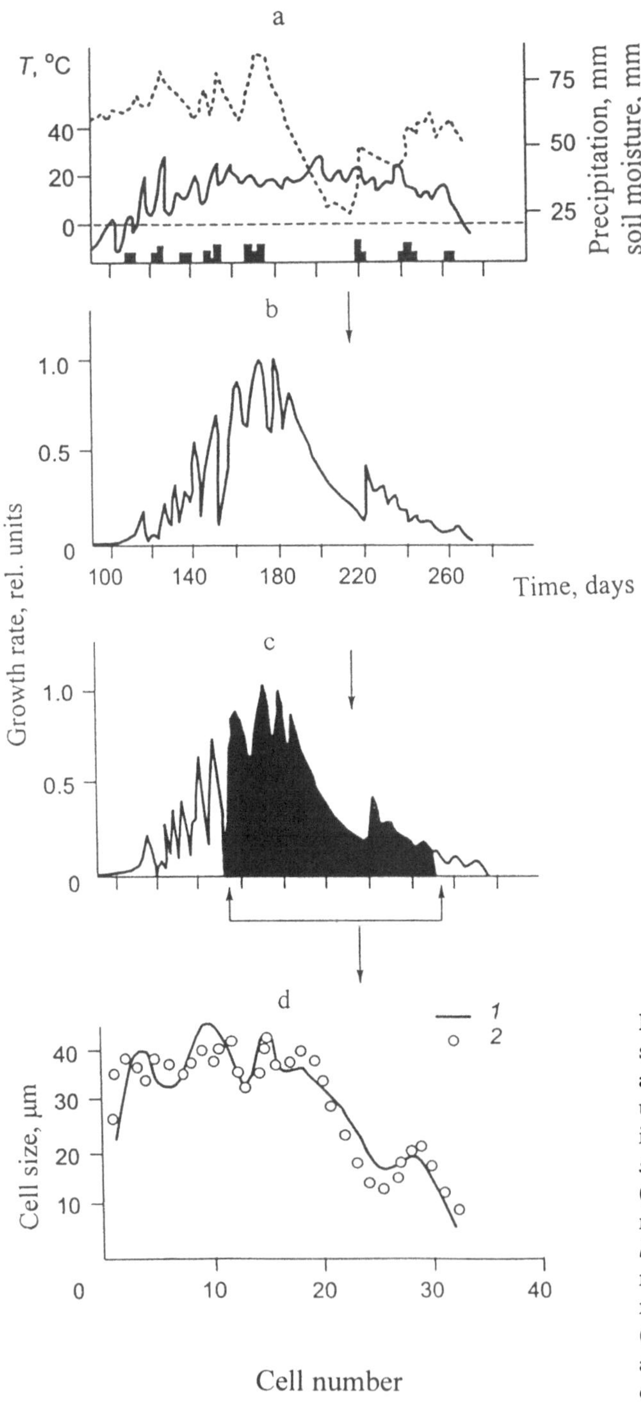

Fig. 7.22. General scheme of calculations according to the VS simulation model. **a** Dynamics of temperature (*T*; *solid line*), precipitation (*P*; *blocks*), and soil moisture (*W*; *dotted line*). **b** Relative growth rate. **c** Relative growth rate for growth period (*shaded*). **d** calculated (*1*) and measured (*2*) tracheidogram

growing season (this example is shown in Fig. 7.23, for the year 1948). It is obvious that, in the steppe environment, the soil water content plays the most important role in tree-ring formation (Shashkin and Vaganov 1993; Andreev et al. 1999). The calculations do show, however, the significant influence of temperature in the first part of the season (date of growth initiation and positive effect when soil moisture is close to optimal) and in the second part of a season (negative effect due to increasing transpiration). The initial value of soil moisture at the initiation of cambial activity is also very important for tracheid production. Conditions affecting initial soil moisture may have significant influence on tree growth. So, high temperature in late autumn (before snow fall) may reduce the soil moisture, which may not be compensated if a dry winter follows. Thus, the basic assumptions of the model are adequate to calculate the intra- and inter-seasonal growth and tree-ring structure variations for conifers growing in the steppe zone environment under temperature and water limitation. The unchanged structure of the model used in more

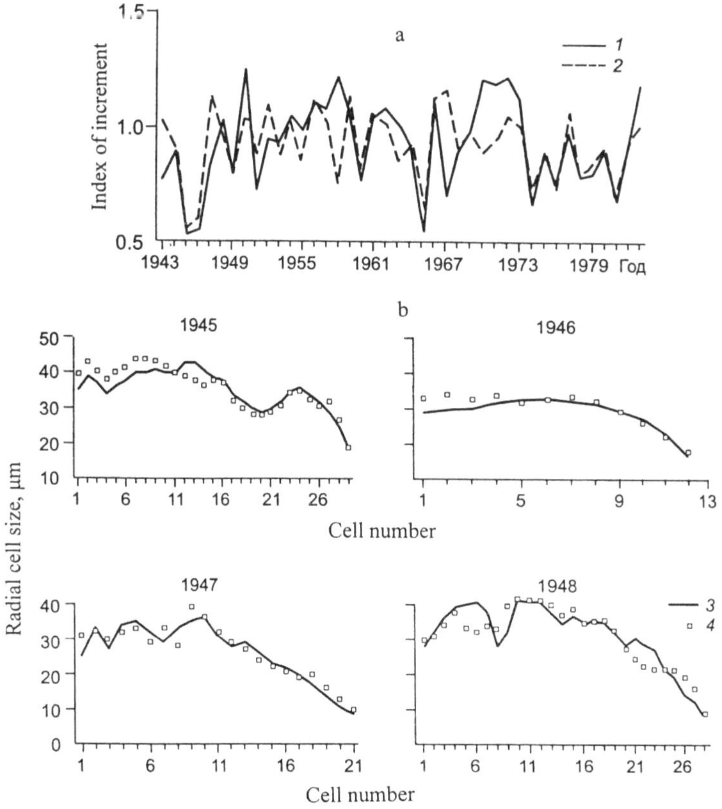

Fig. 7.23. Comparison of measured (*1*, *3*) and calculated (*2*, *4*) growth indices (**a**) and tracheidogram of individual tree rings (**b**) of pine in steppe zone of Khakasia

northern latitudes was applied to simulate growth, which means that the model must reflect some more fundamental mechanisms of xylem growth and environmental control.

7.6.4 Seasonal Growth and Formation of "False Rings" Modeled in Conifer Trees Growing in a Semi-Arid Climate

In extremely dry conditions, numerous "false rings" can be formed in tree rings of conifers, which are characterized by a sharp intra-annual decrease in the radial diameter of tracheids (Fritts 1976; Fritts et al. 1991). We apply the model for calculation of the well known example of ponderosa pine (*P. ponderosa*) growth in the Santa Catalina Mountains near Tucson (Arizona; Fritts 1976, 1990; Fritts et al. 1991). Figure 7.24 shows the temperature, precipitation and soil moisture variation during the period 1963–1967 measured at the experimental site. The characteristic feature of the seasonal dynamics of this environment is the gradual decrease in soil moisture during the dry period (from April to the beginning of July) and then its sharp recovery in July–August due to rains associated with the northern extent of the Mexican monsoon. These within-season changes caused the formation of "false rings"; and in severe drought the reduction of growth is so deep that the "false ring" may not be distinguishable from a real annual ring. Recording dendrograph data show a high rate of growth at the beginning of a season, a gradual decline towards the end of the dry period, and then faster growth in July–August (Fig. 7.25). Summarizing anatomical observations on tree rings formed in 1966, Fritts (1976) compared the variation of cell number in the enlargement and wall-thickening zones with the variations in anatomical characteristics of tracheids (Fig. 7.26). The "false ring" is clearly fixed by a significant reduction in tracheid diameter (the size of these tracheids is equal to typical latewood tracheids) and a slight increase in wall thickness. Only one-third of total xylem annual production is produced by the end of the dry season in late June.

The key process for adequate simulation in this environment, as well as in the steppe zone, is accurate calculation of the water balance in the soil. We used the 1963 data to calibrate some model parameters; and then all calculations were made with constant values of parameters and coefficients. In Fig. 7.27, values calculated in our model and estimated by Thornthwaite and Mather's water balance model for 1964 are compared. There is high level of agreement (R=0.89, n=29, P<0.005). In Fig. 7.28, the simulated and measured data of within-season variation of tracheid diameter and cell wall thickness for 1964 and 1966 are shown. One can see not only good coincidence between modeled and observed values (R from 0.72 to 0.76) but also high coincidence in the character of the variations. This may be seen in the formation of a "false ring" with radial tracheid diameters of about 12–14 μm in the 1964 annual

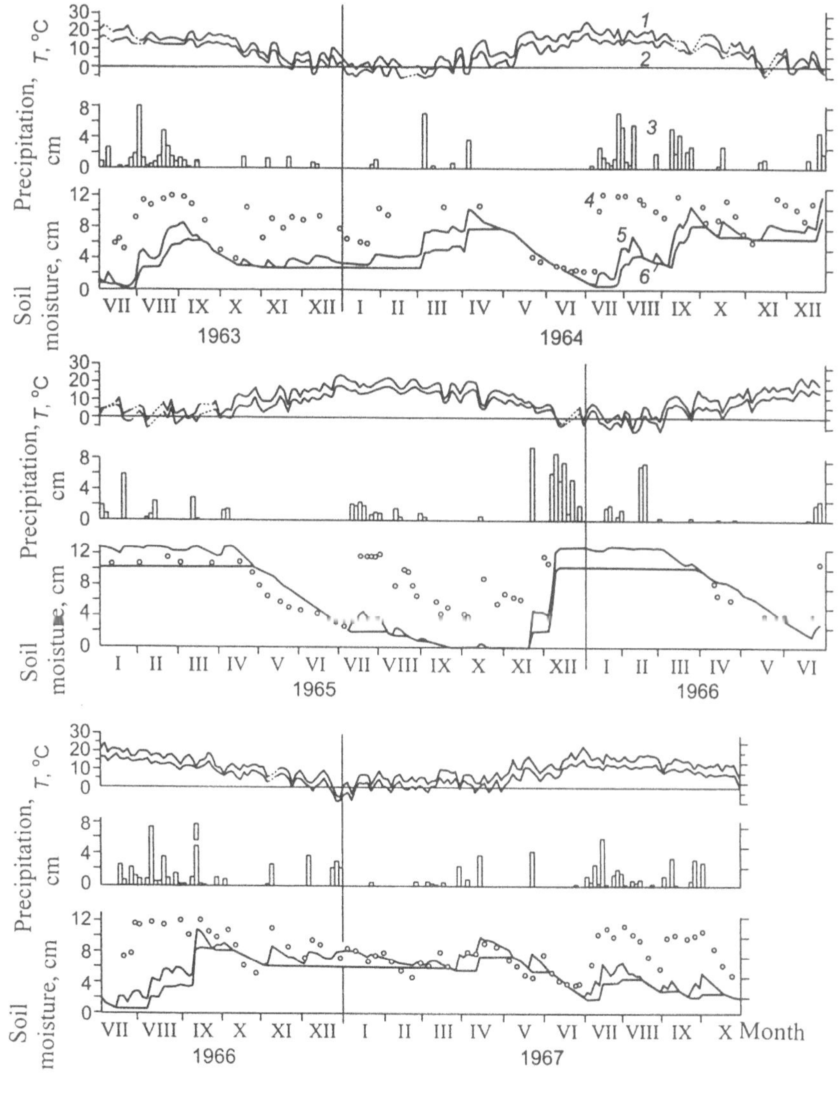

Fig. 7.24. Three-day averages of air temperature, precipitation, and soil moisture measured in the *Pinus ponderosa* study area for 1963–1967, along with calculated upper- and lower-layer water balance (Thornthwaite and Mather 1955). *1* Average day temperature, *2* average night temperature, *3* precipitation, *4* measured and *5*, *6* calculated dynamics of available soil moisture (Fritts 1976)

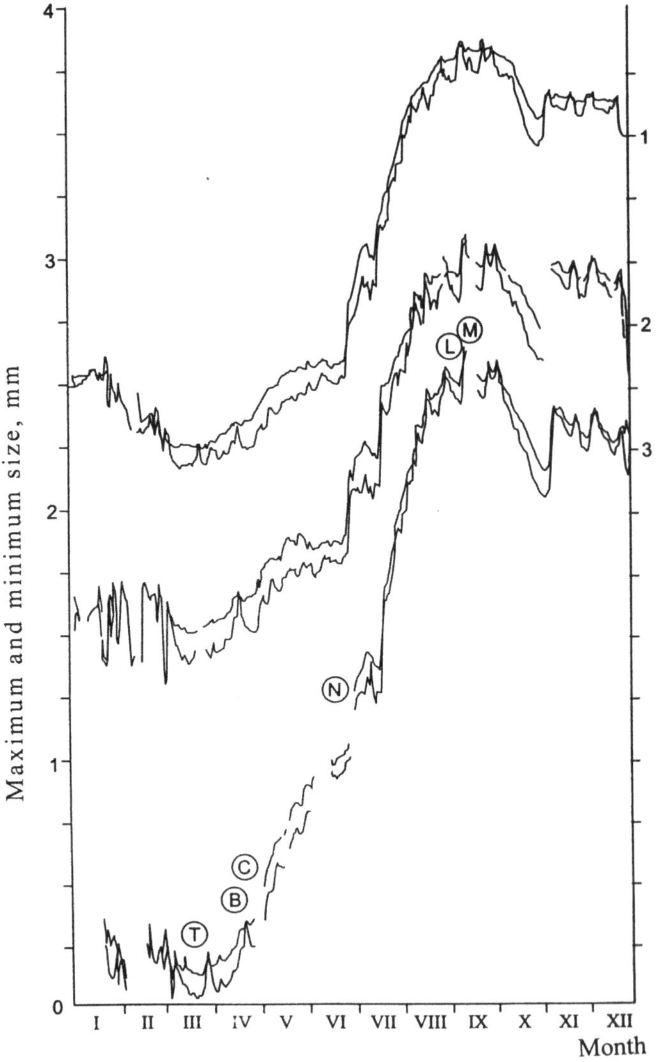

Fig. 7.25. Daily maximum and minimum dendrographic measurements of stem size and phenological observation from one *Pinus ponderosa* for 1966 were made at (*1*) the base dendrograph, (*2*) middle dendrograph, and (*3*) upper dendrograph positions in the main stem and used to check the growth that was simulated. *T* Apical buds swelling, *B* buds opening, *C* cambial activity initiated, *N* needles emerging from bud, *L* lignified latewood cells observed, *M* needles reaching mature size (Fritts 1976)

ring and a local "flush" in tracheid diameter within the "false" ring in the 1966 annual ring. This extreme environment gives a good opportunity to test any kind of simulation model because of the long growing season and strong effect of all climatic factors. We will come back to this example once more in the text when considering the application of the eco-physiological model in Chap. 9.

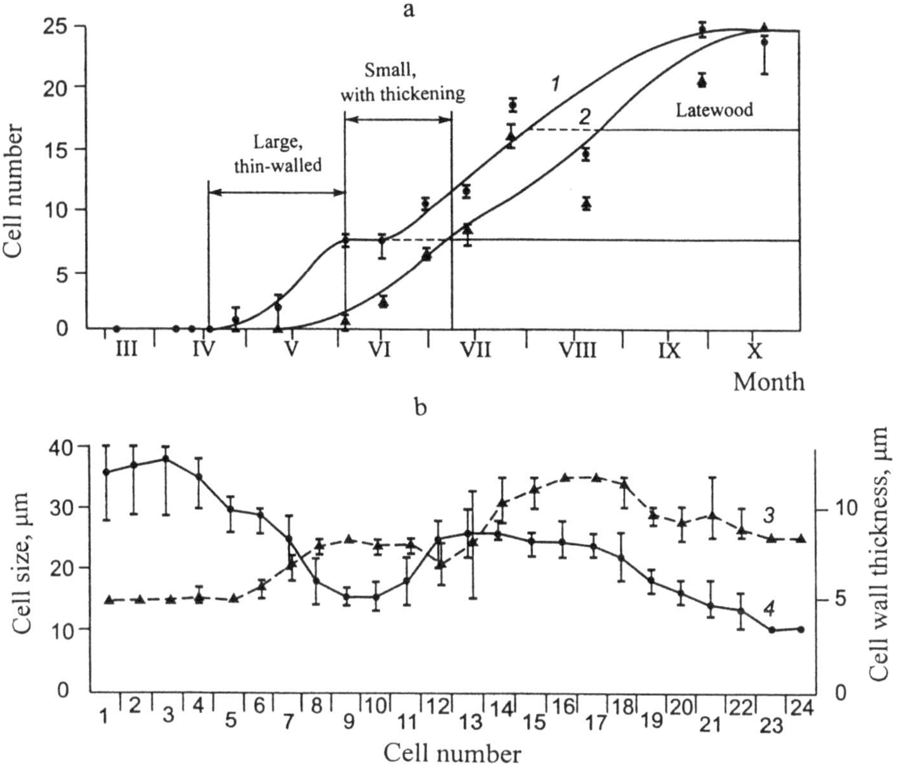

Fig. 7.26. Numbers (*1*) and wall thickness (*2*) of cells differentiated during 1966 in the upper stem of one *Pinus ponderosa* (**a**) and a diagram of observations made from microscopic sections at the end of the growing season (**b**). Less cell-size variation for that year was found at the base of the stem: (*3*) cell size, (*4*) cell wall thickness (Fritts 1976)

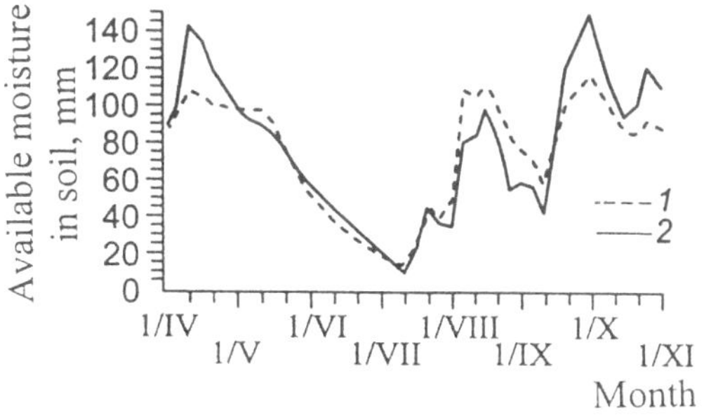

Fig. 7.27. Modeled (*1*) water balance for *Pinus ponderosa* in the Santa Catalina Mountains near Tucson compared to the Thornthwaite and Mather (1955) water balance (*2*). Both dry and wet periods are reproduced well (Fritts et al. 1991)

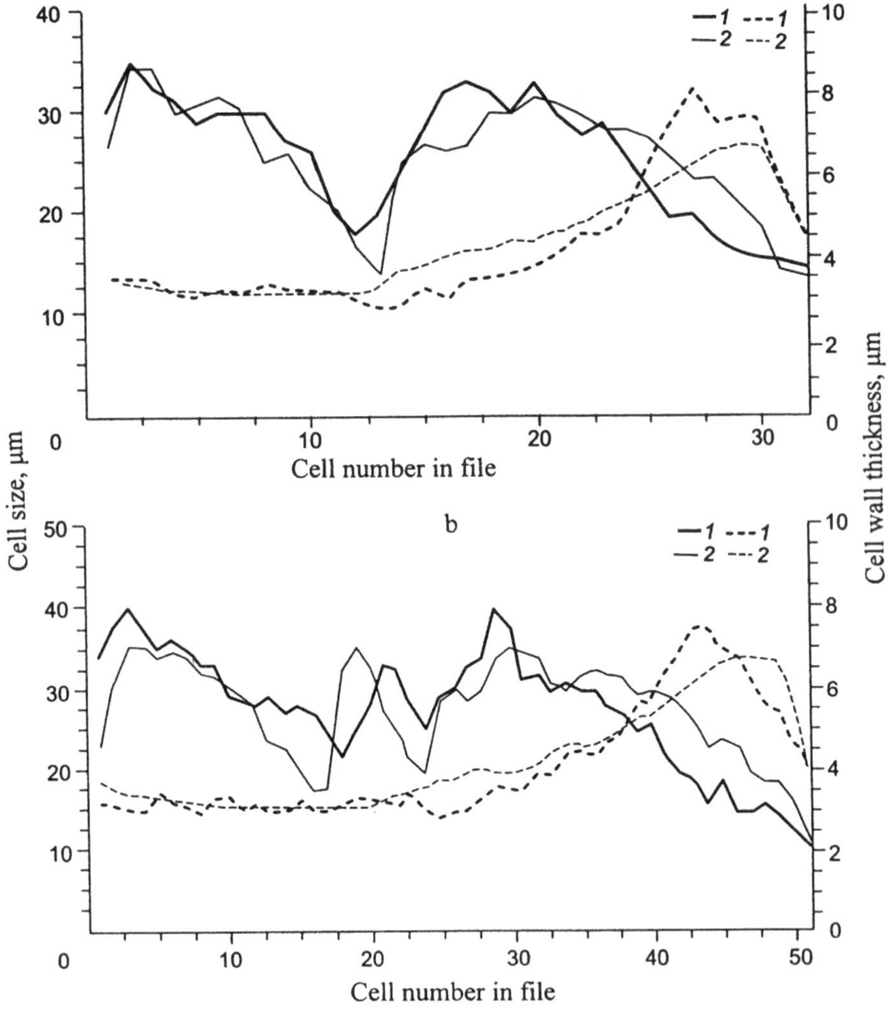

Fig. 7.28. Comparison of measured (*1*) vs modeled (*2*) cell size (*solid lines*) and cell wall thickness (*broken lines*) for the (**a**) 1964 and (**b**) 1966 growth rings from *Pinus ponderosa* in southern Arizona (Fritts et al. 1991)

7.6.5 Modeled Differences in Growth Response to Soil Moisture and High-Temperature Limitation of Conifer Species Growing in a Monsoon Climate

The monsoon climate in South Korea is characterized by a mild winter, long growing season, and specific intra-annual distribution of precipitation: a drier period from mid-May to the end of June and a rainy period from the beginning of July to September (Fig. 7.29; Kostin and Pokrovskaya 1961). The relatively dry period is combined with optimal temperatures and solar irradiation; and there is a limitation by water deficit of conifer growth in many sites (Park 1991, 1993). The high temperature (sometimes reaching 28–34 °C) in July and August results in both a direct and an indirect limiting effect of temperature on the growth rate of trees (Park and Vaganov 1997).

Two conifer species were chosen as objects for experimental analysis and modeling: pitch pine (*P. rigida*) and red pine (*P. densiflora*). Both species are growing on gravel-sandy soil in a mixed uneven-age forest stand located on a southwest-facing slope in Bukhan National Forest (near Seoul). The main differences between the two species growing in the same environment are connected with the anatomical structure of their tree rings (Fig. 7.30b). The pitch pine every year formed annual rings with clear "false rings" which were formed in both the early- and the latewood zones (Fig. 7.30b). The formation of two or three "false rings" during one season is usual. The radial diameter of tracheids in the "false ring" may be as small as 12–14 μm, which is also usual for typical latewood tracheids. In the same years, in red pine stems, annual rings do not show clear "false rings". The intra-annual variations of tracheid diameter and cell wall thickness have a more "normal" structure: a wide earlywood zone with tracheids of large diameter, a clear transition zone, and small or wider latewood. The evaluation of the season's duration was made using wood samples taken periodically (Park and Vaganov 1996, 1997) at about 150–160 days. We used this value to reconstruct the seasonal kinetics of growth rate from measured tracheidograms, using the approach described in

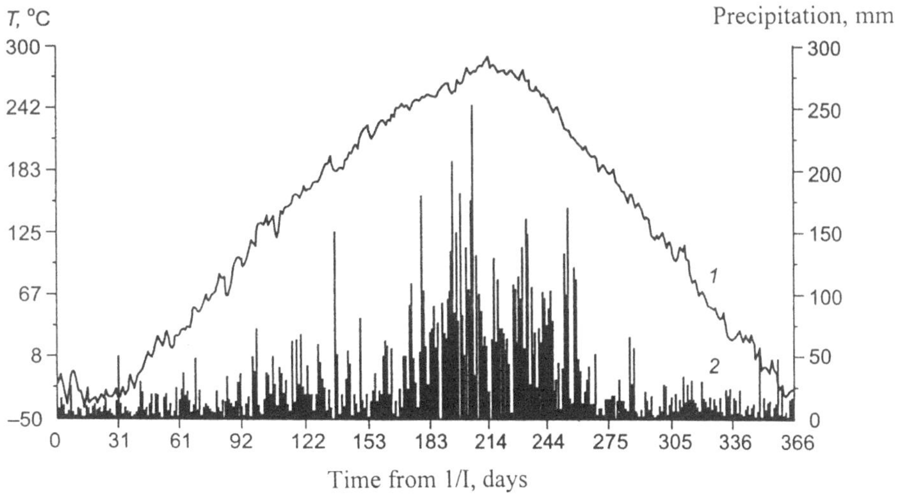

Fig. 7.29. Average long-term data of temperature (× 10) (*1*) and precipitation (*2*) for Seoul meteorological station (South Korea)

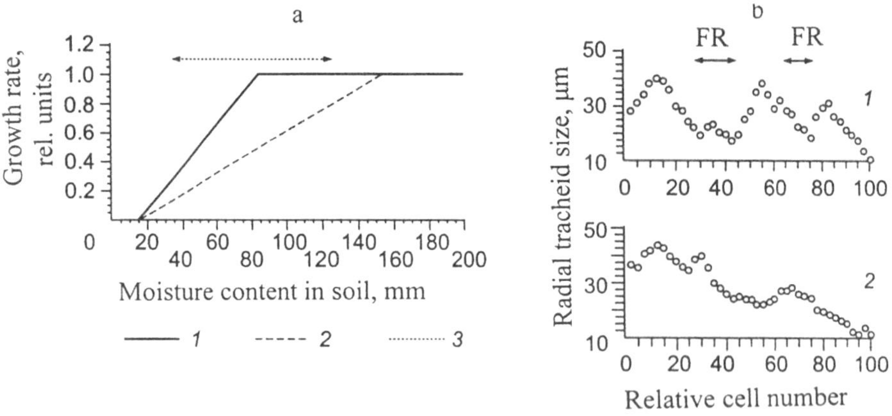

Fig. 7.30. Growth rate reaction of *Pinus rigida* (*1*) and *P. densiflora* Sieb. (*2*) to soil moisture (**a**) and within-season range of soil moisture variations (*3*); and typical tree-ring tracheidograms (**b**). *FR* "False rings"

Sect. 4.5. Examples of the reconstruction are shown in Fig. 7.31. In annual rings of pitch pine and in red pine, a strong decrease in tracheid diameter occurs from the first produced earlywood cells to those which are produced at the end of June under increasing intra-seasonal drought. The rainy season in July was enough to recover the production of typical earlywood cells in tree-rings of pitch pine. False rings do not occur in the red pine. To investigate the reasons for the differences in the anatomical structure of these two pine species we used the model. First, the upper limit of water field capacity for such types of soil was evaluated (Russell 1955; Spurr and Barnes 1980). The maximal daily precipitation into soil was calculated, approximately based on long-term average data and maximal field capacity. On certain days, the amount of precipitation may reach 800–900 mm in this environment, but only a small portion of it was retained by the active soil layer. The low field capacity, mechanical infrastructure of soil, relatively steep slope (20 %), low development of the grass layer, and other factors determined the intensive surface runoff of high daily precipitation.

The calculations with the model were made in three stages. First, we simulated one season of annual ring formation in order to reveal the key parameter that determines the observed differences in anatomical structure of pitch and red pine tree-rings. Then, we simulated seasonal growth for three years with the available measured tracheidograms of pitch and red pine tree rings, and then slightly modified the parameters. Finally, we calculate the inter-annual growth of both species, using long-term daily climatic records to compare simulated tree-ring width indices with the measured ones. This procedure may be considered as the usual iterative procedure used in simulation.

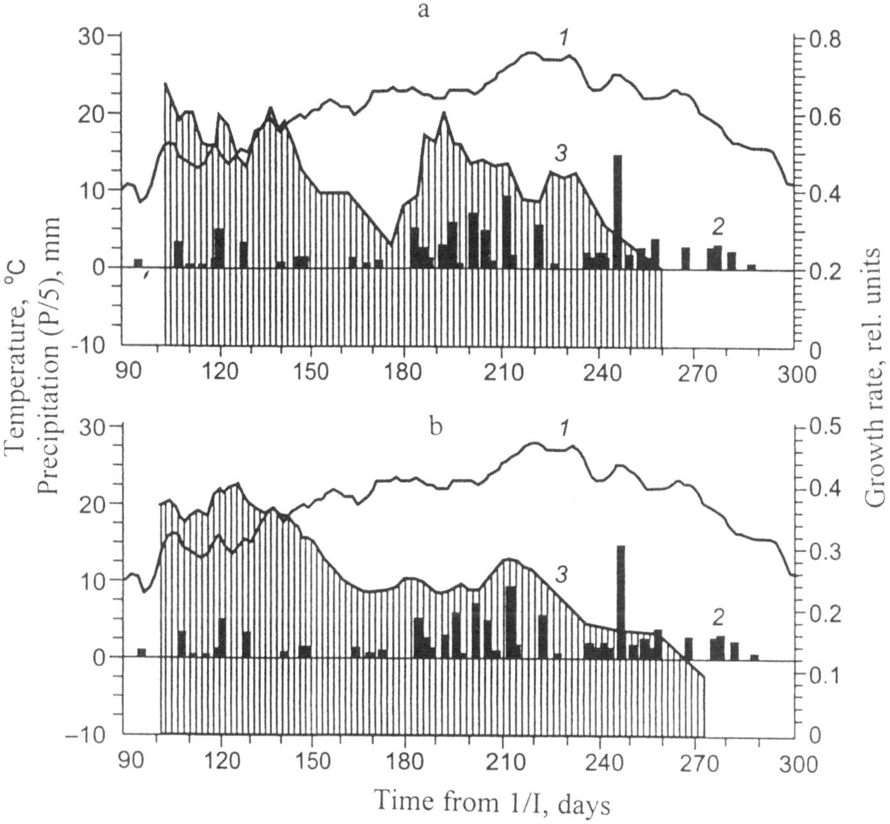

Fig. 7.31. Reconstruction of seasonal growth rate dynamics according to the tracheidograms of *Pinus rigida* (**a**) and *P. densiflora* Sieb. (**b**) in 1983. *1* Air temperature, *2* precipitation (*P*/5), *3* growth rate

The dates of the beginning of the season hardly differed for the two species. For the cambium, initiation occurred at the end of April when the temperature reached 10–12 °C. Growth continued until the end of September, so the duration of a growing season was about 150–160 days. Some 60–70 % of total annual xylem was produced before the onset of the rainy season, so that the first part of the season was characterized by the highest tracheid production rate (Park and Vaganov 1997). The species differences in seasonal growth kinetics and anatomical structure are determined by different response to water content in soil (Fig. 7.30a). Pitch pine shows more sensitivity to lower soil moisture. The red pine is more adaptive to high soil moisture and shows the highest growth rate when the water content in soil is close to optimal. Field capacity reached its maximum at the beginning of the growing season and at the beginning of the rainy season (monsoon). However, in the monsoon period, the temperature also reached its maximum within-season values and limited the growth rate (the temperature was in the range of the decreasing slope of growth response above the optimum; see Fig. 7.12a). Moreover, in July and August, the effective solar radiation also decreased and added an additional limitation to the growth rate. As a result, the growth rate did not reach the same high values in the second part of the season as at the beginning of the season, in spite of optimal soil moisture. The formation of "false rings" in the early and transition wood of annual tree rings of pitch pine was related to minor rainfall events during the long period of intra-seasonal drought (mid-May to end of June). This small amount of water was enough to raise the soil moisture content to a range close to optimal for pitch pine cell growth and production (Fig. 7.30a). Such an amount of water, however, was not enough to provide optimal soil moisture content for red pine growth. The seasonal range of soil moisture is shown in Fig. 7.30a, which clearly illustrates the differences in growth response of both species to water content.

The simulation data were confirmed by direct measurements of seasonal growth. Consider the example of pitch pine growth during 1997. Figure 7.32

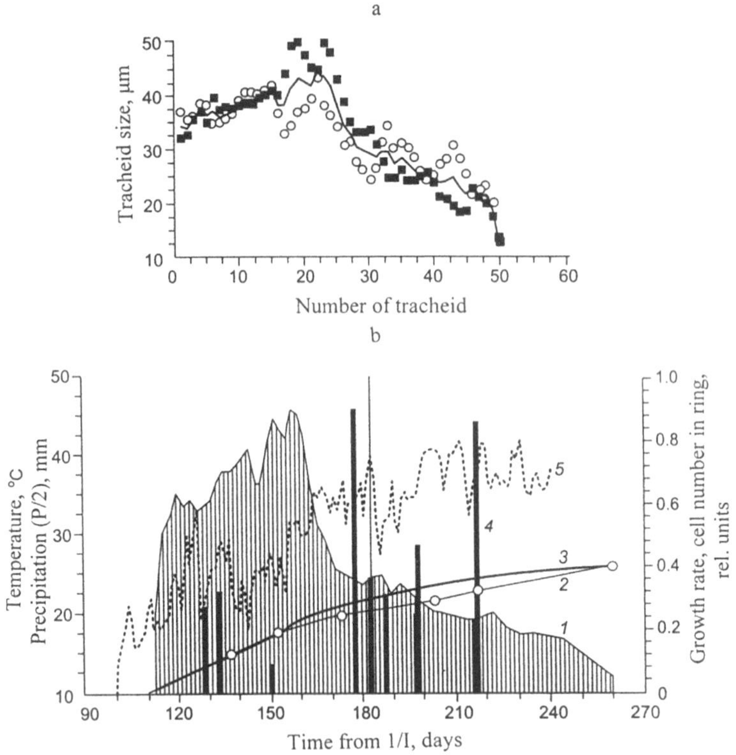

Fig. 7.32. Typical standardized tracheidogram of *Pinus densiflora* Sieb. tree ring of 1997 (**a**). *Circles, squares* Different trees (*n*=5), *line* average. Dynamics of climatic factors and growth rate in 1997 (**b**). *1* Reconstructed growth rate according to the tracheidogram, *2* measured curve of tree-ring growth, *3* calculated curve of tree-ring growth, *4* precipitation, *5* temperature

shows the typical tracheidograms of annual rings formed in 1997, the results of growth rate reconstruction, and measurements. There is good agreement between the measured and simulated courses of tracheid production during the season (R=0.86, n=50, P<0.0001). The highest growth rate occurred in May and the beginning of June under optimal soil moisture, temperature and solar irradiation. From mid-June, the growth rate decreased quite rapidly due to soil drought and did not recover in the second part of a season. This was because the precipitation did not compensate for the decrease in effective solar irradiation and the limitation of growth by high temperatures. The analysis of standard statistical correlation function shows the positive influence of March–May precipitation on the inter-annual variability of red pine tree-ring width, but does not reveal any significant influence on the inter-annual variability of pitch pine tree-ring width (Fig. 7.33). Red pine trees are better able to use the soil water available at the beginning of the season, which is mainly determined by precipitation input before the season (March–May).

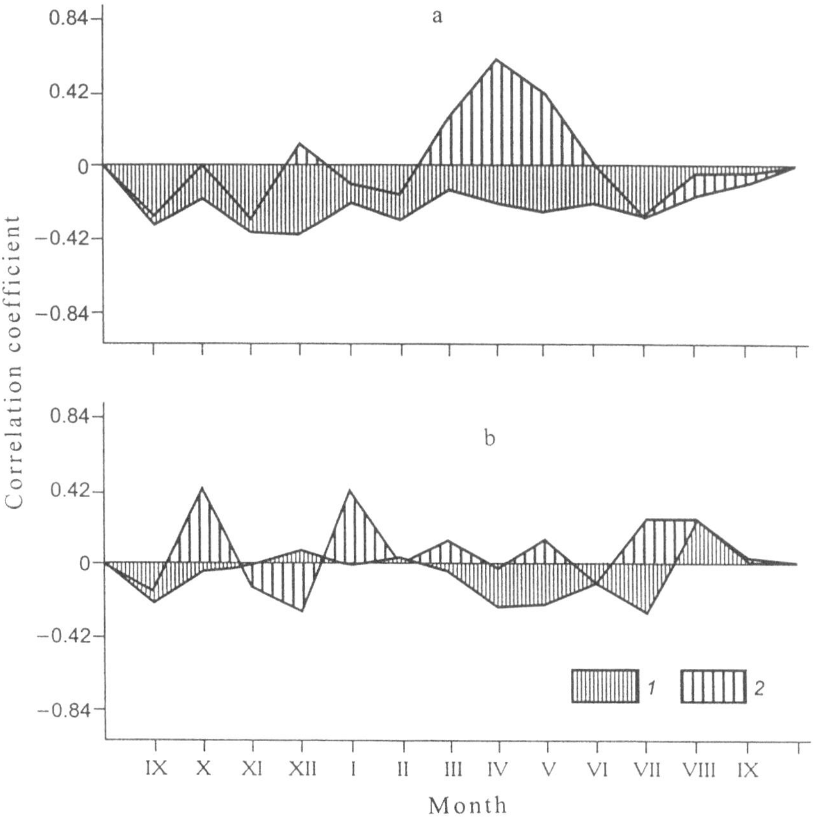

Fig. 7.33. Climatic correlation functions for temperature (*1*) and precipitation (*2*) for tree-ring chronologies of *Pinus densiflora* Sieb. (**a**) and *P. rigida* (**b**)

Pitch pine is better able to use small amounts of additional water in the soil during the season. As a result, it is sensitive to precipitation in March–May and to small amounts of precipitation in May–June. So, high precipitation has a similar response to low precipitation in the case of pitch pine. Finally, the simulation of inter-annual variation of tree-ring width based on fixed model parameters shows good agreement between the calculated and measured curves (R=0.66, 0.62, n=22, P<0.05; Fig. 7.34).

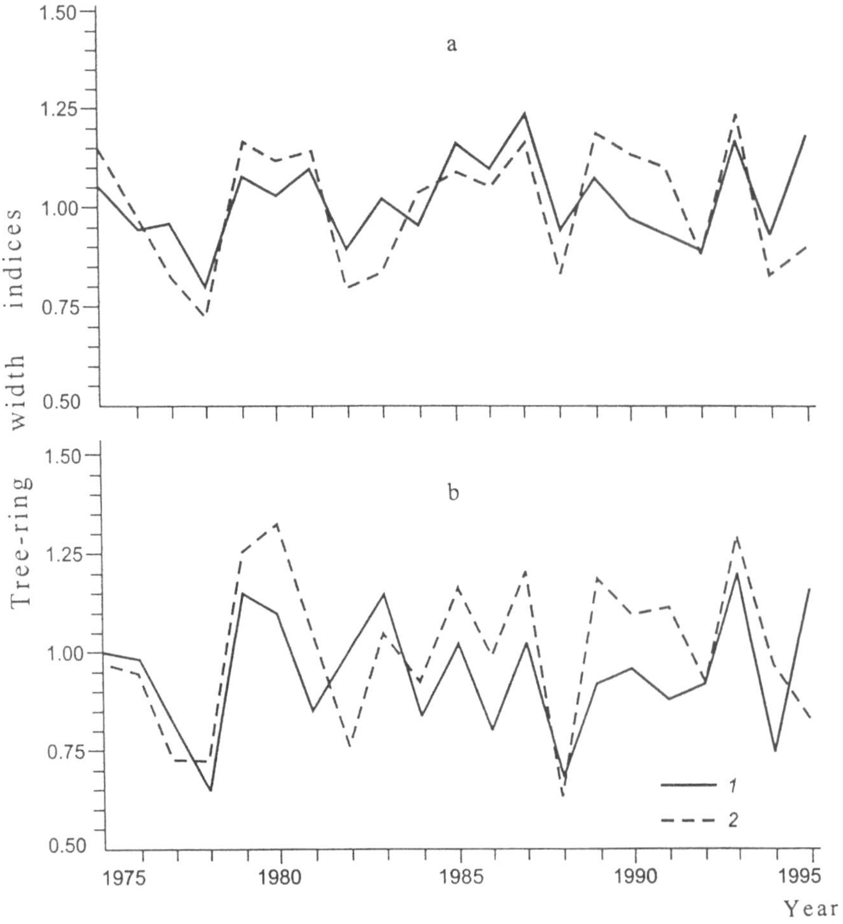

Fig. 7.34. Measured (*1*) and calculated (*2*) dynamics of tree-ring width indices of *Pinus densiflora* Sieb. (**a**) and *P. rigida* (**b**)

7.7 Conclusions and Discussion

The main focus of this chapter has been the Vaganov–Shashkin (VS) model, which permits the simulation of seasonal growth and tree-ring formation, using daily meteorological data. The context for this model was set by summarizing the use of statistical models in dendroclimatology and their strengths and weaknesses. Earlier mechanistic simulation models of tree-ring formation were also discussed and the VS model, the main focus of this book, was introduced. Particular attention was given to the modeling of cell growth within the cambial zone and the production of new xylem cells, as this is the key process through which, in the model, the later stages of tree-ring formation are controlled. The versatility of the model was demonstrated by reference to examples of its application in a wide range of environments, from subarctic to monsoon.

The VS model is novel in its approach to the environmental control of the formation of conifer tree-rings, in that it assumes that the target of environmental control is found in the cambial zone. The examples considered here show that the model works well in different climatic conditions and conifer species and reproduces observed patterns of seasonal growth and ring width. Its theoretical basis is focused on the adequate description of climatically induced variations in tree-ring width and anatomical structure.

The simulations may enhance the interpretation of statistical response functions, in some cases explain the divergences between ecological sense and response function results, and provide additional information about climate effects on tree growth in different environments. Such a simulation model may be considered as the basis for the next step, namely eco-physiological modeling, which can include the mechanisms of interaction between meristems in the whole tree as well as the allocation of assimilates, and the direct and indirect influence of some factors (for example pollution and defoliation) on assimilation tissues etc., which are described in the conceptual scheme of environmental control of xylem differentiation. Some ways to develop an eco-physiological model will be considered in Chap. 9.

The VS simulation model we have described has considerable potential for the prediction of tree-ring growth response under differing climate situations. This might include, for example, the calculation of growth in a former time on the basis of climate scenarios generated by climate models, and the subsequent checking of the actual observed tree-ring record. Similarly, it might be used to predict growth under future scenarios of climate change, although at the moment it is limited by its implicit assumption of constant atmospheric carbon dioxide concentration. In spite of this, the simulation model is free of the restriction that exists for all statistical models, i.e. that it is invalid to use them to extrapolate outside the range of conditions for which they were calibrated (see Sect. 7.2)

8 Simulation of Tree-Ring Growth Dynamics in Varying and Changing Climates

8.1 Introduction

Dendroclimatology faces a number of challenges, some of them as a result of new findings, others as a result of new opportunities. Does the VS model offer the possibility of progress in meeting any of these challenges? If it has this potential, how well has it been realized and what enhancements could improve its utility?

One of the most profound problems in the use of all natural archives of past climate is to know to what extent the assumption of uniformitarianism is valid. Does our conceptual model of the climate control of tree-ring variability take into account all the factors and combinations that have played a significant role over the period whose climate we seek to reconstruct or project? In the particular case of the VS model, is it reasonable to apply it outside the regions of the north temperate and boreal forests for which it was originally developed?

The trees simulated in this model are effectively ageless. Might it be necessary to use different parameters for various age-classes of trees if the aim was to introduce realistic age-structure into the exercise? To what extent is it necessary to change model parameters for each conifer species modeled, or for differing site conditions?

These questions are addressed in this chapter and some opportunites for novel applications of the VS model are discussed. First, apparent time-dependent changes in tree-ring response to climate (Sect. 8.2) and the generality of the VS model (Sect. 8.3) are examined. The dependence of the models on conifer species, tree age, and site characteristics is explored (Sect. 8.4). We then demonstrate the use of the model for investigating the impact of various climate-change scenarios on tree-ring growth, in this case in the boreal zone of Siberia (Sect. 8.5), in temperature- and precipitation-limited cases (Sect. 8.6), for differing species in water-limited conditions (Sect. 8.7), and for longer-term variations (Sect. 8.8). The use of forward and inverse models in climate reconstruction is considered (Sect. 8.9), as is the use of the forward (VS) model in the interpretation of empirical–statistical reconstructions (Sect. 8.10). The prospect of applying the VS model on very large spatial scales is outlined (Sect. 8.11) and conclusions and discussion are presented (Sect. 8.12).

8.2 Time-Dependent Changes in Response of Growth Rate to Climatic Variations

"As with all natural environmental archives, the use of tree rings is based on the principle of uniformitarianism. The assumption is made that the same mechanisms were at work forming the archive in the distant past as in the recent past. Further, it is assumed that the transfer function designed using recent observations takes into account all climate factors that might have a significant effect on ring growth in the reconstructed period. It is also assumed that all combinations of values of the predictor variables (the tree rings) for the reconstruction period are represented in the calibration period" (Hughes 2002).

It is normal practice in dendroclimatology to check whether the statistical transfer function (see Sect. 7.2) to be used to calculate the reconstruction is stable in time (Fritts 1976; Hughes et al. 1982; Cook and Kairiukstis 1990). This test is done for a period not used in the development of the transfer function, but for which instrumental data are available. Any changes in the model between the calibration and "verification" periods may result from a change in tree-ring/climate relationships, or they may simply reflect the uncertainties inherent in the calculation of such a statistical model, in the derivation of the tree-ring chronology, and in the suitability and quality of the meteorological data used. Enough cases of apparent instability in either transfer function or response function have now been reported, especially for high latitudes, to suggest that, in spite of the statistical uncertainties, a reduction in sensitivity or a change in the character of the climate response of tree rings may have occurred in some regions in the last few decades of the twentieth century (Jacoby and d'Arrigo 1995; Briffa et al. 1998; Barber et al. 2000; Lloyd and Fastie 2002; Wilson and Elling 2003; Magda 2003). Some of these changes are probably related to non-climatic factors such as air pollution (Visser and Molenaar 1992; Wilson and Elling 2003). Several others, however, suggest instability of the growth response to climate variables; that is, growth is mainly limited by one climatic factor at one time, and another in a subsequent period.

This instability is well demonstrated by the results of Lloyd and Fastie (2002). They investigated the response of trees growing at the cold margins of the boreal zone in Alaska to climate variations in the twentieth century. During the period from 1900 to 1950, the growth of white and black spruce at and near the alpine and arctic treelines in three regions of Alaska increased, apparently due to increases in the annual and summer temperatures. After 1950, there was a significant decline in tree growth against a background of temperature increase. These inverse growth responses to temperature were more common at sites below the upper forest margin than at sites at the forest margin and were most common in the warmer and drier sites (Lloyd and Fastie 2002). Drought-stress of growth due to increasing temperature has been reported for elsewhere in Alaska (Barber et al. 2000). These results are broadly consistent with those reported by Briffa et al. (1998). They reported a weakening during the late twentieth century in the decadal-scale correlation between summer temperature and maximum latewood density (and, less markedly, ring width) over much, but not all, of the boreal forest in the Old and New Worlds.

Such apparent changes in response are not limited to high latitudes. This may be seen in the results of an analysis of the response function of Douglas-fir (*Pseudotsuga menziessii*) in Idaho (Biondi 2000). By calculating response functions for moving periods of 60 years, a change in effective response from about 1970 on was identified (Fig. 8.1). There was a strong negative response to the current July temperature throughout the period of analysis, but the second-strongest response changed from strongly positive to the current May temperature towards a more mixed response, including a positive response to prior October precipitation. A very similar change was identified quite independently in the same general region by Kipfmueller (2003). A reduction in sensitivity of maximum latewood density to summer temperature since 1970 has been noted in *Abies pindrow* in Kashmir (Hughes 2001).

From the point of view of dendroclimatology, there are two main consequences of these observations, for the study of the future and the past. The applicability of empirical–statistical models to environmental changes that may be going on now, and might continue in the near future, is limited. Their usefulness for the past is also diminished, unless there can be considerable confidence that conditions then were analogous to those of the recent calibration period. This does not mean that empirical–statistical models should be abandoned, but that even greater emphasis be given to such precautions as checking against completely independent types of archive of past climate and, in particular, identifying situations in which nonlinear responses might occur. The VS model could be of considerable help in this latter task.

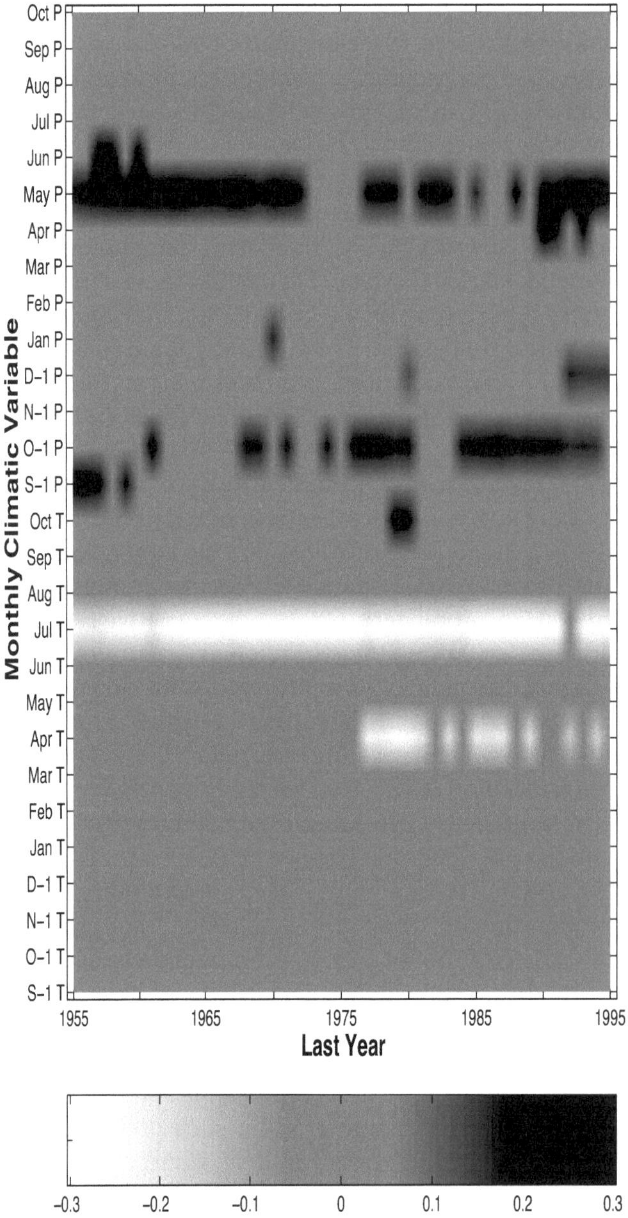

Fig. 8.1. Moving response function (MRF) calculated with divisional climate data and Douglas-fir chronology from NE Idaho, United States, using a moving 60-year period. The size and sign of response function elements are indicated by the gray scale below. *T* Temperature elements from prior fall to growth year, *P* precipitation elements (after Biondi 2000)

8.3 Application of the Vaganov–Shashkin Simulation Model to a Wide Range of Species and Site Conditions

In order to explore the suitability of the VS model described in Chap. 7 for use in dendroclimatology, Evans et al. (2005) examined its application to 208 tree-ring width index chronologies from North America and Russia (Fig. 8.2). No attempt was made to adjust parameters to suit the circumstances of differing species and regions. Rather, a single set of parameters was used. Daily meteorological data from nearby stations were used as input to the model. Their temperatures were partially corrected for the difference between the eleva-

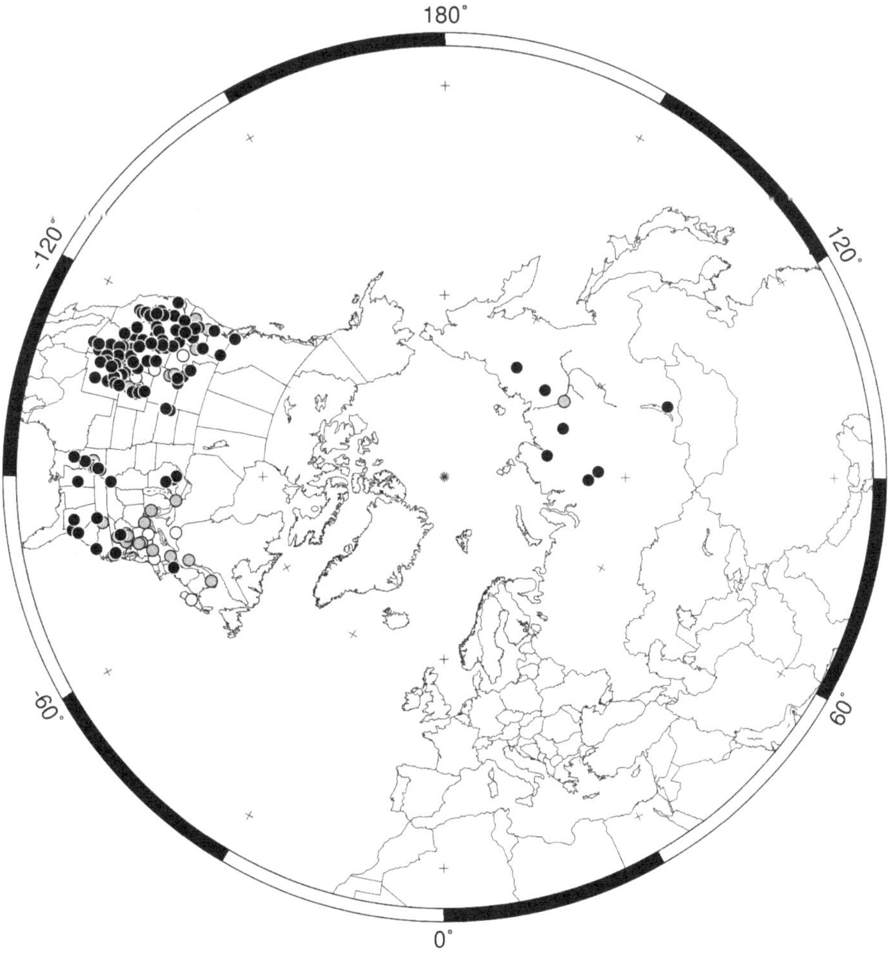

Fig. 8.2. Correlation significance map for the simulation of 218 tree-ring width chronologies from North America and Russia. Significance levels (considering effective degrees of freedom): >99.99 % (*black*), >95 % (*gray*), <95 % (*white*; after Evans et al. 2005)

tions of the tree-ring sites and those of the meteorological stations. We show the results of the intermediate calculations for *Pinus sylvestris* over five years at Ulan-Ude, in a dry region of Siberia southeast of Lake Baikal (Fig. 8.3). It is clear that daily growth rates are controlled by either temperature or water availability, depending on the date, with water availability having much the greater role. Water availability is controlled by the effect of precipitation and temperature on soil moisture. The total number of cells, and hence the ring width, is determined by the growth rates. The Ulan-Ude simulation was not very sensitive to changes in the primary water budget and temperature parameters, such as the temperatures at which minimum, optimum, and maximum growth rates were achieved (T_{min}, T_{opt1}, T_{opt2}, T_{max} in Table 7.2). The sim-

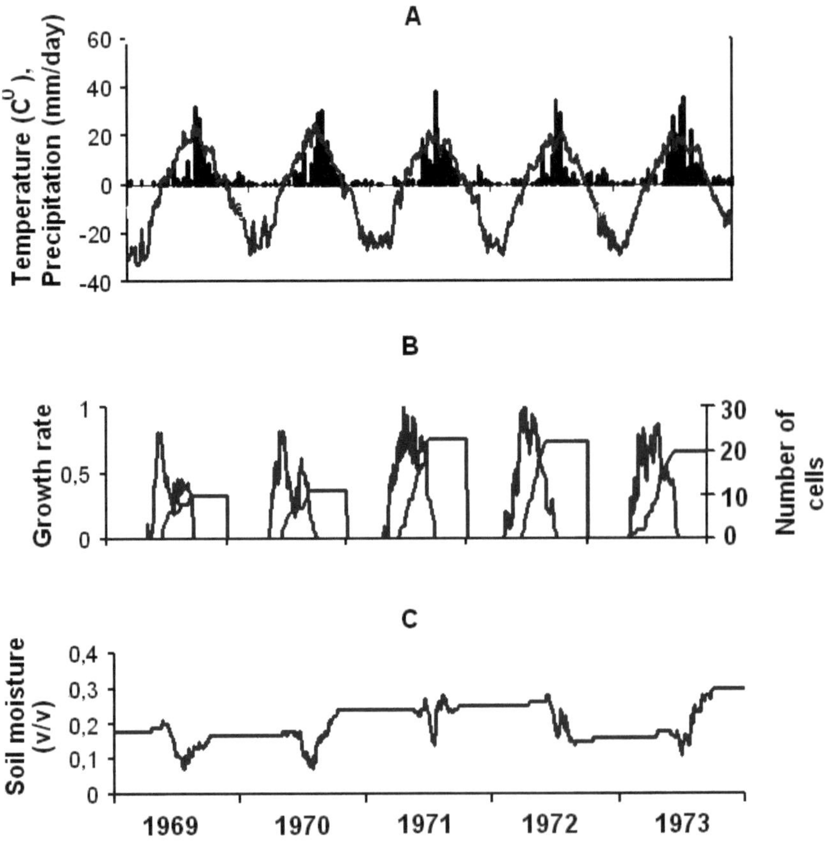

Fig. 8.3. Seasonal soil moisture and temperature controls on simulated tree-ring growth at Ulan-Ude, southern Siberia, 1969–1973. **a** Daily temperature (*gray*) and precipitation (*black*) data used to drive the model. Vertical scale gives both degrees Celsius (temperature) and millimeters/5 days (precipitation). **b** Annual growth response function $G(t)$ (*thin line*) and cumulative number of wood cells per ring (*dark line*). **c** Soil moisture calculated by the water balance component of the model (volume/volume ratio)

ulated series compares well with the actual chronology for Ulan-Ude (Fig. 8.4), not only at inter-annual time-scales (r=0.58, n=65, P<0.01), but also at multi-year time-scales (r=0.81, n=65, P<0.01). The good performance of the model at multi-year time-scales will be discussed later. The results for Ulan-Ude are not unusual. The great majority of the chronologies modeled have highly significant correlations between the simulated and actual chronologies. Most of them, in turn, have similarly significant correlations for decadal averages, rather than annual data, even though far fewer degrees of freedom are available. The strongest correlations between simulated and actual chronologies are concentrated in the interior western United States, the southeastern United States, and Siberia (Fig. 8.2). Most of the weakest correlations are for simulations in the eastern and northeastern United States. It has recently been shown (Anchukaitis et al. 2005) that simulations of chronologies from those regions improve greatly with the adjustment of a single model parameter, K_R, the coefficient for water drainage from soil (see Table 7.2).

Even without this, the results reported by Evans et al. (2005) show that the VS model is applicable in situations as diverse as the Siberian forest-tundra, forest in the mountains of semi-arid western North America and the warm,

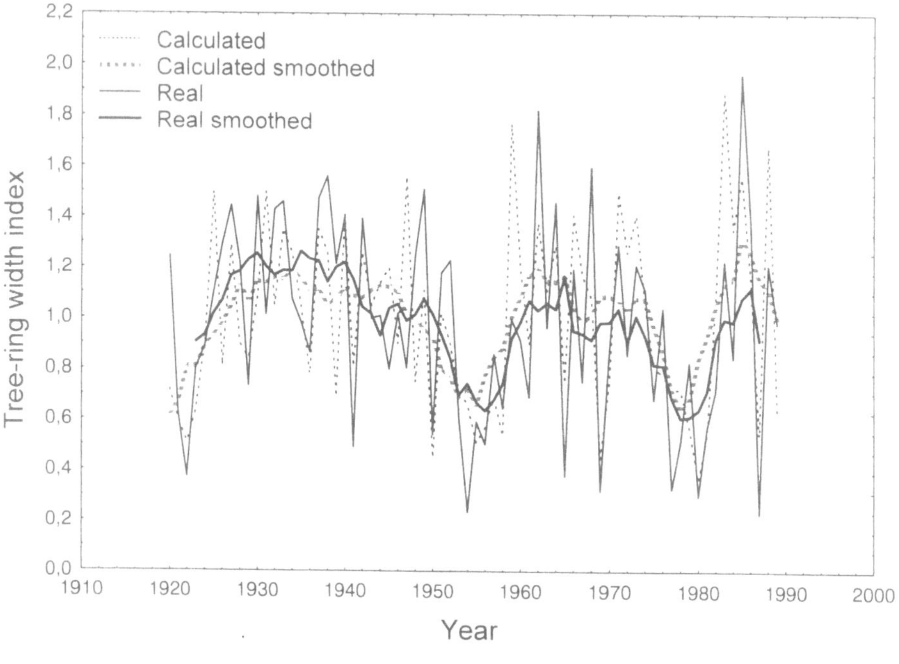

Fig. 8.4. Observed and simulated tree-ring width indices, Ulan-Ude, 1922–1986. Annual (*thin lines*) and 5-year mean (*heavy lines*) correlations are r=0.58 and r=0.81, respectively, both significant at the 99 % level considering effective degrees of freedom (Evans et al. 2005)

moist southeastern United States. The skill they have demonstrated in the VS model comes from three characteristics: (1) its incorporation of nonlinear functional forms, (2) the use of a water balance model to calculate net change in water availability, and (3) the flexibility of response that comes from a daily time-step in the model. We would add a fourth characteristic that we see as particularly important to the skill the model shows at multi-year time-scales. This is its explicit treatment of cambial kinetics, so that the number and size of cells in the cambium at the beginning of each growth season is calculated during the previous, model, growth season.

8.4 Parameters of the Model and Its Relation to Species, Age and Site Characteristics

8.4.1 Age

There are several publications in which an association between age and climatic response has been shown for different species (Szeicz and MacDonald 1994, 1995; Ettl and Peterson 1995; Zarnovican 2000; Carrer and Urbinati 2004). Explanations have been proposed for some of these cases. For example, Sceicz and MacDonald (1994), working with white spruce, suggested that "one possible causal factor of age dependence is that the trees are becoming increasingly stressed with age owing the reduction in the efficiency of water and nutrient translocation mechanism" (p. 20 in Sceicz and MacDonald 1994). This approach has been developed further recently (Carrer and Urbinati 2004) by considering the variation in hydraulic limitation imposed on a tree as its height increases. It is not yet possible to assess how generally applicable this explanation might be.

For example, consider the case of growth in cold environments. Old trees with deeper and more extensive root systems may get some portion of their water from deeper soil layers than young trees. There are several model parameters [depth of root systems, active soil layer available to root system to absorb the water (l), parameters describing the soil thawing (a_1, a_2)] that may influence growth differentially according to age (see Table 7.2). Young trees could be more sensitive to temperature and light variations than older trees, which also is related to competition for light. For example, comparing the response of young and old trees of Scots pine in the same site, Savva et al. (2002b) showed that ring widths of young trees from 25-year plantations have a significant relationship with the temperatures from mid-April to the beginning of May, but old trees only show a significant relationship with May temperatures.

8.4.2 Species and Site Characteristics

To use the simulation model in a particular environment, it is necessary to define the values of the main model parameters (see Table 7.2). Even though the model has been shown to perform well in a wide range of conditions using a single set of conditions (see previous section), it has also been shown that its performance may be considerably enhanced by an appropriate adjustment of some parameters. These could be peculiar to particular tree species and to features of the local conditions (such as soil characteristics, slope, availability of ground water, etc.). Parameterization of the model is the procedure for evaluating model parameters to improve model skill.

The parameters of temperature response (Table 7.2) seem to be mostly species-specific. There is a large range among woody plants in the starting and optimal temperatures for growth (Gates 1980; Kozlowski and Pallardy 1997; Kimmings 1997). In our simulations, different pine species showed different temperature responses. Consider red pine in Korea and Scots pine in Siberia as examples: (1) the temperature for initiation of growth of Scots pine is significantly lower (5 °C) than for red pine (about 10 °C; Sects. 7.6.2, 7.6.5). Optimal temperatures are closer (18 °C, 22 °C, respectively); and high temperatures which limit growth do not differ significantly. A similar situation exists for the minimum temperatures for growth of the main conifers in the northern taiga: the minimum and optimal temperatures for larch are significantly lower (4 °C, 16 °C) than those for spruce (6 °C, 18 °C). However, marked species differences in temperature response may also exist within one species as a result of growth adaptations in certain conditions. Different provenances within a species retain "memory" of the climatic conditions of the place of their origin (Savva et al. 2002a, b, 2003). This "memory" is responsible for a small but significant portion of inter-annual variability of tree-ring width and wood density in the current growth conditions of the provenances. For instance, northern provenances growing in the south taiga and forest-steppe regions in Siberia retain mechanisms to use climatic conditions in the first part of the season more effectively than southern provenances. The southern provenances (in their "native" conditions) use the conditions of the second part of the season more effectively and produce wider rings and denser wood.

Although it is possible to place the variation of temperature-related parameters in a species-specific context and there is some experimental basis for this (Gates 1980; Olszyk et al. 1998; Peltola et al. 2002), the explanation of parameters for water response is more complicated. The first problem is to determine which parameters are defined by species and which by site conditions (mainly by soil type or soil structure). A notable example how water response parameters could be related to species is given in Sect. 7.6.5. Zhang et al. (1997) reported that seedlings of ponderosa pine from a highly drought-tolerant population were particularly sensitive to water availability, so that they can rapidly make use of water and close their stomata in response to water stress. They pointed out the importance of a drought-avoidance mechanism for this adaptation. However, such examples are rare. Probably, if two species grow on heavy soil with a large proportion of clay there will be, first, an increase in water content at the wilting point (W_{min}), because of the higher water retention by clay soils, and, second, an increase in the value of optimal water content (W_{opt1}). According to this, the influence of soil type and structure is more important for the water-limited growth response than the differences in species response (Orwig and Abrams 1997; Peterson et al. 2002).

A larger species-specific effect may be seen for parameters which describe the portion of incoming precipitation intercepted by the crowns of trees and lost to evapotranspiration (k_1, k_2, k_3). If the interception of precipitation by tree crowns is related to tree shape and structure (for some species this may be as large as 20 %; Protopopov 1975), it could be species-specific. The portion related to surface runoff mainly depends on soil structure and understory (grass composition, portion of moss, lichen, etc.). The water loss due to evapotranspiration may be more species-related because of needle structure, frequency and density of stomata, and stomatal conductance as a whole. The importance of the morphology of needles in this process is widely studied and well known (Cregg 1993; Zhang et al. 1997; Olivas-Garcia et al. 2000; Woo et al. 2001; Anfodillo et al. 2002).

8.4.2.1 An Example from an Extreme Environment

Great Basin bristlecone pine (*Pinus longaeva*) grows in high, cool, very dry environments where few other trees can survive. There are some large differences between the typical parameters that may be used with many conifers and those required for skillful modeling of seasonal and inter-annual variation of tree-ring growth in this species. Inferential and empirical–statistical approaches show that inter-annual to multi-century variability in ring-width indices of this species growing close to its lower elevational limit (ca. 2,800 m a.s.l.) may be largely accounted for by variations in available moisture (LaMarche 1974a, b; Hughes and Graumlich 1996; Hughes and Funkhouser 1998), whereas close to the upper distributional limit (ca. 3,500 m a.s.l.) a more complex interaction of temperature and moisture availability appears to control growth (LaMarche 1974a, b; Graybill and Idso 1993; Hughes and Funkhouser 2003). Hughes and Graumlich (1996) did, however, note that a precipitation reconstruction based on bristlecone pine tree-ring width indices from Methuselah Walk in the White Mountains of California reflected all the very dry years of the instrumental period very well, except for five years which happened to have the coolest May in the record. It must be emphasized that this is a very dry region indeed, with average annual total precipitation only reaching 477 mm, even at the nearby very-high-elevation meteorological station (3,800 m a.s.l.) at Barcroft. It was necessary to adjust parameters in order to model the chronology from Methuselah Walk, growing at 2,800 m a.s.l. (Figs. 8.5, 8.6). The best fit of actual and simulated chronologies was achieved with the minimal and optimal temperature parameters (T_{min}, T_{opt1} in Table 7.2) equal to 3 °C and 16 °C, respectively, and when the two coefficients which describe water loss from soil are significantly higher than usual: k_2=0.25 and k_3=0.18. The value of coefficient k_2 (0.25) shows that evapotranspiration at low temperatures is practically twice the typical value for many other species.

We cannot say definitely that these parameters are species-specific to bristlecone pine, because the environment in which these trees grow is unusual in several respects. The lower partial pressure of water vapor at such elevations leads to a higher transpiration rate. The substrate at the Methuselah Walk site is Reed dolomite, which is very white, with very low vegetation density, leading to a high albedo of the ground surface. This could well be associated with higher leaf temperatures than might otherwise be expected at a particular air temperature, for example. The simulation model revealed that, in every year, the growth in the second part of the season is limited by low moisture, but the starting dates and the growth rate at the beginning of the season are completely defined by temperature (Fig. 8.6a,b). This explains Hughes and Graumlich's (1996) observation of the failure to capture drought

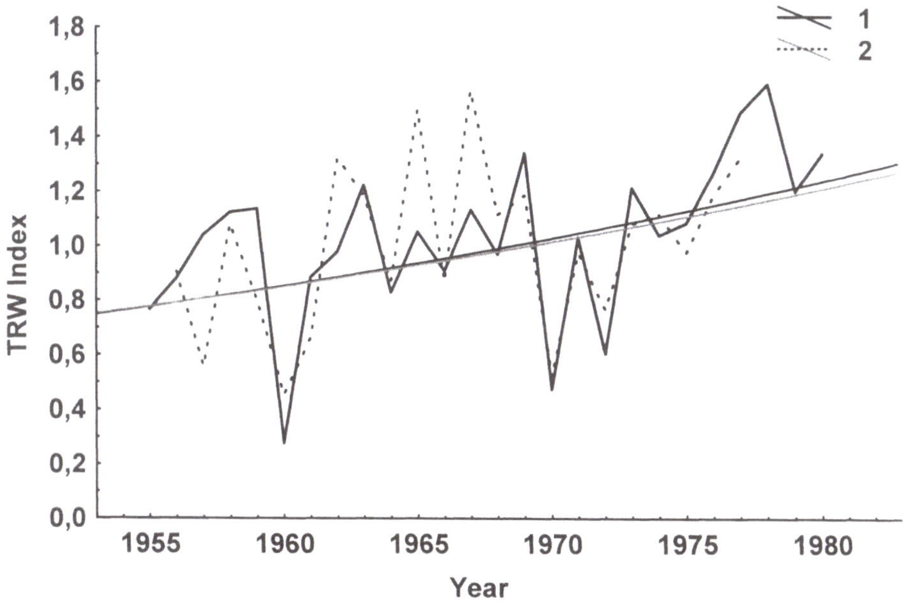

Fig. 8.5. Actual (*1*) and simulated (*2*) fragments of tree-ring chronology of bristlecone pine at Methuselah Walk (White Mountains, Calif.). The long-term trends of actual and simulated chronologies are also shown

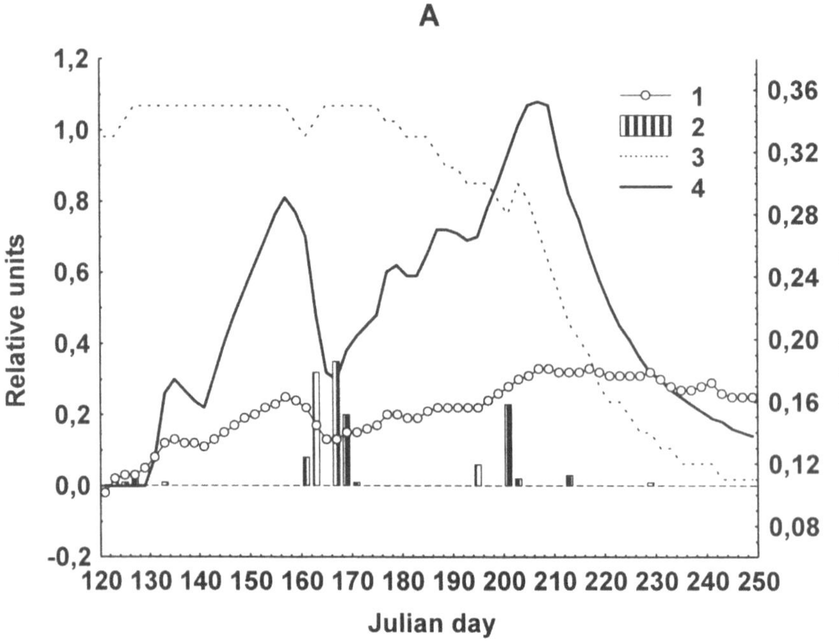

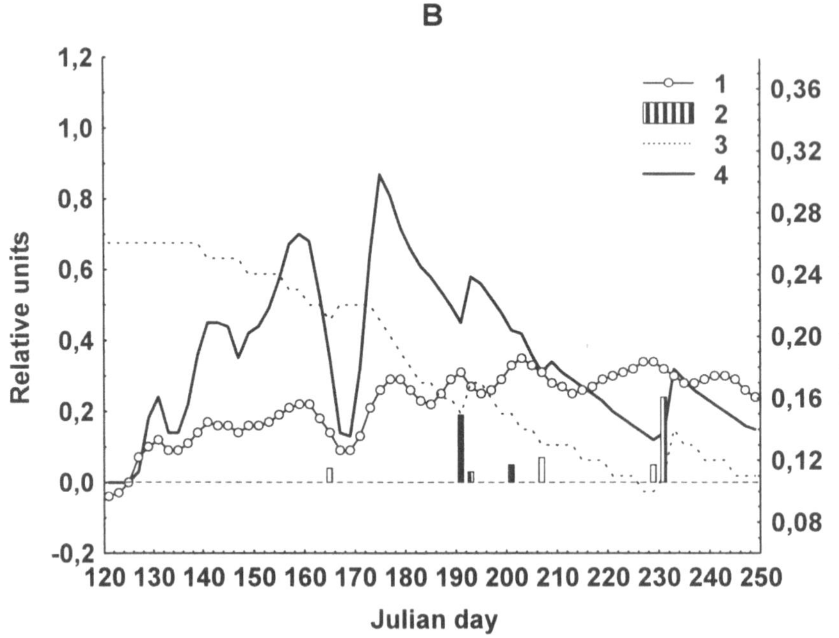

Fig. 8.6. Seasonal climatic conditions (*1* temperature, *2* precipitation, *3* calculated soil moisture) and growth rate (*4*) in more wet (**a**) and dry (**b**) years at the Methuselan Walk site with bristlecone pine

in years with a cold May in their statistically based reconstruction. The initial moisture that might affect growth is mainly defined by the date of the switching of the limit. For example, 1969 was more favorable due to higher initial moisture; and so the growth rate followed temperature variations up to mid-July. Only after that was growth limited by decreasing soil moisture. In 1970, the snow resource and initial water content in soil were very low, which produced the earliest switching from temperature to low soil-moisture limitation (in mid-June; Fig. 8.6b). Note that a short but sharp decrease in temperature in mid-June is associated with a decrease in new cell production, because of the relatively high sensitivity of growth rate of bristlecone pine to temperature at this site.

8.5 Examples of the Simulation of Tree-Ring Growth Dynamics in Temperature-Limited Conditions Under Projected Climate Scenarios

In this and following paragraphs, we consider the results of applying the simulation model described in Chap. 7 to proposed climatic scenarios for different locations and species in the boreal zone. We chose typical sites within the boreal zone of Siberia and basic climatic scenarios for these sites, according to the Intergovernmental Panel on Climate Change (IPCC (2001). We chose sites in three regions of Siberia, for which the IPCC gave the same scenarios: 5 °C increase in temperature by 2100 and 20 % increase in precipitation. Note that we were limited in our choice of sites by the availability of climatic data with daily resolution. The following scheme has been used for all calculations:

1. Climatic scenarios: (a) the period from 1960 until 1990 ("recent") is used for calibration of the model on existing tree-ring chronologies and parameterization of the model, (b) a gradual increase in temperature and precipitation up to the limit of +5 °C and 120 % of precipitation during 31 years (transformation of recent variant in 1960–1990; "gradual"), (c) a sharp increase of temperature and precipitation so that all temperatures are +5 °C greater and all precipitation totals 20 % greater than in the "recent" case ("sharp"). These two projected variants simulate an accelerated increase compared to what is expected by 2100. Our two variants retain the natural variations in climate experienced in the "recent" period (1960–1990), plus the projected variations. They provide an opportunity to see what changes in tree growth response we can expect, first under a gradual increase in climatic variables, and second under a sharp increase. It is hoped that this will cast light on a number of situations trees might face as climate changes. The main model parameters are determined using the subset of years with positive and negative extremes in tree-ring growth

indices. The model is then verified using the whole interval from 1960 to 1990 by comparison of the simulated and actual chronologies.
2. The model is run for the two scenarios and then results are analyzed and compared. In this comparison, we use one of the important features of the model – it calculates the absolute production of cells in the growing tree ring every year and then transforms these absolute numbers into indices by dividing each annual number by the average for the time-interval under investigation (31 years), i.e. standardized absolute values. In the "warming" scenarios, we expect changes (sharp or gradual) in total annual production; and to take this into account we standardized the annual value of cell production by the average for the initial ("recent") case. This was done to retain differences in growth rate associated with each scenario.

The lower Indigirka region in northern Yakutia, and in particular the Chokurdakh meteorological station, was chosen as an example to study tree growth response at the northern timberline to changing climate in temperature-limited conditions. There are good quality long-term chronologies (local, regional) of several tree-ring variables from this region (Hughes et al. 1999; Panyushkina et al. 2003). The data presented above (Sect. 8.3) show that early summer temperature is a main limiting factor for seasonal and inter-annual tree growth in this area. The correlation between real and modeled

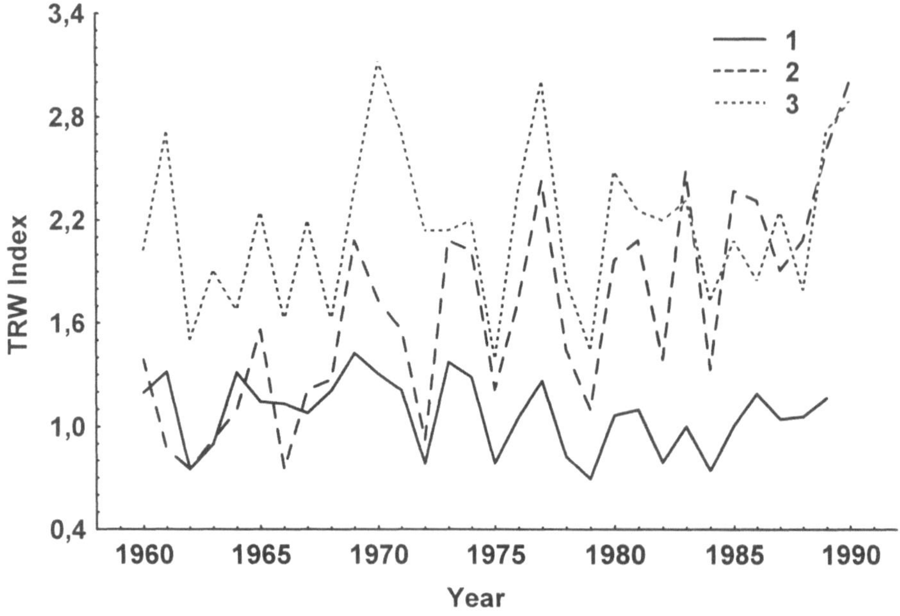

Fig. 8.7. Comparison of recent (*1*) and projected (*2* "gradual", *3* "sharp") scenarios for larch tree-ring growth at the northern timberline, Chokurdakh, Yakutia. The *abscissa* indicates the recent calendar period, but this should only be used as an ordinal scale

chronologies reached 0.79 (n=31, P<0.001). The "recent" chronology for 1960–1990 is compared with the "gradual" and "sharp" scenarios (Fig. 8.7). As expected, there is a gradual increase in tree growth in the "gradual" case. One can see that, in the beginning of the studied period, the "gradual" case resembles the "recent", and at the end of the period, the "gradual" is close to the "sharp" case. The comparison of coefficients of variation (standard deviation/mean) show the following: actual chronology 34 %, modeled "recent" 35 %, "gradual" 35 %, "sharp" 21 %. The significant decrease in variability in the "sharp" modeled chronology is consistent with the relationship shown in Fig. 7.2, because warming is shifting the range of temperature variations and a large portion of these variations is moved to the optimal (non-limiting) range of temperature response.

Some interesting conclusions can be drawn from a comparison of simulation model results with those based on an empirical–statistical model. The June and July temperatures explain more than 60 % of the common tree ring width variability in these conditions (Hughes et al. 1999). The multiple regression model was written as:

$$Y_t = 0.409 + 0.00935T_{JN} + 0.00124T_{JL} \tag{8.1}$$

where Y_t is the tree-ring width index, T_{JN} is the temperature of June, and T_{JL} is the temperature of July. The multiple coefficient of correlation equals 0.82 (adjusted R^2=0.65, F=27.54, P<0.00001).

If the "sharp" scenario was chosen, the calculations of projected tree growth changes must be calculated accordingly:

$$Y_t = 0.409 + 0.00935(T_{JN} + 5) + 0.00124(T_{JL} + 5)$$

or

$$Y_{t\,warm} = 0.409 + 0.00935T_{JN} + 0.00124T_{JL} + 0.0529 = Y_t + 0.0529 \tag{8.2}$$

i.e. in coordinates Y_t–Y_{twarm}, the values of "recent" and "sharp" chronologies calculated from a multiple regression function lie on a straight line (Fig. 8.8).

If we plot the similar values of "recent" and "sharp" chronologies from the simulation model calculations, the final picture is quite different. There is no straight line, the points are dispersed over a very wide range, and some points show that, in certain years, the modeled growth is higher than according the statistical model and, in a few other years, it is less. The simulated and regressed "sharp" chronologies are not well correlated (r=0.42, n=31, P<0.01). This results from the non-linear response of growth to temperature increase and the unequal contribution of the same temperature rise in different parts of the season to growth rate and cell production in the simulation model. In the simulations, a particular temperature increase will have a larger effect on

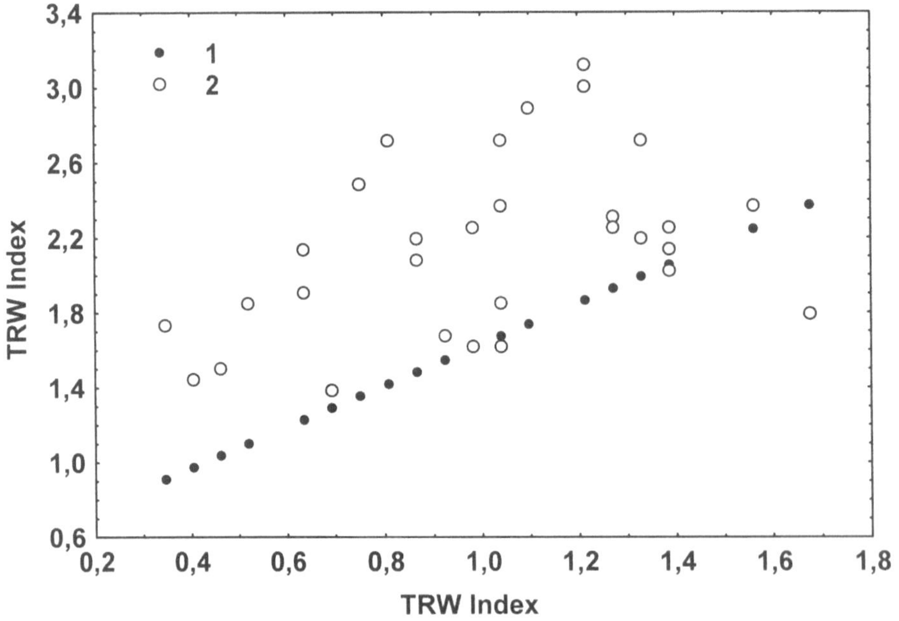

Fig. 8.8. Diagram showing relationship between projected and recent climatic influence on tree-ring indices for larch tree-ring growth at the northern timberline, Chokurdakh, Yakutia, according to typical regression (*1*) and simulated "sharp" (*2*) scenarios. See text for further details

increasing new cell production early in the growth season than later. During a season with a late beginning and a low rate of seasonal temperature increase, the total production will not increase to the same degree. Given the same range of projected temperature changes, it is lower for the mean tree-ring width index multiple regression model than for the VS process-based model. The number of years when simulated and regressed values are approximately equal is six. In the majority of years, the simulated tree-ring index is higher than that from the regression. The significant dispersion of points in Fig. 8.8 indicates that the decadal variations in the modeled chronology will change, particularly in the "sharp" case. Note that if we use the regression model for projection, the multi-year change will be preserved because this is based on a linear transformation (Eq. 8.2).

It has been hypothesized that the dates of snow melt and soil thawing are very important for the time of cambial initiation and total wood production during a season (Vaganov et al. 1999; Kirdyanov et al. 2003). This is especially important in the area of the northern timberline when the seasonal soil thawing reaches only about 50–60 cm. According to long-term observations, it was suggested (Groisman et al. 1994; Groisman and Easterling 1993) that the increasing temperature in April–May in high latitudes will accelerate snow melting. The northern timberline is an area where the beginning of the season is highly variable in time (inter-annual differences can be more than one month) because of the particular patterns of change in temperature and snow dynamics (Kirdyanov et al. 2003). How might this work in scenarios of "gradual" increases in temperature and precipitation? Most precipitation falls during the winter. Two simulations were made: one with the normal block of soil melting and thawing and effects of snow cover, and another without this effect. The results are shown in Fig. 8.9. In the case of a gradual increase in temperature and precipitation, the effect of delay in cambial initiation due to snow melt and the timing of soil-thawing will reduce growth by about 6% every year. Note that this effect does not change the inter-annual or decadal variability. However, there are several years (see the last portion of the chronology) when the differences between thawing and non-thawing variants reach about 20–40%. These are the years with the most spring precipitation (before snow melt) and a low seasonal temperature rise. We cannot predict how frequently such years will occur, but if their frequency continues, the effect of snow and soil melting on cambial activity will be significant in warming scenarios too.

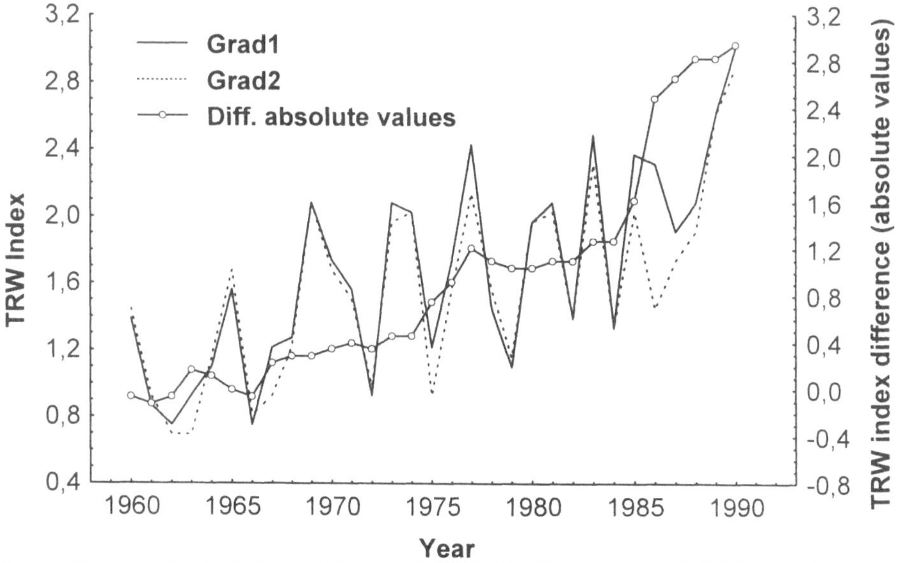

Fig. 8.9. Effect of delay in growth initiation due to later soil thawing in gradual scenario of larch growth in the Northern timberline site. *Diff.* Cumulative difference between two variants (for better explanation, see text)

8.6 Examples of the Simulation of Tree-Ring Growth Dynamics in Temperature-Limited and Low Precipitation Conditions Under Projected Climate Scenarios

Some recent studies of tree-ring growth of Alaskan trees indicated that, in conditions of rising annual temperature, the conifers show growth reduction because of increasing water deficit in the soil (Barber et al. 2000; Lloyd and Fastie 2002; D'Arrigo et al. 2004). It has been proposed that the increased water deficit was produced by increasing evaporation of water from the active soil layer and that this was not compensated by absorption of water from the deeper thawing resulting from increasing temperature. A similar effect has long been known in the conifer forests of Central Yakutia, where high summer temperatures during a short season coincide with low precipitation (Pozdnyakov 1986). We chose this as a typical region for modeling in the most continental part of Yakutia (the Verkhoyansk region), where the lowest winter temperatures (down to −60 °C) and highest inter-annual range in temperature (about 90 °C) coincide with a relatively small annual precipitation (150–350 mm). We parameterized and tested the skill of the model for the "recent" variant against the actual chronology and then calculated the response according the two basic "warming" scenarios – gradual" and "sharp".

In Fig. 8.10a, the measured and modeled chronologies are compared during the "recent" period from 1960 to 1990. The correlation is 0.62 (n=31, P<0.001). Note that the best fitted multiple regression model based on six independent variables (monthly temperatures, precipitation) shows less certain results, probably because of the relatively large number of predictors (six) compared to the number of cases (31): R=0.67, R^2=0.43, F=2.97, P<0.1, error =0.180. The four years with the lowest tree-ring width indices were characterized by a cold and dry early season (1962, 1980), or a very dry June (1972), or mild temperature in the beginning of the season and a low snow depth (1984). The simulations according the two scenarios show that tree-ring dynamics here are more complicated than at the northern timberline. First, the growth increase in the "sharp" case is less than in the northern timberline (2.2, 1.22, respectively; Fig. 8.10b), although it does increase. The corresponding values for "gradual" climate increases are 1.7 and 1.1. The inter-annual variability increases by 30 % in the "gradual" case when compared with the "recent" case. Unlike the situation at the northern timberline, the inter-annual variability increases in both "sharp" and "gradual" warming scenarios, but with a smaller increase in wood production (ring-width index). Thus, the acceleration of growth in warming conditions is restrained by increasing water deficit in many years. In the new warming conditions, the temperature and water content in soil interact in a different way and produce different inter-annual patterns of soil water dynamics. As a result, minima in growth rates occur in different years. They may be seen in 1962, 1972, 1976, 1980, and 1989 in the

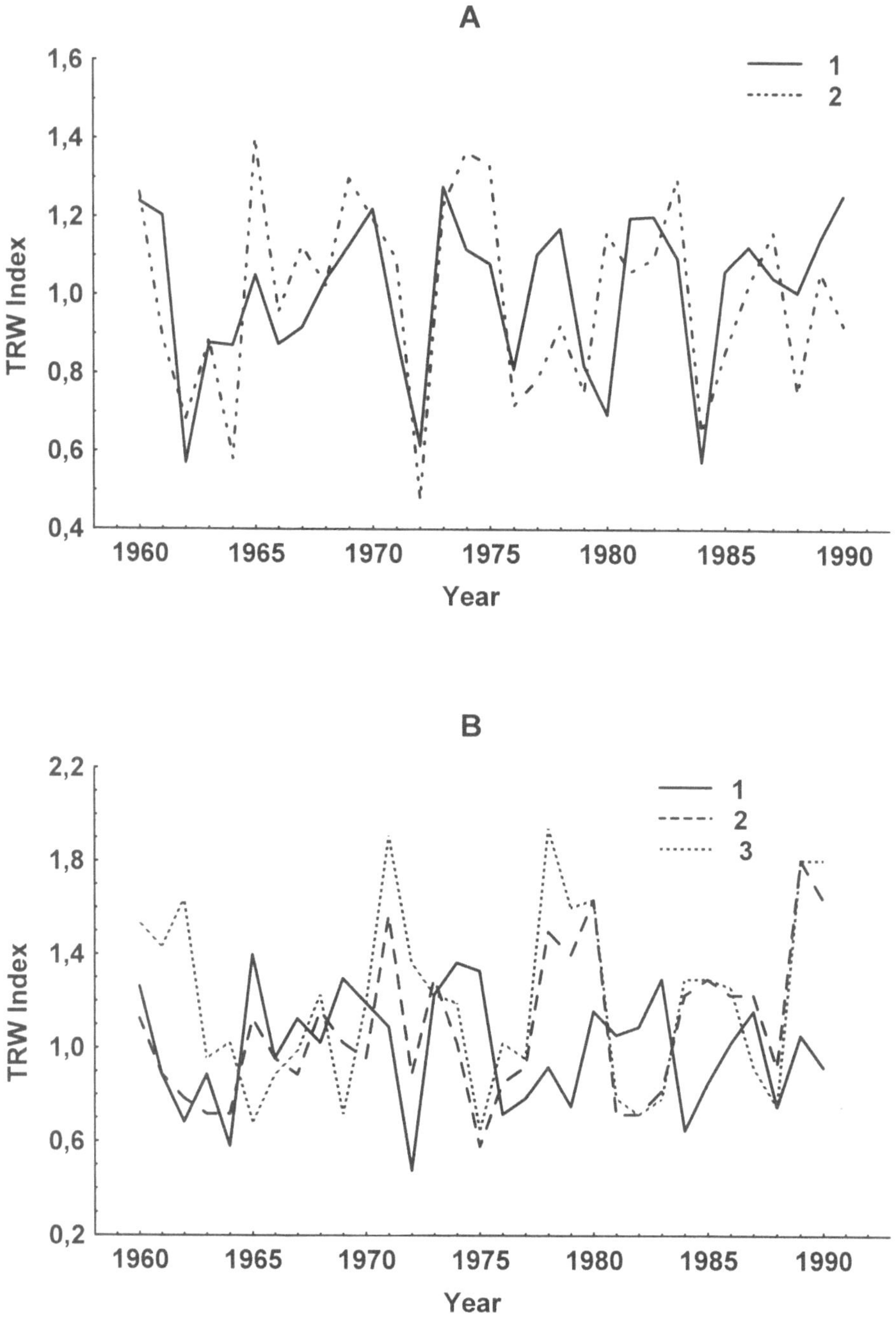

Fig. 8.10. **a** Actual (*1*) and simulated (*2*) "recent" tree-ring chronologies from Central Yakutia (Verkhoyansk region), a temperature- and water-limited site. **b** Simulated chronologies for "recent" (*1*), "gradual" (*2*), and "sharp" (*3*) scenarios

"recent" case, and in 1965, 1969, 1975, 1981–83, and 1988 in the case of "sharp" warming.

The "gradual" warming case is synchronous with the "recent" case during the first 6–7 years, but then becomes synchronous with the "sharp" variant. The switch after 6–7 years means that a 1 °C increase in temperature and a 4 % increase in precipitation were enough to enter new inter-annual and decadal growth dynamics. If the number of extreme (lowest and highest index) years in initial and both "warming" scenarios is retained, the decadal variations are completely different. It also can be shown by correlation: "recent"–"sharp" has $R= -0.25$, "recent"–"gradual" has $R=0.09$, "gradual"–"sharp" has $R=0.74$. The new regime of decadal variations of tree-ring indices commences, and after the temperature reaches +1 °C and the precipitation becomes more than +4 %, this new regime becomes more or less stable, even with further changes of temperature and precipitation. At the same time, both "warming" scenarios show the increase in growth despite the increasing water deficit, because this region is characterized by temperature limitation. The increasing temperature in the first part of the season (at the beginning of cambial activity) compensates for the loss in productivity due to water deficit in the second part of the season (increases of 10 % in the "gradual" variant, 20 % in the "sharp" variant). Even so, +1 °C warming and +4 % increasing precipitation are sufficient for a transition to different decadal tree-ring variations in this region.

These peculiarities of the "warming" chronologies may be seen on the graph of $Y_{twarm}-Y_t$ (Fig. 8.11). The number of years when simulation model calculates the tree-ring index higher than multiple regression model is reduced (from 19 to 12). The dispersion of points is no less than in the case of the northern timberline. Simulated estimates are smaller than statistical estimates for seven years. In some years, there are intervals when growth rate is limited by high temperatures coinciding with low soil moisture content. Typical of such years is 1975, with high temperature (up to 23 °C) for about ten days (Fig. 8.12). The increasing growth rate caused by higher temperature is compensated for by low soil moisture. In some other cases, the high temperatures limit the growth rate directly because these temperatures are higher than optimal. Such cases depress the new cell production more than slow changes in temperature or soil moisture and lead to a larger reduction in total wood production during a season (Fig. 8.10).

In the multiple regression model, the highest partial correlation of the "recent" chronology is with June precipitation ($R= +0.39$) and the correlation is less with early summer temperatures (May, June). The "sharp" chronology shows the highest partial correlation (R= +0.48) with May precipitation. This may be explained by the earlier initiation of cambial activity and production of new wood caused by commencement of growth in May and by the importance of May precipitation for soil moisture at the beginning of the season.

Considering the results discussed in this and previous paragraphs, it is interesting to point out one common and important conclusion. The "sharp"

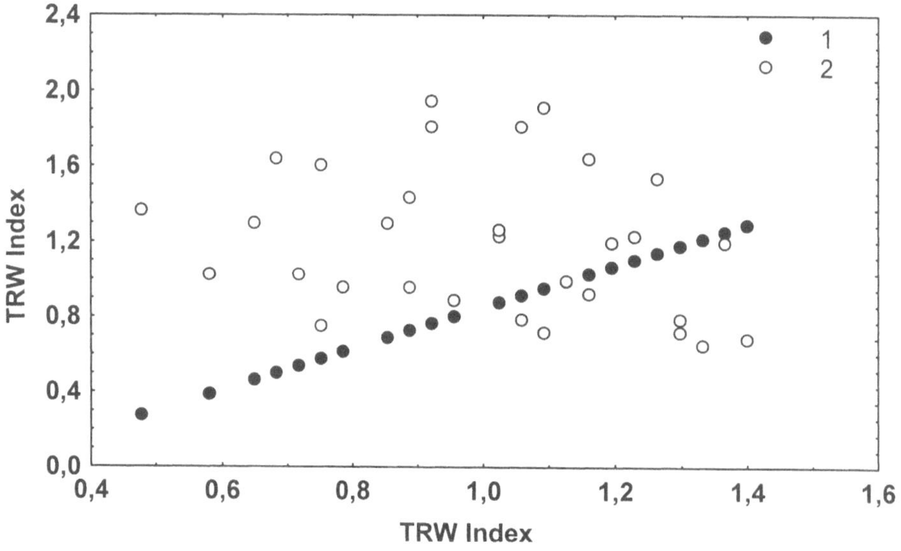

Fig. 8.11. Graph of relationship between recent and projected scenarios from Central Yakutia (Verkhoyansk region) for typical regression (*1*) and simulated "sharp" (*2*) scenarios

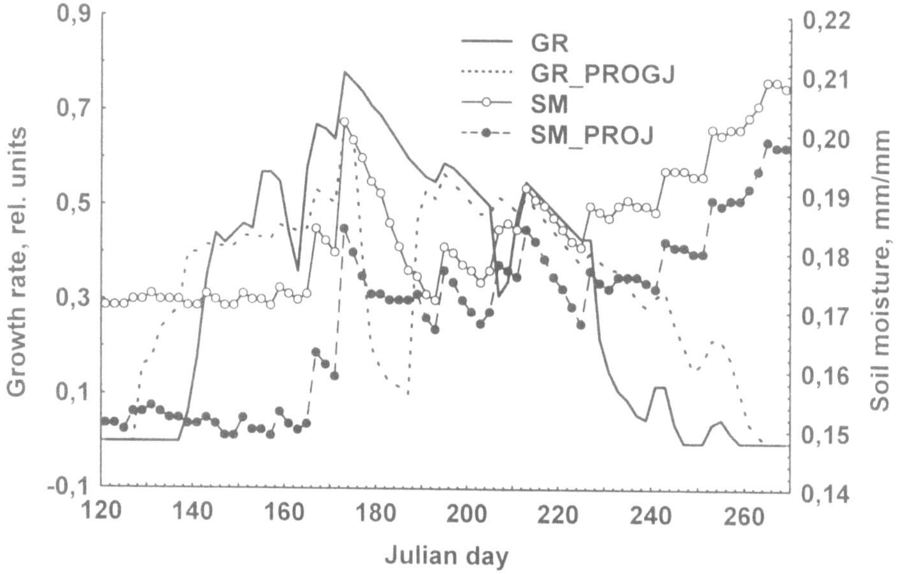

Fig. 8.12. Seasonal dynamics of growth rate and soil moisture according to recent and projected scenerios for 1975 in Central Yakutia (Verkhoyansk region). *GR*, *SM* Growth rate and soil moisture for "recent" scenario, *GR_PROJ*, *SM_PROJ* growth rate and soil moisture for "sharp" scenario. All given as relative units

warming, even where growth is strongly limited by temperature, leads to significant change in long-term variations in the tree-ring series. This is related to the non-linear response of tree-ring growth to temperature and soil moisture. If we apply this conclusion to the past variations in tree-ring series, it implies that, in periods with higher temperatures, the contribution of high frequency variations is reduced compared to cooler periods. This permits the use of such statistical characteristics as standard deviation (or concordance coefficient) as additional variables to reveal the long-term temperature signal in tree-ring width series (Naurzbaev and Vaganov 2000).

Modeling of the Verkhoyansk chronology gives another example of the non-linear response of tree-ring growth to climatic variables. Temperature plays a double role: (1) as a direct effect accelerating or depressing growth, (2) as an indirect factor switching limitation of growth from temperature to soil moisture because of a water-deficit in the hottest interval of the season. Clearly, the simulation model is better suited to detect such effects than are empirical–statistical models.

8.7 Examples of the Simulation of Tree-Ring Growth Dynamics in Water-Limited Conditions Under Projected Climate Scenarios: Comparison of Two Species

Two chronologies were chosen for further analysis: pine and larch. The region is characterized by dry conditions (steppe zone, southern part of Krasnoyarsk region) with low annual precipitation and a longer vegetation season than in the examples discussed in previous paragraphs. The sites where the trees were sampled are located on a gently sloping southwest-facing slope and the soil type is clay–detritus. The larch trees are mainly located close to the bottom of the slope in richer and wetter soil, but the pine trees occupy the upper part of the slope. The measured chronologies show moderate correlation with each other (R=0.38, n=31, P<0.01). The model was tested as described earlier and the results show that the simulated chronologies have significant correlations with the actual ones (for larch R=0.62, n=31, P<0.001; for pine R=0.68, n=31, P<0.001). The model captures both long-term variations and inter-annual fluctuations. The main difference of larch from pine is related to the response of evapo-transpiration to temperature: the larch needs to transpire slightly more water at the same temperature than pine. The reason for this is that pine has needles with a thicker cuticle, which reduces water loss as compared to larch. This allows pine to occupy the driest sites in the southern forest border in Siberia as well as elsewhere. There are also small differences between the parameters of the model for these two species. The first temperature optimum, T_{opt1}, is 18 °C for pine and 19 °C for larch. Pine reaches its optimal level of soil moisture ($W=_{opt1}$) at a lower level than larch.

These small differences in temperature and soil moisture response are reflected in the modeled chronologies. So, the average cell production in "recent" chronologies for larch is lower than for pine (33 vs 41 cells annually), but the coefficient of variation is very close in value (27 %, 26 %). In all "warming" scenarios, the average annual production of wood (and the tree-ring width index) is less than in the "recent" case, which means that warming leads to reduction of total growth and reduction of seasonal growth rate. In the case of "gradual" warming, the inter-annual variation of the larch chronology increases (30 %), but for pine it remains about the same level (26 %). In the "sharp" cases, the total production of wood is reduced when compared to the "recent" cases: larch lost 30 % and pine 25 %. The larch "sharp" chronology shows a higher inter-annual variability than pine (34 % vs 30 %). Changes in long-term variations can be seen in the "sharp" chronology, although they are not as evident as in the case of the Verkhoyansk chronology. Indeed, the "recent" and "sharp" chronologies are well correlated (R=0.74, n=31, P<0.001; Fig. 8.13).

The "gradual" chronology helps us identify the temperature change needed to render irreversible its depression relative to the "recent" chronology According to Fig. 8.14, this was approximately 2.0–2.5 °C. A comparison of the

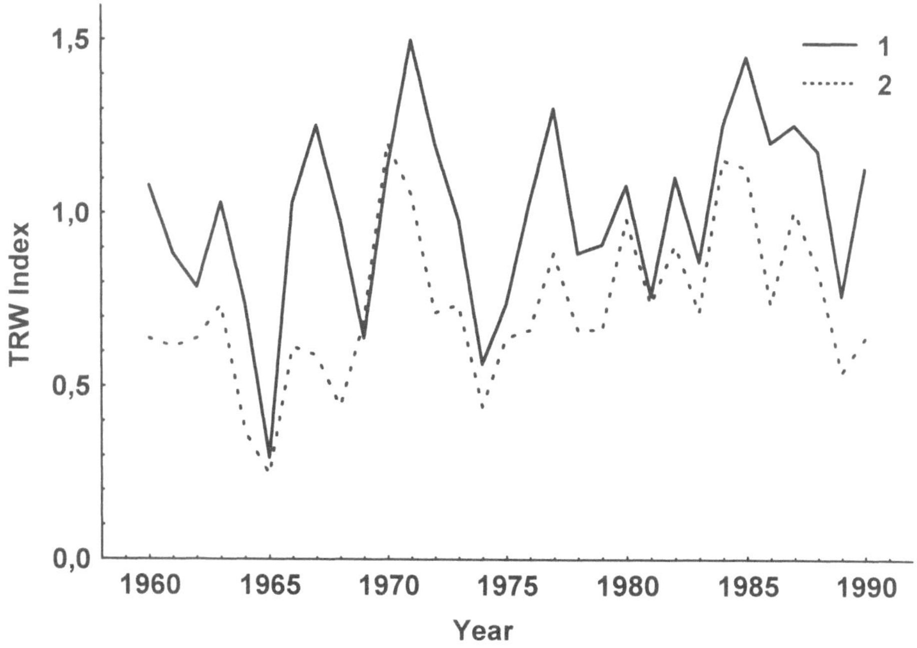

Fig. 8.13. Simulated "recent" (*1*) and "sharp" (*2*) pine chronologies from the steppe zone (south of Krasnoyarsk region)

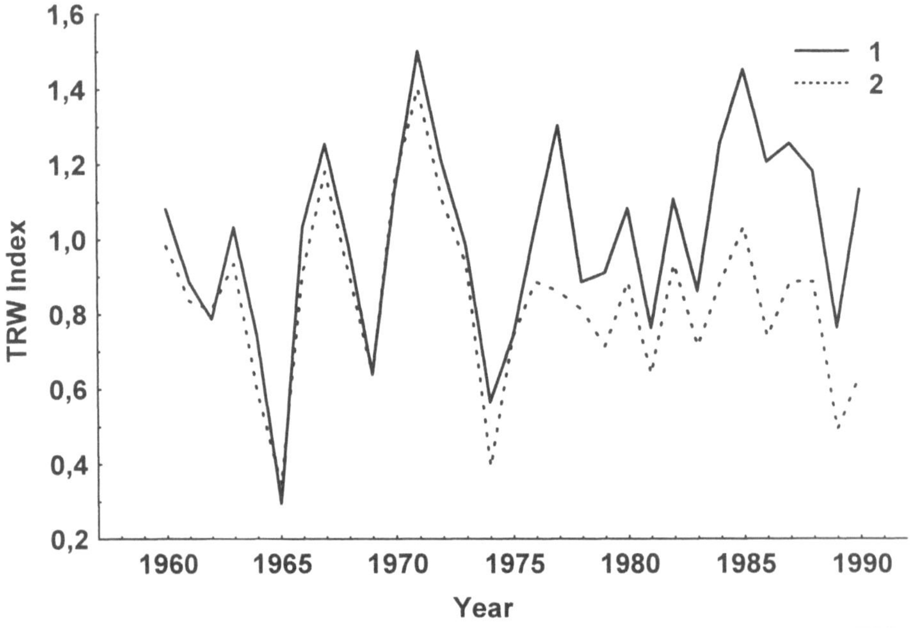

Fig. 8.14. Simulated "recent" (*1*) and "gradual" (*2*) pine chronologies for the steppe zone (south of Krasnoyarsk region)

differences in indices between "recent" and "gradual" chronologies and the corresponding temperature increase (Fig. 8.15) shows that:

1. The growth response to increasing temperature is non-linear.
2. A 7 °C temperature increase stops pine growth (a smaller increase is needed for larch). This could be the critical level of warming that these two species could survive in this region.
3. The non-linear response means that the negative effect of temperature on the growth of tree rings increases with temperature.

In these simulations of larch and pine tree-ring growth in these conditions, there are several years with direct limitation of growth rate by very high temperatures, mostly in the mid-season. This effect is new for the boreal zone. In most cases, high temperatures (up to 30 °C) have an indirect effect on seasonal growth rate through the lower soil water content. We have already modeled a direct negative effect of high temperatures on growth rate in the seasonal and inter- annual growth of conifer tree rings in a monsoon region. In those conditions, however, the highest temperatures coincided with very rainy seasons, mitigating the effect. In southern Siberia, high temperatures practically always coincide with low soil-water content, causing the conditions to become closer to those seen in semi-arid regions like on the southwest United States (see Sect. 7.6.4). High temperatures and low soil moisture lead to temporary

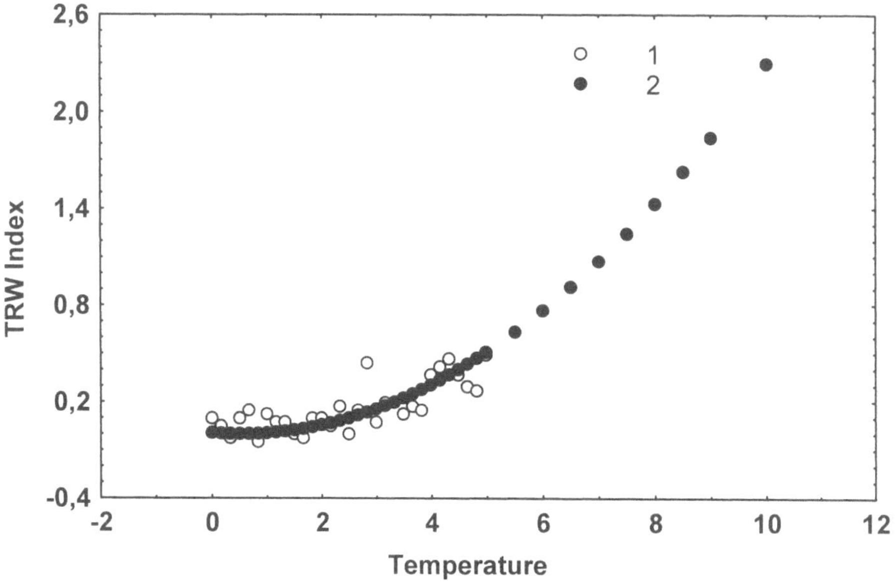

Fig. 8.15. Difference in tree-ring width index of "sharp" and "recent" chronologies from the steppe zone, south of Krasnoyarsk region, plotted versus increase in temperature (*1*) and to fit exponential curve (*2*)

cessation of cambial activity. The simulation allows cambial activity to recover when the conditions become better. In reality, there may be other requirements for the reactivation of cambium and apical meristem after such a hiatus. Such requirements may be related to photoperiod (reactivation, for example, needing an increasing photoperiod) or other internal or external factors. If conifer species growing in this area are conservative in reactivation of growth, they could stop growth in the first interval of cessation and then not resume. In that case, the depression of growth under the projected "sharp" scenarios will be greater and the total reduction could reach 38 % for pine and 42 % for larch. Such intra-annual cessations of growth lasting more than one week are seen in approximately one-third of all years in the simulation. Such intra-seasonal cessation of growth without its resumption leads to great changes in long-term variations in the projected chronologies when compared with the "recent" one. The main conclusion of the simulation is that, with increasing temperatures, the tree-ring growth of conifers in the southern part of Siberia becomes more and more similar to the semi-arid type of growth, with the intra-seasonal cessation of growth and production of "false" rings.

8.8 Simulation of Longer-Term Variations in Tree-Ring Dynamics in Dry Conditions

In previous paragraphs, we came to the conclusion that decadal to multi-decadal variations in temperature-limited conditions are significantly changed in a warming climate. In the steppe zone, this warming may or may not greatly change the variability of tree-ring series on these longer time-scales, but 30 years are not enough to make such a statement. This is why we applied the model to the growth response to warming scenarios for longer meteorological and tree-ring series from another part of southern Siberia, in the steppe zone near Lake Baikal (Buryatia). This region has a well replicated multi-century chronology of *Pinus sylvestris* and one of the longest meteorological records with daily data (Ulan-Ude station; Sect. 8.3). In Fig. 8.16, the modeled and "recent" chronologies are compared. Both "gradual" and "sharp" variants show a reduction in tree-ring growth (80 % for the "gradual" compared with the "recent", 60 % for the "sharp"). The coefficient of variation scarcely changes: 36 % for "recent", 37 % for "gradual", and 33.5 % for "sharp" chronologies. The "sharp" chronology has a few years (five out of 70) with intra-seasonal growth cessation, which means that, in this large steppe area, this factor does not become critical within the projected range of temperature

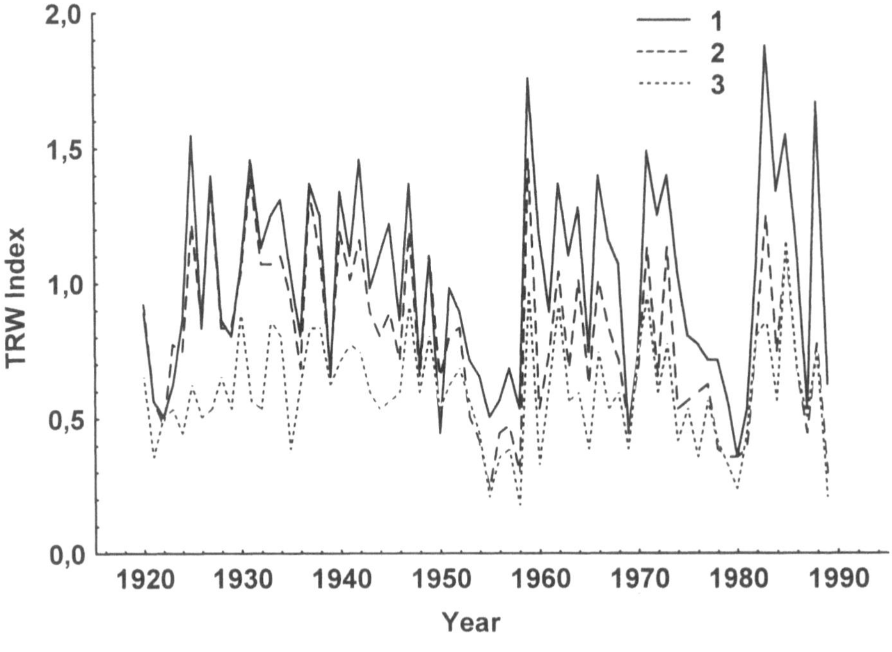

Fig. 8.16. "Recent" (*1*), "gradual" (*2*) and "sharp" (*3*) simulations of tree-ring chronologies of *Pinus sylvestris* in a dry site (Buryatia, Russia)

increase. The most remarkable result of the simulation is that both "gradual" and "sharp" chronologies keep the same long-term variations in tree-ring growth as in the "recent" variant ("recent"–"gradual" (R=0.83, n=70, P<0.0001), "recent"–"sharp" (R=0.76, n=70, P<0.0001). Such stability in long-term (and even inter-annual) variation means that the increasing temperature has practically no direct effect on tree-ring growth, but works indirectly through an intensifying water limitation. Under warming conditions, the water balance in soil decreases, limiting growth more strongly, leading to lower wood production (narrower tree rings).

The stability of long-term variations in tree-ring series in water-limited conditions could change greatly with a change in the seasonality of precipitation. If the most of the rainfall occurs in May–June rather than July–August, there will be wetter conditions in the first part of the season and less limitation of the growth rate by soil moisture during the most productive interval of the season. The production of wood and tree-ring width will increase and, in combination with temperature changes, long-term variations will change.

8.9 On the Use of Forward and Inverse Models in Climate Reconstruction

Dendroclimatology based on empirical–statistical models has been most successful, in terms of generating reliable reconstructions of past climate variability that are consistent with independent records and physical models, when certain guidelines have been followed. These involve the selection of species, regions, individual trees, and tree-ring variables so that a single seasonal or annual climate factor is the dominant influence on the inter-annual variability of the chosen tree-ring variable. Hughes (2002) wrote "Site chronologies from near the limits of the ecological range of a species commonly respond to a single seasonal climate factor, unlike those from the center of the range where the response to climate may be more complex. Thus tree-ring widths and densities from high-elevation or high-latitude sites without significant moisture stress commonly have a strong, consistent, linear relationship with surface temperature in part or all of the growing season (e.g. Parker and Henoch 1971). Those from near the lower forest border in Mediterranean or semi-arid regions have a similarly strong and simple relationship to, for example, winter and spring precipitation (e.g. Fritts et al. 1980; Till and Guiot 1990; Touchan et al. 1999). Many factors are known to influence the variability of tree-ring features; and so it is remarkable that many hundreds of tree-ring chronologies have been shown to have strong, linear correlations with climate variables such as seasonal mean temperature or total precipitation, the linearity of the tree-ring/climate relationships is in part the result of the careful selection of sites, trees and tree-ring variables, based on a century of experience and scientific progress by dendroclimatologists."

There are, however, some significant limitations to the use of the empirical–statistical approach, as discussed in Chap. 7 of this work. It is based on the assumption that the combination of climate conditions and tree-ring responses found in the calibration period covers all situations that might have occurred in the period to be reconstructed. Not only might no-analogue conditions exist outside the instrumental range of combinations of climate variables and tree-ring responses, but the relationships may well not be linear outside that period. Moreover, other factors such as the composition of the atmosphere might come into play. This is as much a problem for the reconstruction of the past as for the anticipation of the future.

In response to this problem, Hughes (2002) wrote: "Ideally, transfer functions would be mechanistic, rather than statistical-empirical. Process-based models that include the biology and physics responsible for creating the tree-ring record would be inverted so as to provide estimates of past climate variables. These reconstructions would not be based on extrapolation to conditions outside the range recorded during the instrumental period, unlike statistical transfer functions." In this, he was following the argument of Verstraete and Martin (1994). In the present empirical–statistical approach to dendroclimatic reconstruction, we take time-series of tree-ring measurements – the common pattern in trees at each site – and convert them to time-series of climate values. We assume that there are linear, constant relationships between them and that all factors that might be relevant are included. Then we derive regression models to calculate past climate from tree rings.

Verstraete and Martin (1994) proposed a different approach, using the slightly different response of each tree at one site. A model capturing the essential features of the process controlling the tree-ring variable, e.g. width, is established. It is applied to calculating one year's ring growth in each of n trees subject to the same climate conditions. To do this, we need only some measurable, unchanging parameters for each of the n trees and the microsite of each (e.g. optimum temperature for growth, field capacity of the soil) and that year's values for the climate variables. Therefore, this forward model could be used to calculate ring growth for each tree for any year for which we have climate data. Thus it could use climate to specify tree rings in many years in many trees.

This model could be inverted for a year in which the climate variables were not known. The same climate produces the differing ring growth values for each of those n individual trees. Starting with a guess of the climate values for that year (for example, mean summer temperature 12 °C, total winter precipitation 400 mm), the ring-growth value for each tree is calculated using the forward model. This is repeated with slightly varying climate values until the closest possible match to the actual set of ring-growth values from the n trees is found. These values are the most probable estimates of the climate variables in that year. Hence, the inversion enables the specification of climate in one year from tree rings in many trees.

The advantage of this is that process-based understanding of climate/ring growth relationships is used rather than "blind, empirical" statistical models. The model described in Chap. 7 may have the necessary characteristics for such use, especially with a relatively small number of parameters which are in turn capable of being measured experimentally or in the field. The major drawback of the inversion approach is that, unlike the regression case, the problem is ill-defined. Any one combination of tree-ring responses in a particular year may result from a number of climate scenarios. In order to constrain these possibilities, it is important to have markedly different climate responses to the same climate variability clearly associated with measurable differences in the parameters of the forward model.

Some recent developments in the application of geospatial techniques open up a possible opportunity to apply this approach. In the high, dry, cold environments of subalpine trees in the mountains of the semi-arid western United States, closely neighboring trees may have markedly different responses to the same climate (e.g. Villalba et al. 1994; Bunn 2004; Bunn et al. 2003, 2005a,b). These markedly diverse responses to the same climate amongst close neighbors are produced by the rugged topography. Whilst one tree may be growing in a small depression, another, a few meters away, may be on a small ridge. Such minor differences in micro-site will have major effects on the availability of moisture in such dry environments. Similarly, in these relatively cloud-free, high-elevation locations, differences of slope, aspect, and shelter may produce large differences in plant and soil temperatures. The effects of these differences on tree-ring growth would be hard to detect in an environment more favorable to growth, because the stand would be much denser. Interactions between neighbors of the same or differing species would have large impacts on tree-ring growth, masking the effects of topography. Here, however, trees such as bristlecone pine (*Pinus longaeva*, *P. aristata*), foxtail (*P. balfouriana*), and limber pine (*P. flexilis*) may be found in mono-specific, very open stands in severely cold and dry conditions. It is reasonable to expect that interactions between trees are minimal in such situations and that the effects of the topographic modification of tree-ring growth response to climate may be detected.

This has been shown to be the case in research on foxtail pine (*P. balfouriana*), a close congener of the bristlecone pines, in Sequoia National Park, Calif. There are strong associations of proxies for soil moisture (topographic convergence) and solar radiation (potential relative radiation) with ring-width patterns (Bunn et al. 2005b). Put another way, trees within a few meters of one another on the same hillside may have very different tree-ring series, but these differences may be reliably predicted from their placement on the detailed landscape of bumps, slopes, and hollows. In the foxtail pine study, topographic convergence and potential relative radiation were calculated from a digital elevation model (DEM). Topographic convergence index (TCI), a function of the upslope contributing area and local slope, measured the tendency of water to collect on the landscape (Moore et al. 1991; Urban et al. 2000; Bunn et al. 2003). The index had high values in depressions or drainages and lower values on well drained areas, such as ridge tops. The potential relative

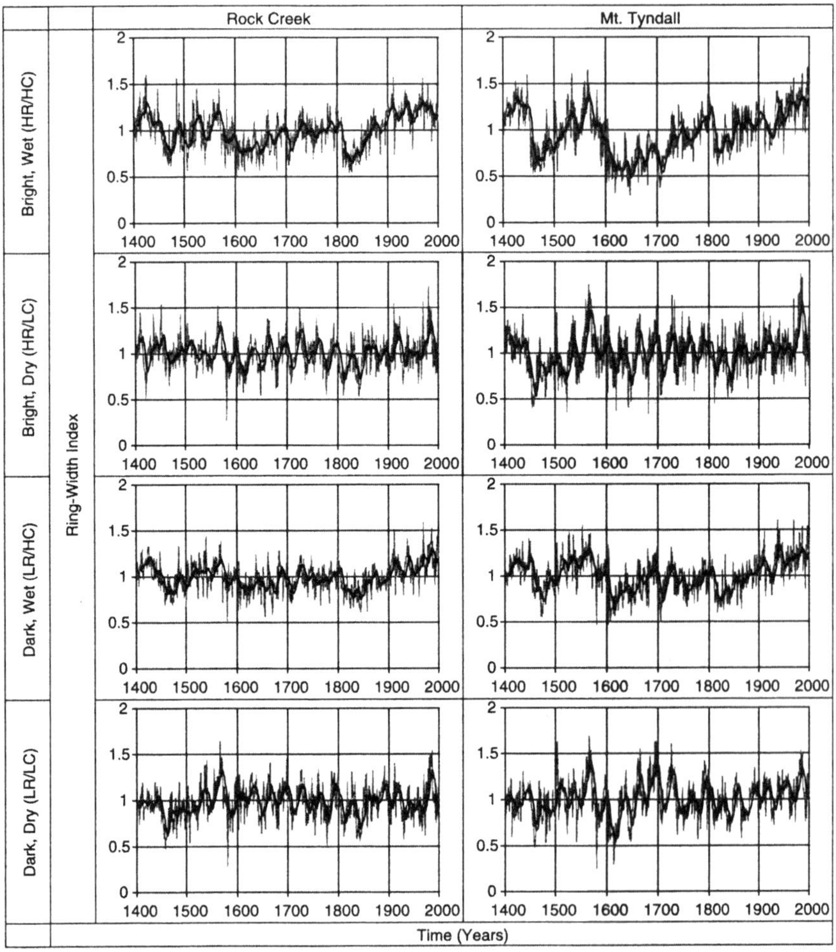

Fig. 8.17. Topographic modification of tree-ring response to climate in foxtail pine (*Pinus balfouriana*) in the Sierra Nevada, Calif. Four tree-ring chronologies from different physical settings (*rows*) are shown for two sites (*columns*). *X-axis* Years, *Y-axis* tree growth shown as a dimensionless ring-width index. *Gray line* Raw data, *thick black line* 10-year running average. All chronologies run from 1400 AD to 2000 AD. Marked differences among the chronologies by biophysical setting are clearly visible. For instance, the bright and wet plots (*HR/HC*; *top*) have different modes of temporal variation than the dark and dry plots (*LR/LC*; *bottom*) and the wet plots (*HC*) show increased growth in the twentieth century that is characteristic of many temperature-sensitive tree-ring chronologies (From Bunn et al. 2005b)

radiation (PRR) index used measurements of the sun's position over the course of the year and the shadow-producing effects of hills to measure the relative amount of sunlight that a particular point receives (Bunn et al. 2003). Ring-width chronologies from the different topographic situations showed different characteristics and correlated with different climate variables. Thus, the chronologies from wet plots correlated most strongly with temperature and showed a strong low-frequency variability (i.e. hundreds of years), while ring-width chronologies from the dry plots correlated with moisture availability and showed strong variability on multi-decadal scales (Fig. 8.17). These situations were characterized by demonstrably strong topographic modification of the response of tree-ring growth to climate, the availability of the necessary digital elevation models, and precise location information on the individual sample trees. This may permit an attempt to derive paleoclimate information by inverting a forward model of tree-ring growth, such as the VS model discussed in Chap. 7.

A further case that might permit inversion of the forward model involves data sets where several variables that are simulated by the model have been measured in each ring, such as earlywood width, latewood width, total ring width, density variables including maximum latewood density, and cell dimensions across the ring. To the extent that these variables respond independently and differently to climate, an inversion of the forward model may be possible.

8.10 On the Use of the Forward Model in the Interpretation of Empirical–Statistical Reconstructions

Even in the many situations where the conditions for a useful inversion of the forward process-based model are not met, it may be of value in assessing the robustness and fidelity of statistical–empirical reconstructions. First, as discussed in Chap. 7, the process-based model provides a better method for determining the nature of the climate signal stored in the tree rings, when the necessary daily meteorological data exist to allow its use. The benefits of this may be seen in the case of bristlecone pine at Methuselah Walk, Calif., as discussed in Sect. 8.4.2.1, where low spring temperatures may reduce a drought response, even in a very dry region. One of the process-based model's strengths is the explicit treatment of non-linear responses that a linear statistical method cannot provide. For example, the response to climate warming obtained by simulation will significantly differ from that reconstructed using regression models, as discussed in Sect. 8.5 and illustrated in Fig. 8.8. If the range of temperatures in a season extends above that where growth is temperature-limited to that where temperatures are optimum, there will be a stronger association between temperature and the tree-ring variable of interest at low temperatures than at higher, optimal temperatures. As a consequence, the fidelity with which lower temperatures are recorded by the tree rings will be greater than that for higher temperatures, resulting in a negative bias of higher temperatures as recorded by the tree rings. Such a bias could have a number of consequences for empirical–statistical reconstructions, including a negative bias over a larger part of the calibration range than that directly due to the "optimum effect" as a simple result of the forced linearity of the fitted relationship and a reduction in the apparent range of tree-ring response over and above the loss of variance due to the regression effect. Given the extent to which present estimates of hemispheric temperatures in recent centuries depend on tree rings (Hughes and Diaz 1994a, b; Mann et al. 1998, 1999, 2000; Briffa et al. 2001; Esper et al. 2002; Bradley et al. 2003), it is important that the extent of any such distortions be known.

The first step in detecting whether there is any such bias in a particular case is to make a scatter plot of the tree-ring variable against the appropriate seasonal temperature to see whether there is any decline in sensitivity and a correlation with increasing temperature. If there is, then it may be necessary to revise the interpretation of the reconstruction. Even if such a change is not identified in the instrumental period, it may have occurred under warmer conditions in the pre-instrumental past and so there may be unknown cool biases in the reconstructions. Thus, for both these reasons, it is necessary to develop techniques to detect and correct any such biases. It should be possible, for example, to identify combinations of values of several tree-ring variables associated with the highest and lowest seasonal temperatures, rather than depending on a linear relationship with a single tree-ring variable. These combinations could be determined using the instrumental record. They might also be determined for a much wider range of conditions using large numbers of process-based model runs for different seasonal progressions and the resulting sets of values for the tree-ring variables.

For instance, the warmest summers may be characterized by an early start to growth, the longest duration, and a warm autumn. These conditions lead to the formation of a wide ring with high earlywood cell diameter, complete formation of latewood cells, and high latewood cell wall thickness. So, the combination of ring width, tracheid dimensions, and density parameters may be used to indicate the warmest seasons. In contrast, the coldest years may be characterized by late dates of growth initiation, a slow rate of temperature increase, low mid-summer temperature, and a cold autumn. The narrow rings so formed would have small diameter earlywood tracheids and thin latewood cell walls. The seasonal progressions of climate and the size and internal structure of the tree-rings might be organized by some classification or ordination scheme. If these schemes show a strong relationship to one another, it should be possible to infer the seasonal progression of climate from the set of tree-ring values (width, density trace, tracheidogram, etc.). Other seasonal climatic progressions might result in similar or distinct ring structures. If it is possible to classify the seasonal temperature regimes, or organize them along some measurable gradients, it may be possible to identify the seasonal progression of a particular year in the past from the set of tree-ring variables.

It is possible to go further than this. The tracheid dimension and density profiles may be used to reconstruct seasonal growth rate and the seasonal growth curve (Vaganov et al. 1985, 1992). This is one of the important consequences of the simulation of intra-seasonal growth (see examples from Chap. 7). The only problem in this case is determining the exact time-scale for the seasonal growth curve. This can be solved in a stepwise manner: (1) definition of the class of seasonal temperature regime by multivariate tree-ring structure characteristics, as discussed in the previous paragraph, (2) determination of the date of the growth initiation from this, (3) use of the tracheidogram to reconstruct the seasonal growth rate, (4) using the reconstructed and time-scaled growth curve to reconstruct the course of temperature during the season. At least such a procedure can work in temperature-limited conditions. This could permit the detection of particularly warm summers that might be missed using the empirical–statistical approach.

8.11 Simulation From Local and Regional to Hemispheric Scale: Projections For the Future

Among the strengths of the model described in Chap. 7 are that it has a relatively small number of parameters, that they have clear biological or physical meaning, and that in fact very few of these need to be varied to simulate satisfactorily ring-width chronologies from a vast range of temperate and boreal conditions. As a result, the model is potentially suitable for the repeated application that characterizes spatial applications. We have already discussed (Sect. 8.9) the possibility that the model might be used in the exploration of the topographical modification of tree-ring responses to climate on spatial scales of a few meters. It could also be used for simulating large spatial fields of tree-ring response to climate, for example, continental or hemispheric, if fields of the necessary daily meteorological variables are available. There are three possible sources of such fields of daily data – gridded conventional meteorological data, the output of the reanalysis of observed data such as the United States National Centers for Environmental Prediction Reanalysis data set, and the output of spatially explicit climate models. Thus, conventional data might be used to test to what extent the observed large-scale patterns of tree-ring growth could be explained in purely climatic terms for the instrumental period, and to what extent there are systematic departures between the actual and modeled tree-ring data that might be caused by factors such as the changing composition of the atmosphere or, indeed, various treatments of the primary tree-ring data.

The reanalysis data could be used in a similar way, with the advantage that these data sets deal explicitly with the subalpine elevations from which many tree-ring records come. The existence of runs of coupled ocean–atmosphere models forced by the best available estimates of the main candidate forcings (explosive eruptions, solar receipts, greenhouse gases for recent centuries, plus orbital forcings for recent millennia) makes possible the simulation of "age-free" spatiotemporal tree-ring records driven by the output of these models. These may be tested against the observed chronologies for the regions of interest, so as to explore the limits of the climate information that might be extracted from a perfectly sampled, age-free tree-ring network.

In order for any of this to be possible, it will be necessary to assign appropriate parameters to the process-based model. Two main approaches might be considered. In the first, parameters from the literature and experimental evidence might be assigned. This could be done site by site, or for some aggregated version of the tree-ring data set, for example, a gridded version. In the second, the parameters for the modeling of each chronology could be determined by some empirical scheme designed to optimize the simulation of the actual tree-ring network. In either case, the chosen values of the parameters must be physically and biologically defensible. Clearly, a great deal of work remains to be done if the application of the process-based tree-ring model to spatial fields is to be accomplished.

8.12 Conclusions and Discussion

Questions concerning the applicability of the VS model and its potential for novel applications, have been discussed in this chapter. First, apparent time-dependent changes in tree-ring response to climate and the generality of the VS model (Sect. 8.3) were examined. The dependence of the models on conifer species, tree age and site characteristics was explored. We then demonstrated the use of the model for investigating the impact of various climate change scenarios on tree-ring growth under diverse conditions. The use of forward and inverse models in climate reconstruction was considered, as was the use of the forward (VS) model in the interpretation of empirical–statistical reconstructions. The prospect of applying the VS model on very large spatial scales was outlined.

The Vaganov–Shashkin (VS) model described in Chap. 7 is well suited for the simulation of tree-ring growth dynamics in varying and changing climates. It possess a considerable degree of generality, performing creditably with only minimal adjustment of parameters in several major forest types, including forest-tundra, taiga, mixed conifers in semi-arid regions, and monsoonal conditions. It captures most inter-annual and multi-year to decadal variability in these diverse situations. It is particularly suited for the study of the effects of varying and changing climates on tree-ring growth because it is a process-based model. Hence no complex schemes of empirically-determined, time-varying parameters are needed. It is capable of capturing the role of changing and varying climate and so may be used to identify this, so as to help assess any impact of other factors such as increasing atmospheric concentration of carbon dioxide, which are not dealt with in the model.

This model has considerable potential for use in the study of past climate. It may be used to improve understanding of the formation of the natural archive, the tree ring, and so enhance the evaluation and interpretation of paleoclimate reconstructions based on the empirical–statistical approach. This may be done not only at the level of the single site tree-ring chronology, but also a various spatial levels, from local to regional and even hemispheric, if the appropriate model parameters are assigned. This opens up the possibility of combining the model with spatial information so that the topographic modification of the recording of climate signal in tree rings might be dealt with explicitly, or that the climate forcing of continental-scale patterns of tree-ring growth may be modeled.

Finally, there may be situations in which it is possible to reconstruct climate by inverting the forward model of climate/tree-ring relations described in Chap. 7 (the VS model). This could provide, for example, a means of disentangling confounded temperature and precipitation signals in trees from cool, dry places such as the high mountains of continental interiors.

9 Eco-Physiological Modeling of Tree-Ring Growth

9.1 Introduction

As discussed in Chap. 7, plant growth may be modeled empirically or using process-based models, which may also be described as "physiology-based" or "dynamic" models. Models in this second group are designed to explicitly represent relationships and variables based on the biological processes that determine plant growth. Typically, it is necessary to assign a large number of variables and parameters to process-based models (Reynolds and Thorley 1982; McMurtie 1985, 1990; Valentine 1985; Landsberg 1986; Makela 1986; Running and Coughlan 1988; Wang and Jarvis 1990; Hunt et al. 1991; Running and Gower 1991; Thornley 1991; Weinstein and Beloin 1991; Deleuze and Houller 1995; Friend 1995; Chen et al. 2000). They include models of the growth of the crown, branches, assemblages of roots, and individual trees in a stand, as well as the stand as a whole. The particular modeling approach used will depend on the spatial and temporal scales of interest.

Interest in the operation of the global carbon cycle has grown greatly in recent years, largely due to the link between anthropogenically increased atmospheric concentrations of optically active gases such as carbon dioxide or methane and enhanced global warming. The terrestrial vegetation cover plays a key role in the dynamics of the global carbon cycle; and so it is necessary to model the climatic control of its growth. Photosynthesis takes a central place in such models, which must deal with the relationship of photosynthesis to environmental characteristics, such as irradiation and temperature, to leaf age, and to the behaviour of stomata.

For the purpose of studying the carbon cycle, it is necessary to estimate absolute values of photosynthetic rates. This requires some additional information on crown structure, leaf area, and biomass as well as quite precise calculations of absorbed solar energy and diffusion rates of carbon dioxide in chloroplasts (McMurtie 1990; Rauscher et al. 1990; Wang and Jarvis 1990; Running and Gower 1991). Such information, however, is usually not available for long time-periods, limiting the possibilities of model verification by comparison with denrochronological data. Nevertheless, eco-physiological models for the simulation of the effects of climatic variation on stem carbon accumulation are used. Commonly, net primary production (NPP) is calculated and a particular pattern of allocation of carbon to the stem is used. Hunt et al. (1991) used the FOREST-BGC model to simulate annual increments of stem carbon in a ponderosa pine stand. They assumed that a constant part of each year's NPP is accumulated by the stem. Using a model of photosynthesis adapted from Farquhar et al. (1980) jointly with a soil water balance model, significant correlations were found between simulated NPP and observed tree-ring width index for pine (*Pinus sylvestris*) and larch (*Larix cajanderi*) stands in northern Siberia (Benkova and Shashkin 2003). In the MAIDEN model developed by L. Misson (Misson 2004; Misson et al. 2004), the pattern of carbon allocation is more complex. The value of carbon accumulation in the stem is a constant fraction of the current (daily) NPP and the reserve carbon pool. Moreover, accumulation occurs only during the phenological phase when the growth of foliage and roots has finished. The MAIDEN model shows that spring-time utilization of reserve carbon is of major importance for leaf and root growth; and it accounts for the high level of autocorrelation of stem increment in deciduous species. Carbohydrate-reserve dynamics, phenology, and stem growth are less connected in evergreen species modeled by MAIDEN. Even so, a major part of all tree-ring variance is accounted for by MAIDEN, not only in the case of an oak (*Quercus petraea*) stand in Belgium but also in a set of Aleppo pine (*P. halepensis*) stands in southeastern France.

The eco-physiological model described here resembles a number of these other models in that it simulates daily NPP, but it differs from them in that carbon allocation to the stem (wood production) is controlled by explicitly modeled cambial activity. The cambial cells and their differentiated derivates are a sink for the current products of photosynthesis and carbohydrates from reserves. The cambial block enables our eco-physiological model to simulate not only tree ring width, but also cellular structure and carbon accumulation in tracheids.

The model described in Chap. 7 (the Vaganov–Shashkin or VS model) has a stronger emphasis than the eco-physiological model and the others discussed above for describing tree-ring response to climate variability. Dendrochronological data based on relative values, such as standardized chronologies or normalized tracheidograms, are used for its verification. Thus it is not necessary to estimate absolute values. As a result, the number of parameters in the VS model is considerably reduced and the use of simple functions, which approximate experimental data on photosynthesis and water balance, is justified. It is assumed that the functional relationship of relative growth rate/weather conditions has the same character for all the trees in given habitat and does not vary with time or with factors such as the atmospheric concentration of carbon dioxide. Comparison between simulated and observed chronologies has shown that this assumption holds true for a wide range of climatic conditions in the case of the VS model (see Sect. 8.2).

However, it is well known that the growth and formation of tree rings is a rather complex process. It involves substrate production by photosynthesis, transformation of photosynthetic products to biomass, and a complex system of control of this transformation. (see Chap. 1). A significant portion of photosynthetic production is spent on plant cell and tissue maintenance (maintenance respiration). This varies with temperature and may use as much as one-third of the product of total annual photosynthesis (Linder and Axelsson 1982; Ryan and Waring 1992). Furthermore, there is competition for substrate and priority of growth between different plant tissues. In an individual tree, the biomass of stem, branches, roots, and needles is strongly correlated with stem diameter, although this correlation varies within and between species (Utkin 1982; Usoltsev 1988). Changes in allometric relationships between stem biomass and diameter within any species may reflect the influence of environmental factors on tree form (Kofman 1986; Gower et al. 1987). The share of the total amount of carbon fixed by photosynthesis that is used in wood increment does not remain constant. Water and temperature stress-induced carbon reallocation "in favour" of needles and roots has been observed (Waring and Pitman 1985; Gower et al. 1992, 1995). In contrast, a larger share of carbon has been shown to be used for wood increment after fertilization (Gower et al. 1992; Snowdon and Benson 1992).

One of the main objectives of the eco-physiological model described in this chapter is to make a first step to link the growth rate (G) with the photosynthetic rate and substrate available for growth and with the regulation of cambial activity. In order to do this, it is necessary to consider that xylem is a tissue with functions of water conduction, mechanical support, and storage, and that the control of tree-ring growth rate must be connected in some way to these needs under particular environmental conditions. This involves considering what is the biological meaning of the growth rate (G) as calculated in the VS model and how such a complex net of processes and control loops is incorporated in such an apparently simple set of modeled relationships? What, for example, is the relationship between this quantity and processes such as photosynthesis?

The eco-physiological model, in which an attempt is made to incorporate these links, will necessarily be more complicated than the VS model described in Chap. 7. It may, however, be useful for the description and analysis of growth changes under the influence of a wider range of environmental conditions, including non-climatic factors such as insect defoliation, air, soil pollution, etc. (Fritts and Shashkin 1995; Fritts et al. 1999). One of its main differences from the VS model is that the eco-physiological model permits the checking of not only the simulated characteristics of the tree ring, but also the intermediate quantities such as the rates of photosynthesis and transpiration and the stable isotopic composition of wood, foliage, and root tissues.

In this chapter, we describe an eco-physiological model based on the VS model (Sect. 9.2) and the determination of quantitative values of its coefficients and parameters (Sect. 9.3). Some examples of its application to several different problems are given (Sect. 9.4) and conclusions drawn (Sect. 9.5).

9.2 Description of the Eco-Physiological Model

A. Shashkin and H. Fritts (with further modification by G. Downes and D. Hemming) have developed a model designed to calculate absolute ring-width sizes (Fritts and Shashkin 1995; Fritts et al. 1999; Hemming et al. 2001). It is based on the VS model, as described in Chap. 7. The main part of the eco-physiological model is the cambial component of the VS model. The eco-physiological model, called TREERING, resembles the VS model in that it describes tree radial growth as a response to climate factors (temperature, precipitation, solar irradiation), considers other external factors influencing growth (such as soil properties, nutrition, competition in stand, etc.) to be constant, and is "age-independent". It is assumed that tree height and the mass of foliage and roots are constant. The growth of phloem and bark is not considered.

As in the VS model, tree radial growth is the result of cambial activity, cell enlargement, and cell wall synthesis, which are simulated in daily steps. The rates of all these processes depend on three values: air temperature, the water-status of the tree, and photosynthate concentration, which are under hormone regulation. This dependence is described by a function analogous to $G(t)$. Therefore (for example), temperature affects the growth of cells in the cambial zone either directly or indirectly, i.e. through changes in photosynthesis, respiration, and transpiration rates. Hormone regulation is included in the model indirectly through the phenomenological relationship between crown growth and xylem formation. This particular model attempts to describe the processes governing the growing living cells in the cambium, the zone of enlargement, and the zone of maturation in the main stem. It keeps track of the tracheid elements that die and become part of the ring structure in the wood, along with the new mass of cells in the crown and roots. Thus it provides detailed information on the structure of individual cells making up the annual growth ring, as does the VS model.

The eco-physiological model consists of four interrelated blocks: the microclimatic block, the block dealing with photosynthesis and the photosynthate allocation and utilization by the foliage, stem, and roots of the tree, the water balance block, and the cambial block (Fig. 9.1).

In the microclimatic block, climatic data (daily maximum/ minimum air temperatures, precipitation) are transformed into values directly influencing physiological tree processes: daily mean temperature, water deficit of air, and daily solar irradiation.

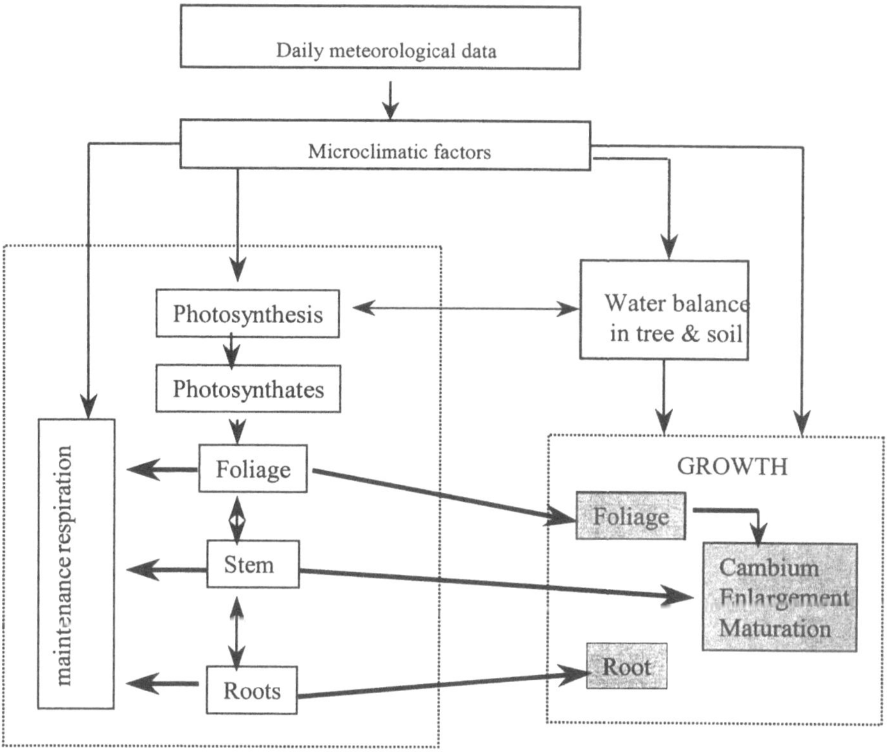

Fig. 9.1. Block diagram of the "TREERING" model

The next block simulates daily photosynthesis, photosynthate distribution between foliage, stem, and roots, and photosynthate uptake by foliage, stem, and roots for maintenance respiration and growth respiration.

In the third block, soil water balance is described, depending on daily precipitation and transpiration. The tree water balance is calculated depending upon the potential rates of absorption, the loss of water due to transpiration, and the resistance of the leaves. The water status of the tree is described through the resistance change due to water balance maintenance.

In the cambial block, the dynamics of cell numbers in the cambium, enlargement, and maturation zones are calculated.

9.2.1 Microclimatic Data

Usually the available meteorological data consist of daily maximum temperature, daily minimum temperature and daily precipitation. These are not adequate for calculating the rates of photosynthesis and transpiration. The calculation of the necessary microclimatic values is based on the known climatic principles and algorithms (see Running et al. 1987; Hungerford et al. 1989). The model is driven by daily meteorological data. Maximum and minimum temperatures and total daily precipitation are read directly from input data files. Average daytime temperature is calculated from the maximum (T_{max}) and minimum (T_{min}) temperatures (Running et al. 1987):

$$T = 0.606T_{max} + 0.394T_{min} \tag{9.1}$$

Vapor pressure deficit and absolute humidity are calculated from empirical equations (Murray 1967) based on the data by Bristow and Running (Running et al. 1987; Bristow 1992). The latter show the high correlation between minimum air temperature and dew point temperature: $T_{dew}=a+bT_{min}$. The vapor pressure (e) in the air is:$e=e_s(T_{dew})$, where $e_s(T)$ is the saturation vapor pressure at the given air temperature, calculated as (Murray 1967):

$$e_s(T) = 6.1078 \cdot \exp\left(\frac{17.269 \cdot T}{237.3 + T}\right) \tag{9.2}$$

The absolute humidity (ρ, kg/m^3) is:

$$\rho = 0.622\frac{\rho_a}{P_a}e \tag{9.3}$$

where ρ_a is the air density (1.226 kg/m^3 at 15 °C), P_a is the atmospheric pressure (101.325 kPa at 15 °C).

The vapor concentration deficit ($\Delta\rho$) is:

$$\Delta\rho = 0.622\frac{\rho_a}{P_a}(e_s(T) - e) \tag{9.4}$$

To estimate incoming solar radiation on a horizontal surface, the algorithm based on general meteorological concepts by Nikolov and Zeller (1992) has been used. Daily potential radiation (E_0) is a function of latitude, solar declination, sunrise/sunset hour angles, and the day of a particular year (Klein 1977):

$$E_0 = \frac{s_c}{\pi}(1 + 0.033\cos(\frac{2\pi}{360}j))(\cos\varphi \cdot \cos\delta \cdot \sinh s + \frac{hs}{57.296}\sin\varphi \cdot \sin\delta) \tag{9.5}$$

$$\delta = \arcsin\{0.39785 \sin[4.868961 + 0.017203j + 0.033446\sin(6.224111 + 0.017202j)]\} \tag{9.6}$$

$$hs = \arccos(-tg\varphi \cdot tg\delta) \tag{9.7}$$

where E_0 is the solar radiation received on a horizontal plane at the top of the earth's atmosphere (W/m^2), s_c is the solar constant (1,360 W/m^2), j is the Julian day of the year, ϕ is the latitude (degrees), δ is the solar declination (degrees), and hs is the day length (degrees).

The amount of solar radiation received by the earth's surface is less than that at the top of atmosphere because of scattering and absorption within the atmosphere. This attenuation of radiation has been expressed as a linear function of undepleted solar radiation at the top of the atmosphere and the average cloudiness (Bristow and Campbell 1984; Hungerford et al. 1989):

$$E = E_0(1.0 - e^{-0.003 \cdot \Delta T}), \qquad (9.8)$$

where $\Delta T = T_{max} - (T_{min} + T_{max}) / 2$.

The results of calculations of microclimatic values for the middle taiga region in Siberia (Eniseysk, Russia) and for a semi-arid zone (Chiricahua Mt., Ariz., USA), carried out according to these algorithms, are shown in Fig. 9.2.

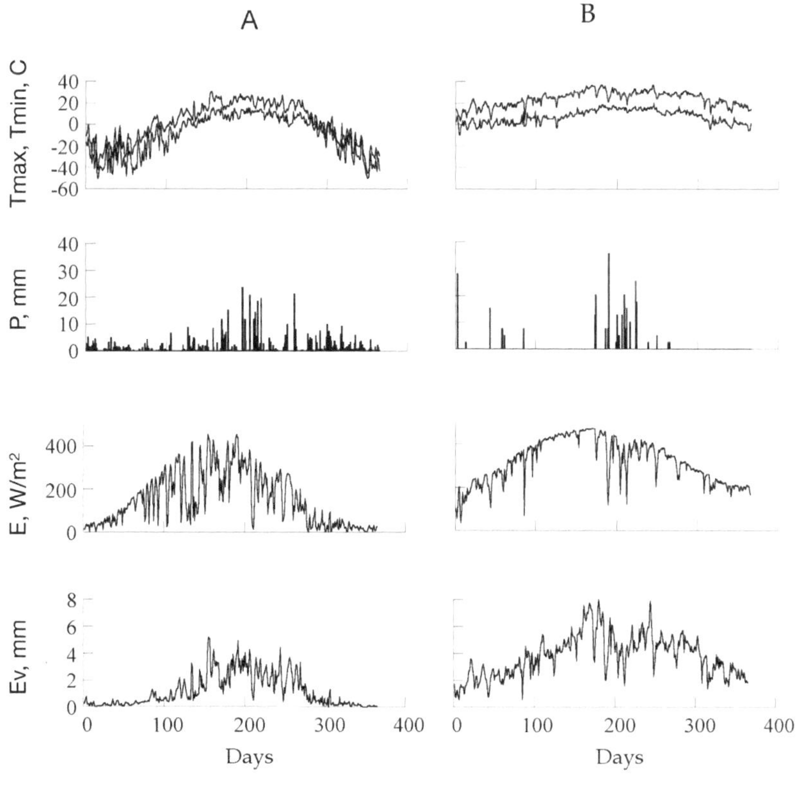

Fig. 9.2. Seasonal dynamics of climatic factors (daily maximal and minimal temperature, T_{min}, T_{max}, and precipitation, P) and calculated solar irradiation (E) and transpiration (E_v) for: **a** middle taiga (Siberia, Russia), **b** semi-arid zone (Arizona, USA)

9.2.2 Soil Water Balance

"The processes governing water movement through the soil, into the roots, through the plant and into the atmosphere are highly inter-related (Penman and Schofield 1951; Zahner 1955, 1968b; Noble 1974; Kramer and Kozlowski 1979; Gates 1980; Zimmermann 1983; Boyer 1985; Johnson et al. 1991; Ellsworth and Reich 1992). Transpiration is the dominant factor because evaporation of water produces the water potential gradient in the plant that drives the water movement. It controls the rate of absorption and produces diurnal water deficits in the leaves as well as throughout the entire plant. This influences the water status of guard cells on the leaf surface and the stomates change in size, often closing midday, reducing the flux of carbon dioxide into leaves and the rate of photosynthesis in the plant. Water deficits can also affect enlargement of cells in the cambium and cell division can be reduced or even stopped" (pp 27–30 in Fritts and Shashkin 1995).

Soil moisture dynamics are described in this model in the same way as in the VS model (described in Chap. 7) but the rate of tree transpiration, *Ev*, is described more precisely:

$$\frac{\Delta W}{\Delta t} = f(P) - Ev - k_R W, \qquad (9.9)$$

where ΔW is the change in water content in soil of a defined thickness per time unit, $f(P)$ is precipitation per time unit (day), *Ev* is the water loss in soil by transpiration, and $k_R W$ (=*Q*) is runoff.

Transpiration is basically the physical process of evaporation (Gates 1980). It is affected by plant factors such as leaf area, leaf structure, the difference between leaf and air temperatures, and the behavior of stomata. The rate of diffusion of water vapor from the leaf to the air depends upon resistances in the diffusion pathway and the vapor gradient from the free water surface in the leaf to the air outside the plant, according to Fik's law (Gates 1980; Kramer and Boyer 1995):

$$Ev = \frac{\Delta\rho}{R} \qquad (9.10)$$

where $\Delta\rho$ is the vapor concentration difference between the sites of evaporation in the leaf and the outside air (kg/m^3) and *R* is resistance in the diffusion pathway of water vapor (s/m; pp 27–30 in Fritts and Shashkin 1995). Commonly, the vapor concentration difference ($\Delta\rho$ in Eq. 9.10) is not equal to $\Delta\rho$ in the air (Eq. 9.4), because the leaf temperature could differ from the air temperature. Therefore, the Monteith equation based on the leaf energetic balance (Monteith and Unsworth 1990) is often used for the estimation of the transpiration rate. To use this equation, some additional parameters and an estimate of the energy balance of the leaf are required. Conifers are well aerated, so estimates of the transpiration rate by Eq. 9.10 show a high correlation with observed data (Jarvis and Stewart 1979; Whitehead and Jarvis 1981).

The resistance to water diffusion from leaves is determined by boundary layer resistance, stomatal resistance, cuticular resistance, leaf interior resistance, and mesophyll wall resistance. The largest part of the water diffuses through the stomata, so that stomatal resistance is the most important control of transpiration rate and tree water balance. The resistance, R, can be changed from minimum values (R_{min}) when the stomates are open to maximum values (R_{max}) when the stomates are closed. The values R_{min} and R_{max} are specific for various species and perhaps for particular sites. They are mostly the result of genetic selection and consequent adaptation to environmental conditions operating at the population level for the particular species or genotype. The three major environmental influences on stomatal conductance at the leaf and canopy levels are the available light, the vapor pressure deficit in the air, and the water status of the leaves (Schulze and Hall 1982; Kramer and Boyer 1995). There are empirical relationships between stomatal conductance and photosythetic rate (Wong at al. 1979; Cowan 1982; Farquhar and Sharkey 1982; Ball et al. 1987). As for stomatal conductance being controlled by vapor pressure deficit, there does not seem to be a direct effect, but rather an indirect effect through the water balance in the tree and through the "soil–plant" hydraulic conducting system (Buckley et al. 2003). In the eco-physiological model, the rate of photosynthesis and the daily balance between transpiration and the water supply to leaves from the soil determine the stomatal resistance. The available soil moisture per volume unit is: $\omega_{max} - \omega_{min}$. The potential absorption rate of water by roots of the tree ($J_{r\,max}$) depends on the roots density m_r (kg/m^3) and on soil moisture ω (v/v):

$$J_{r\,max} = q \cdot f(w) \cdot m_r v_r dlen \tag{9.11}$$

where $J_{r\,max}$ is the maximum rate of water absorption by roots (g/day), v_r is $s \times l$ [where s is $1/LAI$, which is equal to the soil surface (s) per unit of foliage area (LAI is the leaf area index), and l is the root depth], q is a constant, $f(w)$ is the normalized function of ω describing a trapezoid-shaped curve (see Chap. 7), and *dlen* is day-length (s).

The stomatal resistance (R) is determined by both transpiration and the availability of water. If the stomatal resistance R is not limited by photosynthesis, then the potential transpiration Ev^p (kg/m^2, daily) is determined as:

$$Ev^p = dlen \frac{\Delta\rho}{R_{min}} \tag{9.12}$$

It is assumed that the balance between transpiration and water supply is maintained during a day. The balance is achieved by variations in stomatal resistance, according to the following rule:

$$J_r = Ev = \min(Ev^p, J_{r\max}) \tag{9.12a}$$

and:

$$R = R_{\min} \max(1.0, \frac{Ev^p}{J_{r\max}}) \tag{9.12b}$$

In other words, if the potential transpiration is greater then the maximum value of water absorbtion, the stomatal resistance decreases. Under such conditions, the resistance is inversely proportional to the water vapor deficit, in agreement with experimental data (Cowan 1982).

If the resistance is changed under the influence of photosynthesis and reaches the value $R^c > R_{min}$, then:

$$J_r = Ev = \min(\frac{R_{\min}}{R^c} Ev^p, J_{r\max}) \tag{9.12 c}$$

and:

$$R = R_{\min} \max(1.0, \frac{R_{\min}}{R^c} \frac{Ev^p}{J_{r\max}}) \tag{9.12d}$$

Then, the dynamics of the water content in the soil is calculated by Eq. 9.9.

9.2.3 Photosynthesis

There are several approaches to modeling whole-leaf photosynthesis. Some of them are based on the empirical relations between the photosynthesis and solar irradiation, temperature, and carbon dioxide concentration (see Gates 1980; Jarvis 1985; Running and Coughlan 1988; McMurtie 1990; Wang and Jarvis 1990). There are the mechanistic models of photosynthesis based on its biochemistry (Farquhar and von Caemmerer 1982; Friend 1995). In the present eco-physiological model, the empirical approaches described in the book by Gates (1980) were used. In the model it was assumed that the crown structure did not change and the efficiency of utilization of solar energy was constant.

The rate of photosynthesis (A) is described as a function of light radiation (E), carbon dioxide concentration inside the leaf (C_i) and leaf temperature (T; Gates 1980):

$$A = A_{\max} f_c(C_i) f_T(T) f_E(E) \tag{9.13}$$

where A is the photosynthetic rate (expressed per second as mmol CO_2/m^2), A_{max} is the maximum potential photosynthetic rate (expressed per second as mmol CO_2/m^2) and f_c, f_T, and f_E are the functions described below.

The dependency of the photosynthetic rate on CO_2 [$f_c(C_i)$] is a saturation curve. We used the simple approximation:

$$\begin{cases} f_c(C_i) = 0, & if \quad C_i \le a \\ f_c(C_i) = \dfrac{C_i - C_a}{b - a}, & if \quad a \le C_i \le b \\ f_c(C_i) = 1, & if \quad b \le C_i \end{cases} \tag{9.14}$$

where a and b are constants. At the small scale, Eq. 9.14 gives an incomplete describtion of the response of A to the intercellular CO_2 concentration, C_i, since such responses are invariably observed as being biphasic (Farquhar et al. 1980). However, with increasing scale, spatial heterogeneity in carboxylation processes masks the nature of this response. The dependency on light intensity is described as the Michaelis–Menten equation (Buwalda and Smith 1990):

$$f_E(E) = \frac{E}{E + E^*} \tag{9.15}$$

The dependence of net photosynthesis on temperature is the same as described in Chap. 7 for the growth rate.

The supply of CO_2 for photosynthesis is governed by the rate of CO_2 diffusion from external air to the leaf interior and can be written:

$$A = \frac{C_a - C_i}{1.6R}, \tag{9.16}$$

where C_a is the concentration of CO_2 in air, R is stomatal resistance. The factor 1.6 in Eq. 9.16 is the approximate ratio of binary molecular diffusion coefficients for water vapor and CO_2 in air. Equations 9.13, 9.16 respectively represent the demand and supply of CO_2. They are sufficient to define the tree's photosynthetic rate. A simultaneous solution of Eqs. 9.13, 9.16 is given by:

$$\begin{cases} A = A_m \dfrac{C_a - a}{b - a + A_m 1.6R}, & if \quad A_m 1.6R \ge C_a - b \\ A = A_m, & if \quad A_m 1,6R \le C_a - b \end{cases} \tag{9.17}$$

where $A_m = A_{max} f_T(T) f_E(E)$.

The daily photosynthesis is calculated as A_{dlen}, where $dlen$ is the day length (s).

When the rate of photosynthesis is limited by temperature or light intensity to a level when $A_m 1.6R \le C_a - b$, the CO_2 concentration in the leaves reaches or exceeds the saturation level b. Under these conditions, it is assumed that the stomatal resistance increases and thus the CO_2 concentration inside the leaf at level b is maintained:

$$R^c = \max(\frac{C_a - b}{1.6A_m}, R_{\min}) \quad , \qquad (9.18)$$

This equation shows the linear dependence of stomatal conductance ($1/R_c$) on photosynthetic rate (Farquhar and Sharkey 1982). Thus, decreasing soil moisture could limit the assimilation rate through increasing stomatal resistance and, vice versa, limitation of photosynthesis by light or temperature could cause transpiration to decrease.

9.2.4 Growth Dynamics

The total living tissue in a tree is composed of tree compartments: a crown (foliage, M_l), a stem (M_s), composed of the living cells of xylem and phloem, and roots (M_r). It is assumed that "living" mass growth of all tree compartments during the growth season is compensated by cell mortality. The foliage and root growth was included in the model only for the estimation of assimilate utilization. The total foliage was separated into several different age classes $M_{l0}, M_{l1}, M_{l2}...M_{ln}$ ($\Sigma M_{li} = M_l$). Foliage growth rate (μ_l) is described by:

$$\mu_l = \mu_{lo} M^*_{lo} (1 - \frac{M_{lo}}{M^*_{lo}}) F_l(S,T,w) \qquad (9.19)$$

where M_{l0} is the current foliage mass, $M_{l0}{}^*$ is the potential foliage mass, μ_{l0} is a constant and F_l is a normalized function which determines the relations between growth rate and the amounts of assimilates (S), temperature (T), and soil moisture (w).

Root growth was simulated analogously:

$$\mu_r = \mu_{r0} M_r F_r(S,T,w) \qquad (9.20)$$

Stem growth occurs by the formation of a new tree ring during the growth season. It was assumed that the number of initial cells is proportional to the number of cells or the living cell mass in phloem and xylem (M_S). Cell growth and differentiation in the cambial zone are presented in the model as the cambial block described in Sect. 7.4.2. In this case, the function $F_c(S,T,w)$ was used instead of the function $G(t)$. Tree-ring growth integrates the division and differentiation of cells in the cambial zone, cell growth by elongation, and the formation of a secondary cell wall. All stages of cell differentiation and maturation depend on three variables: the amount of assimilates in the stem, temperature, and soil moisture. This dependence is also described in the model by the normalized functions $F_e(S,T,w)$ and $F_m(S,T,w)$. All functions $F_i(S,T,w)$ are the same and differentiated only by their coefficients:

$$F_i(S,T,w) = \frac{S_i}{S_i + S_i^*} f_i(T) f_i(w), \qquad (9.21)$$

where i takes the values l, r, c, e, m (foliage, roots, cambial zone, elongation and formation of cell wall), S_i* is a parameter, S_i is the photosynthate content in foliage l, stem s, and root r, $f_i(T)$ and $f_i(w)$ are the piecewise linear functions of the influence of temperature and soil moisture on growth (analogous to the functions g_T and g_w; see Sects. 7.4.1, 7.4.2.). As is evident from the earlier chapters of this book, quantitative knowledge of the growth processes of tracheid sizes and the synthesis of their walls is limited and largely qualitative. Adequate quantitative models describing these processes are absent. Studies of cell growth in seedlings and roots have shown two growth phases: a relatively short-term phase of "acceleration", when the growth rate increases to a maximal value, and the "slowed-up" phase, when the growth rate gradually decreases (Pritchard 1994; see also Fig. 4.3). In the model, environmental conditions maximally influence the growth of cells in cambial zone and the rate of elongation at its initial stage. In accordance, this cell elongation is described in the model as:

$$\frac{d}{dt} D(t) = V_e = k_e (D_p - D(t)) \qquad (9.21a)$$

where, D(t) is the radial size of cell at position j at time t (μm), V_e is the growth rate (μm/day), and $k_e = k_{e0} F_e(S_s, T, W)$.

Cell size is determined during differentiation in the cambium zone (D_p is potential cell size) and during elongation when the growth rate is changed by the environment. The potential cell size D_p is a function of the distance ($y = \sum_{i=1}^{j} D_i$) from the initial cell when cell j started enlarging:

$$D_p = D_{max} - (D_{max} - D_{min}) e^{b5(b6-y)}$$

where b_5 and b_6 are parameters and D_{max} and D_{min} are the maximum and minimum cell sizes. The smaller the value of b_5, the greater the effect of distance (y) across the cambial zone on maximum cell size. If $b_5 = 0.1$ and cambial growth is slow with few dividing cells present (y is small), the potential enlargement is reduced. This would occur in the case of a mid-summer drought. As the number of cells in the dividing zone declines, the ability for cells to enlarge is correspondingly reduced. Parameter b_6 is the critical width of the cambial zone at which y begins to influence the potential cell size. Below that width, the size of the dividing layer has no effect on potential cell size. Cell walls enter maturation when the growth rate becomes less than the critical value V_{ecr}.

The dynamics of cell wall synthesis and wall thickening are similar to those controlling cell enlargement. The potential cell wall thickness (L_{wpot}) is a function of size of the j-th cell:

$$L_{wpot} = L_{w\max} - (L_{w\max} - L_{w\min})\exp(-a_5(D_j - D_{\min}), \quad (9.22b)$$

where D_j is the cell size. The cell wall cannot be bigger then $D_j(1-L_{\min})/2$, where $L_{\min}$ is the minimum lumen size expressed as a percentage. In fact this function describes the boundary of late- and earlywood cells. Wall synthesis irreversibly stops when V_m is less then V_{mcr}. Note that the equations for wall thickening are the same as those for enlargement. At optimal conditions, the wall thickness of the j-th cell is: $L_w = L_{wpot} - (L_{wpot} - L_{w\min})\exp(-k_m T_m)$. The duration in the wall-thickening stage, T_m, is: $T_m = \frac{1}{k_m}\ln(\frac{k_m}{V_{mcr}}(L_{wpot} - L_{w\min})$. Latewood with thicker-walled cells forms when the foliage mass is higher than a certain critical value, which is taken as one of the parameters in the model (Vaganov and Terskov 1977; see also Fig. 2.12). From Eq. 9.21a, which the processes of cell enlargement and cell wall synthesis follow, it can be seen that the duration of cell growth (or wall synthesis) is longest for the biggest cells. The decrease in the rate of growth through coefficient k gives a smaller cell size and a decreased duration of growth. This model of cell growth treats cell size as being determined not only by cambial activity but also by the environment's direct influence on expansion and maturation.

9.2.5 The Allocation of Assimilates

The distribution and utilization of assimilates by different organs, tissues, and cells are a complicated and not entirely known part of the carbon cycle in a tree (De Wit et al. 1970; Monsi and Marata 1970; Patefield and Austin 1971; Hesketh and Jones 1975; Sheehy et al. 1979; Gifford and Evans 1981). Following the assumption made by Thornley (1970), the assimilates translocate along the gradient of their concentration. The volume of each compartment (foliage, stem, roots) is proportional to its dry weight (M_l, M_s, M_r. So, if s_i is the concentration, then the content of assimilates in a compartment is $S_i = s_i M_i$. The translocation rate is high enough so that equilibrium along a tree is established during a day (Canny 1973).

The assimilates are utilized for growth and respiration. The daily rate of respiration, Rm_i (mM CO_2/kg), exponentially depends on the temperature (Amthor 1989; Buwalda 1991; Kramer and Kozlowski 1979):

$$Rm_i = \beta_{0i}\frac{s_i}{s_i + s_i^*}e^{\beta_{1i}T}M_i \quad , i=l, s, r \quad (9.23)$$

In the model, the utilization of assimilates for the growth of the leaves and the roots is proportional to their growth rates:

$$GR_l = \alpha_l \mu_l \tag{9.24}$$

$$GR_r = \alpha_r \mu_r \tag{9.25}$$

The utilization of assimilates for the growth of cambial (c) cells and cells in the enlargement (e) and maturation (m) stages is given in the model as:

$$GR_c = \alpha_c \cdot \rho_w \cdot N \cdot L_{w0} (\sum_{n_c} V_c + 2 \cdot D_t \cdot D) \tag{9.26}$$

$$GR_e = \alpha_e \cdot \rho_w \cdot N \cdot L_{w0} \sum_{n_e} V_e \tag{9.27}$$

$$GR_m = \alpha_m \cdot \rho_w \cdot N \sum_{n_m} V_m \tag{9.28}$$

where α_i is the amount of assimilates used to produce a unit of dry weight (mM CO^2/kg), ρ_w is the cell wall density (kg/μm^3), L_{w0} is the cell wall thickness of cambial and enlarging cells, n_c, n_e, n_m are the numbers of cells in the cambial, enlargement and maturation zones, and N is the number of initial cells per mass of stem (M_s).

9.3 Determination of Quantitative Values of Coefficients and Parameters

The selection of parameters is one of the most difficult problems in modeling. The present eco-physiological model is no exception to this. All the parameters and coefficients used in the model can be separated into two groups. The first includes the so-called physiological constants. Their values can be found in the literature (see, for example, Gates 1980; Larcher 1980; Jones 1983). Such physiological constants include: the relationship of the photosynthesis rate to temperature, CO_2, and irradiation, the needle resistance relationship to the diffusion of water vapor and carbon dioxide (R_{min}, R_{max}), the maximum value of photosynthesis (A_{max}), the soil-water availability, etc. The values of many parameters and coefficients of quite clear physiological meaning are selected by trial and error during simulation. Observed tree-ring chronologies are used for model verification. Some parameters and coefficients are common to all growth conditions, but others are unique to each habitat (for example: soil properties, optimal temperature, distribution of biomass between stem, crown, roots, etc.). Some assumptions have been made, as follows.

Foliage area was considered to be constant. The relative mass of stem and root living cells was obtained from allometric relationships between the biomasses of the foliage; and the sapwood volume of the stem and the fine roots, taking into account that the percentage of living cells, is about 7 % from sapwood mass (Grier and Waring 1974; Gholz 1980; Callaway et al. 1994; Kramer and Boyer 1995; Ryan et al. 1996). The number of initial cells in the cambial zone (i.e. the number of radial files) was estimated from the mean cambial cell sizes. The density of cell walls was taken as 1,500 kg/m[3] (Evans 1994). The functions describing relationships between growth rates and the amount of assimilates, temperature, and soil moisture content were assumed to be identical in shape. It was taken into account that the coefficients in the temperature–respiration relationship could vary slightly for different plant tissues (Ryan et al. 1996).

9.4 Examples of Model Applications

For a simple simulation of the effect of climate variability, the model was tested against a chronology of pine (*Pinus ponderosa*) from a semi-arid location in the Chiracahua Mountains of southeastern Arizona (USA). This location was chosen for several reasons. First, complete climate data for almost 50 years were available. Second, the observed chronology was known to be sensitive to climate variability. Third, tree rings formed at this site had a quite complicated anatomical structure. Almost every tree ring had a "false" ring forming in the middle part of the growth season. This was a response to intra-seasonal drought, caused by an almost total absence of rainfall between the end of May and the end of June. Furthermore, researchers from the Laboratory of Tree-Ring Research of the University of Arizona had been conducting a phenological and ecological–physiological study of seasonal tree growth in this place for several years. Therefore, some coefficients and parameters needed for the model had been determined experimentally and others were selected by trial and error. The similated data for tree-ring increment agree well with observed data (R=0.64, n=42, P=0.05; Fig. 9.3). Simulated tracheidograms, shown in Fig. 9.4 (top row), show some similarities with the measured tracheidograms for a number of trees. There are discrepancies in the position of the false rings (small cells) within the ring, especially between the simulated and measured tracheidograms. In most cases, the position of the false ring is due to the number of cells produced before the false ring, resulting from the time when ring growth started. For the first two years (1965, 1966), the discrepancy is related to the indeterminate initial conditions when the model is started. Clearly, further development of the model is needed to simulate more precisely the cellular structure of the rings.

In simpler environmental conditions with less complex rings, the model better simulates tree-ring growth. For example, data for model verification were derived from the growth of Scots pine (*P. sylvestris* L.) growing in a homogenous stand of fire-succession forest about 200 years old. The stand was located on the edge of the West Siberian Plain (Zotino, Russia, 60°45' N, 89°23' E) on gently undulating, alluvial sands with no underlying permafrost.

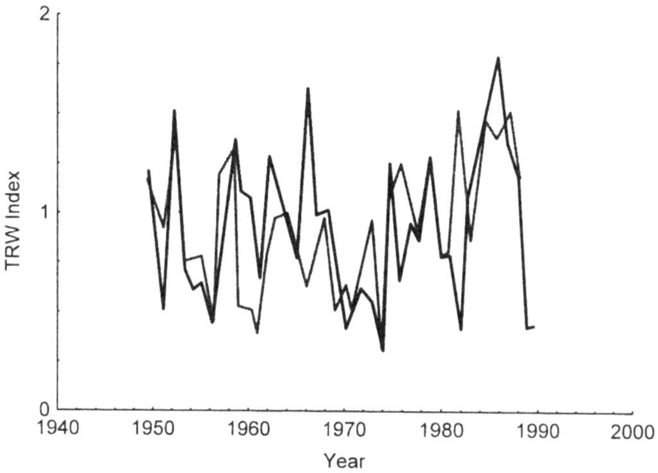

Fig. 9.3. Tree-ring chronology (*gray line*) and calculated index of cell number (*discontinuous line*) in semi-arid conditions (*Pinus ponderosa* in southern Arizona, USA)

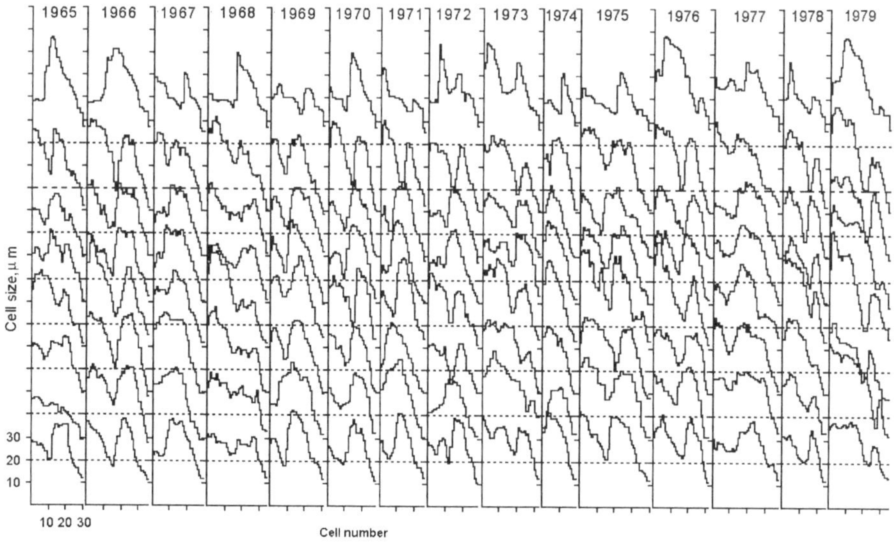

Fig. 9.4. Simulated (*top row*) and measured data of annual (1965–1979) variation in tracheid diameter in the rings of seven trees (*P. ponderosa*, Arizona, USA). Tracheidograms are normalized to the average measured cell number for each year

Seasonal patterns of diurnal variations of climatic data, photosynthesis, transpiration, soil moisture, and the cellular dynamics of tree rings were simulated (Fig. 9.5). Given the complexity of the model, it should ideally be validated not only against a final tree-ring chronology, but also against internal variables like photosynthetic rate, for example. Unfortunately, these data are

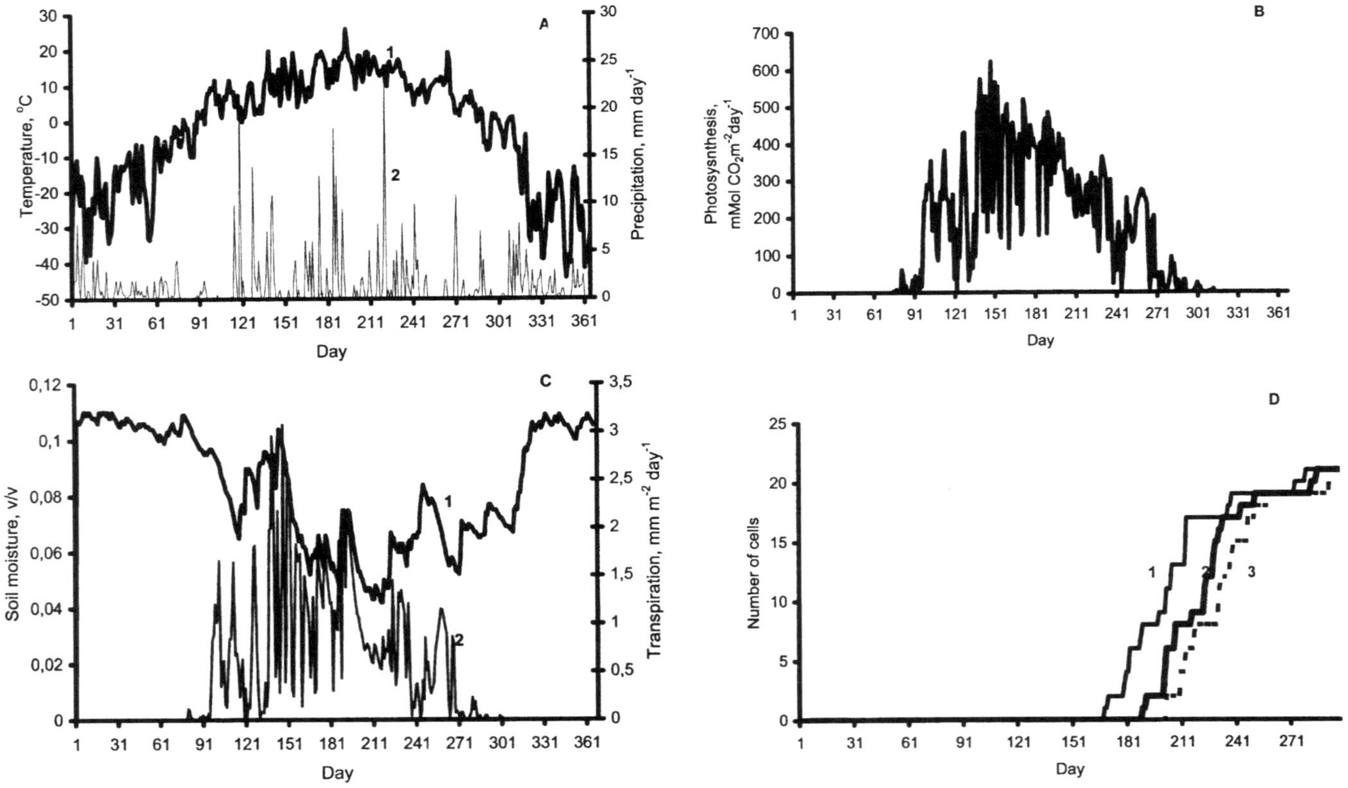

Fig. 9.5. Simulated and measured data for *Pinus sylvestris* at Zotino, Russia: **a** climatic data for 1997 (*1* temperature, *2* precipitation), **b** simulated dynamics of photosynthesis, **c** soil moisture (*1*) and transpiration (*2*), **d** cells in the ring (*1*), maturing and matured cells (*2*), and matured cells only (3)

rarely available. So far as can be determined, the simulated dynamics at Zotino correspond to the actual cellular structure (Figs. 9.6, 9.7).

The stable isotope composition of wood material, especially tree-ring cellulose, is now widely discussed, although rarely used (Hughes 2002) as an environmental proxy and indicator of tree response to environmental fluctuations. Although the theoretical basis linking the isotopic composition of C_3 plant material to environmental and physiological changes is reasonably well understood (Farquhar et al. 1989; McCarroll and Loader 2004), hitherto no attempt has been made to integrate this understanding with a model of daily tree growth and ring formation. The present eco-physiological model has provided this opportunity (Hemming et al. 2001). Using model estimates of the average daily concentration of CO_2 inside the leaves, C_i, the basic equation of carbon isotope fractionation in C_3 plants has been used to model the carbon isotope composition ($\delta^{13}C$) of the photosynthate (Farquhar et al. 1982):

$$\delta^{13}C_p = \delta^{13}C_a + a \times \left(1 - \frac{c_i}{c_a}\right) + b \times \left(\frac{c_i}{c_a}\right) \tag{9.29}$$

Daily estimates of new photosynthate designated for storage are added to leaf, stem, and root pools, whose bulk isotopic composition is subsequently modified by the new additions.

"Modelled daily $\delta^{13}C$ compositions of stem whole wood vary between –18‰ and –28‰ throughout the growing season. These values are in general

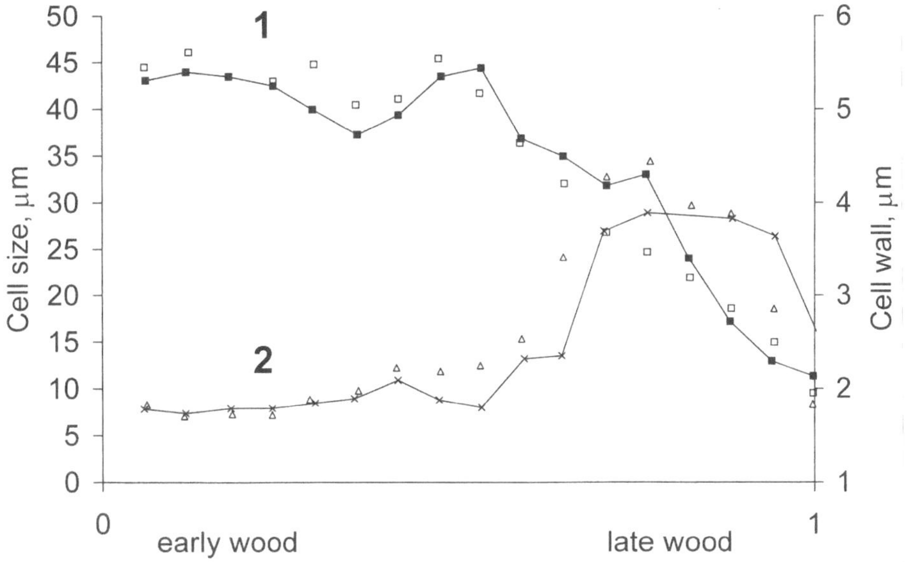

Fig. 9.6. Cell size (*1*) and cell wall thickness (*2*), showing a comparison of measured (*open squares, open triangles*) and calculated (*filled squares, crosses*) data (*P. sylvestris*, Zotino, Russia). Calculated cell size and wall thickness are connected by *solid lines*

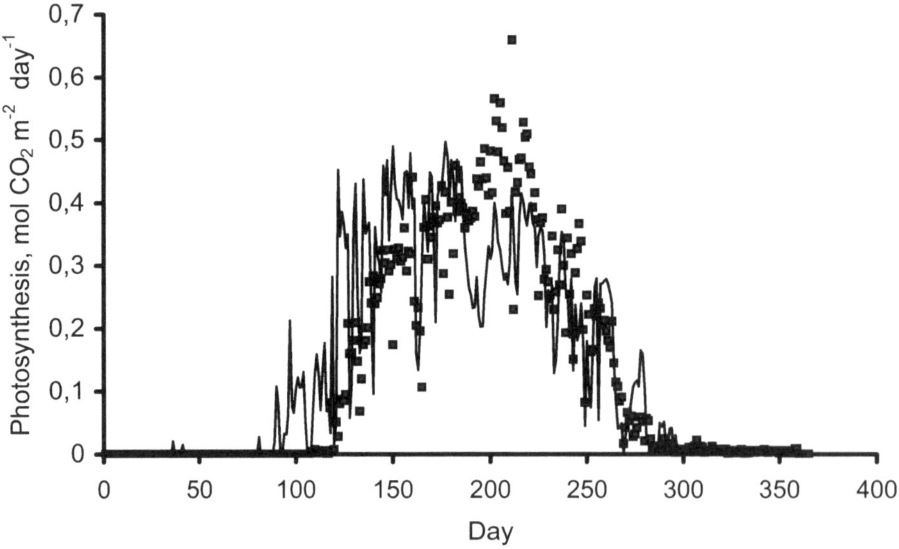

Fig. 9.7. Comparison of the simulated rate of photosynthesis (*continuous line*) and experimental data (*dots*) by Lloyd et al. (2002)

agreement with actual measurements from the species (*Pinus arizonica* Engelmann) and location (Santa Catalina Mountains near Tucson, Arizona) for which the model is tuned. A significant feature of the tree rings of this species in this region is the presence of a false-latewood band associated with a mid-growing season hyperarid period and subsequent summer monsoonal precipitation. This is apparent in the modeled $\delta^{13}C$ composition by a decrease in discrimination linked with the reduced moisture availability, followed by an increase in discrimination, presumably resulting from the rapid increase in precipitation and coincident decrease in tree water stress. From these preliminary results it is clear that the $\delta^{13}C$ component of the model is sensitive to intra-annual climatic variations and simulates reasonably well the absolute as well as intra-annual variance in the $\delta^{13}C$ composition of tree-rings" (p. 23 in Hemming et al. 2001). The model was also tuned for Siberian mid-taiga conditions (Zotino), where it is possible to compare simulation data of the distribution of $\delta^{13}C$ along the ring with measured $\delta^{13}C$ (Wirth, unpublished data; Fig. 9.8). We used only one year for tuning (1997), for which most of the needed data are available. The model of the discrimination of stable isotopes in the different plant tissues needs further development. In the case of the $\delta^{13}C$ calculation, it is important to know how and what kind of carbon is used for cell wall synthesis – whether it is the current products of photosynthesis or carbon from the reserve compartments. The eco-physiological model is sensitive to the pattern of carbon allocation among different tissues and the pattern of carbon exchange between labile pools of carbon. In fact, the model can

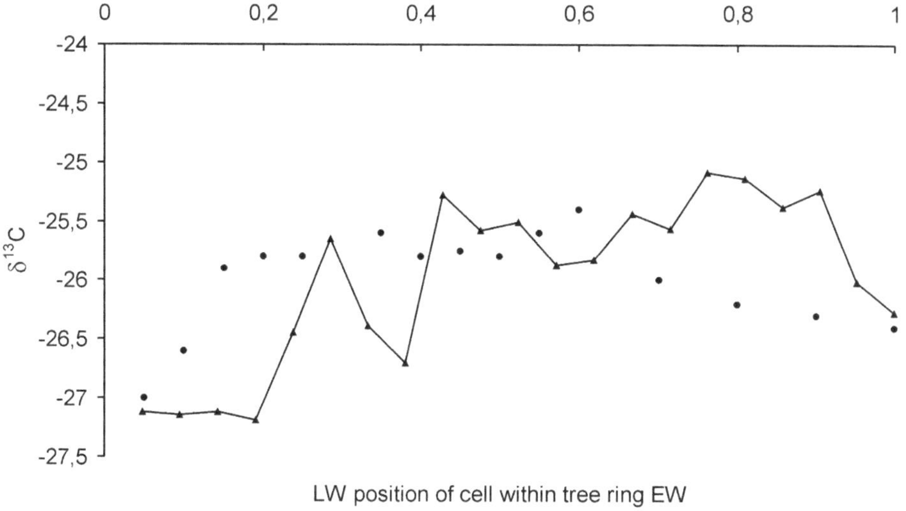

Fig. 9.8. Isotopic composition ($\delta^{13}C$) of cell wall across a tree ring at Zotino, Russia. *Unconnected diamonds* Measured $\delta^{13}C$, *connected triangles* simulated $\delta^{13}C$

be used as a tool for the investigation and estimation of the different processes of carbon discrimination and allocation.

The prediction of radial growth decline and ring structure for trees subject to pollution stress is a good demonstration of the capabilities of the eco-physiological model. Several mechanisms are known for pollutant influence on the morphological–physiological characteristics of trees, dependent on the sources of pollution (Eckstein et al. 1974; Innes 1993; Kharuk et al. 1996a, b). These include stomatal damage (dust emission), soil acidification (emission of sulfur dioxide), necrosis of needles, damage of apical meristems (high ozone concentrations), etc. The influence of pollution on growth is often complex, affecting both above- and below-ground plant parts.

Attempts have been made to model the growth of rings of larch trees grown in a zone of industrial emissions near the Norilsk metallurgical plant (Simachev et al. 1992; Ivshin and Shiyatov 1995; Kharuk et al. 1996a, b). Long-term daily climatic data from the Turukhansk weather station were used. Two mechanisms of pollutant influence were considered: (1) stomatal damage from both dust emission and sulfur dioxide, which resulted in a declining activity of stomatal guard cells and reduced stomatal conductance, (2) gradual acidification of soil from the deposition of oxides (sulfur, carbon, nitrogen), which resulted in a reduction of root ability to absorb water (Matyssek et al. 1995). The first pollutant influence was described in the model as a linear increase of stomatal resistance versus time. The second was simulated as a gradual reduction of coefficient q (see Eq. 9.11). It produces a disturbance of water balance and a limitation of photosynthesis by stomatal conductance;

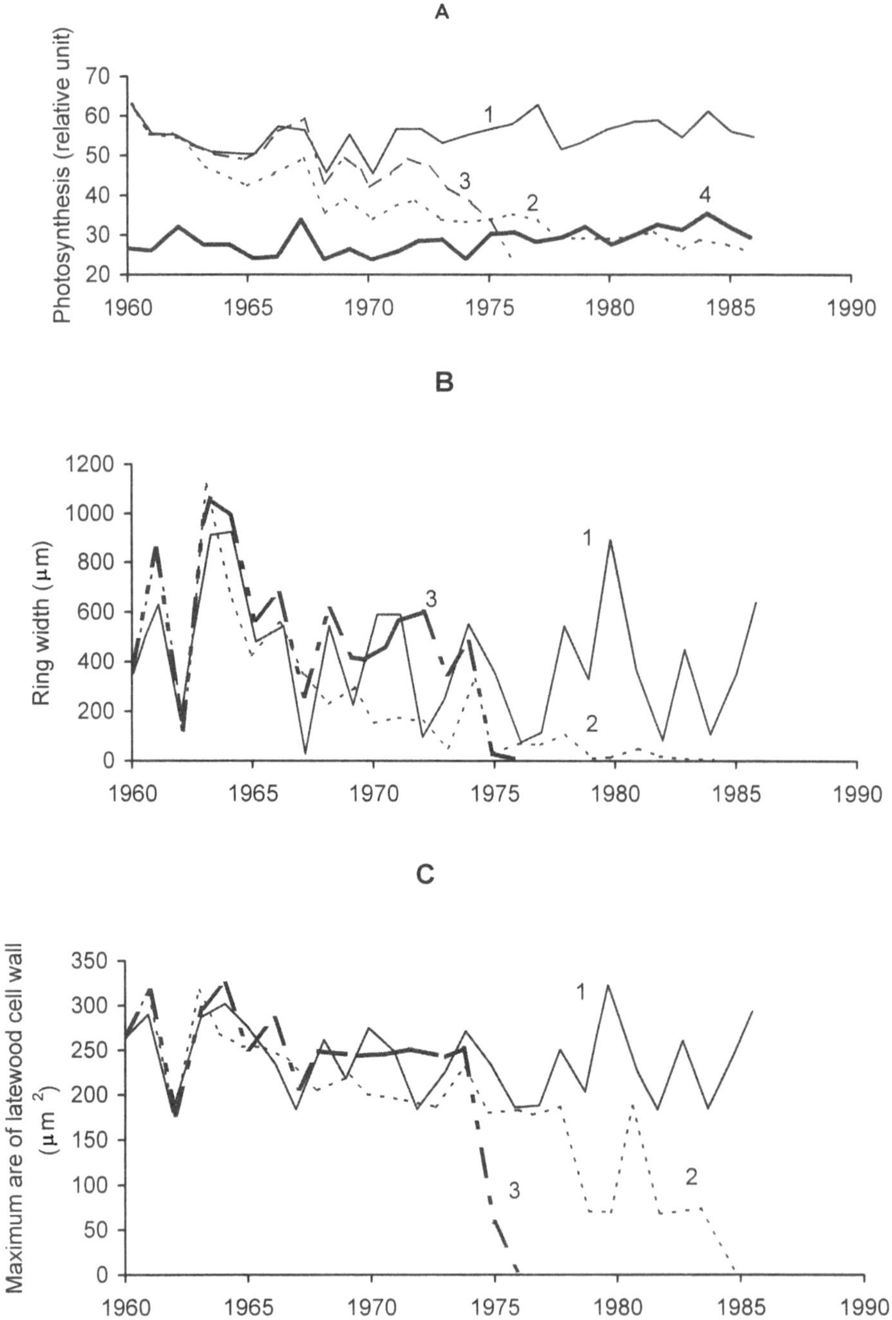

Fig. 9.9. Dynamics of the rate of photosynthesis (**a**), variability in tree-ring width (**b**), and maximal area of cell wall in latewood (**c**) with: *1* climatic change, *2* stomatal damage, *3* decrease in root water-absorption capacity, *4* respiration

and the latter decreases as a result of poor water supply. The results of the simulation of these two mechanisms were compared to the results of a purely climatic model (control), in which only temperature, precipitation, and irradiation were used to determine the radial growth dynamics (Fig. 9.9). In the calculations, it was assumed the pollution began its influence in 1960, as a matter of convention.

In the case of stomatal damage, photosynthesis started to decrease in comparison with the control value after 2–3 years, so that, by 20–25 years, the difference between photosynthesis and respiration became negative. In the case of soil acidification, photosynthesis slightly differed from the control (i.e. from the photosynthesis in a "pollution-free zone") until 1970, when it started sharply to be reduced and, during the next 5 years, it became less than respiration. For both mechanisms mentioned as well as the control, the radial increments differed slightly from one another within the first 5–7 years. Later on, for the first mechanism, the increment steadily declined relative to the control, up to the minimum which resulted in tree death. For soil acidification (the second mechanism), the reaction of the tree differed from this pattern: the increment kept level, close to the control during quite a long period of time; and then the tree growth was discontinued sharply. Similar patterns were seen in the formation of latewood cell walls (Fig. 9.9 c). It should be particularly emphasized that the dynamics of annual growth differ from the dynamics of total annual photosynthesis. This is partly derived from the fact that assimilates are used not only for growth but for respiration too. The nonlinear relationship between tree-ring width and the mass of cell wall material in this ring is more important, as the latter directly depends on the amount of assimilates used. The simulated results agreed with the tree-ring chronologies that were obtained for the control conditions and for stands under pollution pressure (Fig. 9.10). The difference between the two curves (i.e. between the tree ring widths of the control and experimental trees) is clearly visible; and it testifies that, since the early 1980s, the radial growth of the trees under pollution pressure has been suppressed. Detailed examination of the wood microanatomy has shown that in dead trees, as a rule, only the outmost two or three rings are unusually narrow (Schweingruber and Voronin 1996). Wide outermost rings would mean that the tree died suddenly. Analysis of the results of dendrochronological studies of the influence of elevated sulfur dioxide concentrations on woody plant radial growth (Fischer 1993) has shown that: (1) genetically identical plants (clones) react to pollution characteristically, (2) there are species-specific responses, (3) the seasonal dynamics of pollution exposure influence the tree response, (4) morphological damage is not always correlated with radial growth and tree-ring structure. Therefore, simulation modeling provides a tool with whose help it is possible to connect internal changes in the growth processes under pollution influence with anatomical changes. Within the framework of the eco-physiological model, it becomes possible not only to analyze the causes of individual and specific

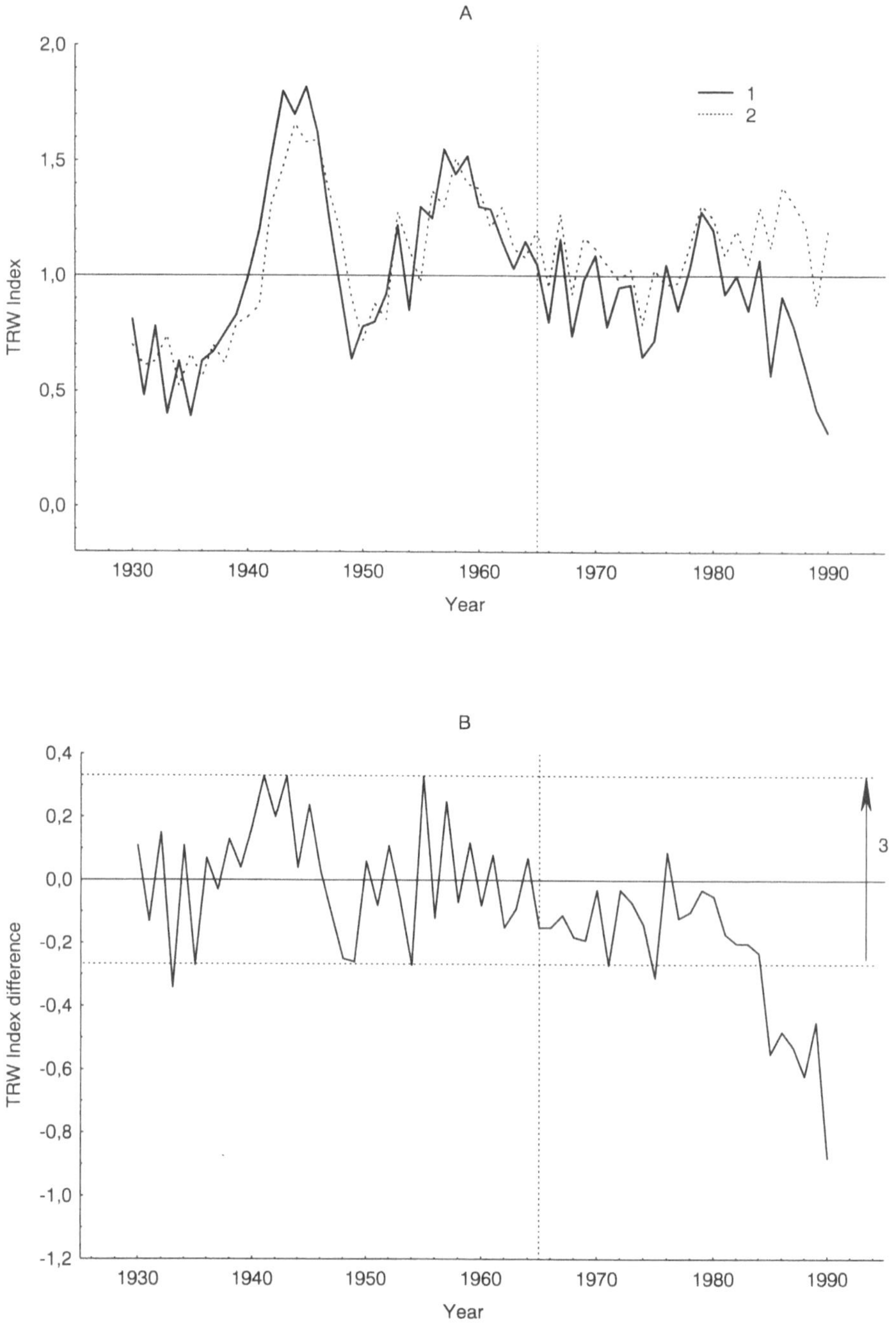

Fig. 9.10. Changes in the dynamics of radial increment in larch trees near Norilsk, Russia. **a** Control (*curve 1*) versus experiment (*curve 2*). **b** The difference between *curves 1* and *2*, showing the influence of damage

peculiar responses to pollution, but also to define those critical seasonal periods of greatest pollutant influence for a tree.

9.5 Conclusions and Discussion

An eco-physiological model has been described that is based on the Vaganov–Shashkin (VS) model. The determination of quantitative values of its coefficients and parameters has been discussed, examples of its application to several different problems given, and conclusions drawn.

The eco-physiological model is a modification and extension of the VS model described in Chap. 7. It differs from the VS model in having an explicit treatment of photosynthesis, respiration, and the partitioning of assimilates. As a result, it may be used to calculate absolute as opposed to relative values of a number of quantities, including tree-ring widths, cell dimensions, and the ratios of the stable isotopes of carbon. It is intended as a step towards linking the growth rate with the photosynthetic rate and the substrate available for growth with the regulation of cambial activity. Although it needs further development, it provides a valuable tool for a number of kinds of investigations including, for example, the diagnosis of the impacts of pollution on tree-ring size and structure.

10 Epilogue

Our aim has been to introduce a way of thinking about the environmental control of tree-ring variability that is expressed clearly in the title of this book: "*Growth dynamics of conifer tree rings: images of past and future environments*". In particular, each ring contains an image of the time when the ring formed, projected onto the ring's size, structure, and composition. The lens through which this projection occurs is the vascular cambium, the site of the development of each year's ring, and it is on its dynamics that we have focused. Our particular perspective comes from our chosen task – the extraction of an image of past environments, especially climate variability, from the natural archives that tree rings offer.

The objective has been to simulate the inter-annual and decadal variability of conifer tree rings as it is driven by climate variability. This has been achieved to a considerable degree in the Vaganov–Shashkin (VS) model described in Chap. 7, by focusing strongly and uniquely on the direct environmental control of cambial activity, without any explicit treatment of photosynthesis, respiration, and transpiration. We justify this approach by its success. The VS and eco-physiological models described here do capture the main features of the behavior of interest, climate-driven tree-ring variability, remarkably often. They not only reproduce the patterns of inter-annual and decadal variability seen in empirical, well dated and replicated tree-ring chronologies from many climates and regions, they also capture the characteristic persistence added to the climate signal in such records. Of course, it may be that the model is effective because of some unplanned correlation with a more direct causal link. That is a common circumstance for simulation models that may only be disposed of by extensive testing of the intermediate as well as final predictions of the model.

Even so, several studies have appeared in recent years whose results support the possibility of some degree of direct, local, environmental control in the cambium and identify the specific molecular agents of control (see Sect. 4.6). If such evidence continues to accumulate, the decision to focus on the cambium will be further justified. However, even if this does not occur, the fact that the model works as well as it does, as an effective "mimic" that not only captures inter-annual variability but also simulates the persistence of tree-ring series faithfully, makes it a valuable tool for the examination of the formation of this widely used natural archive of past environments. It may also be of use in diagnosing the ecosystem effects of changing climate variability. As more is learnt about the strong temporal texture of climate variability, such as the widespread effects of the El Niño–Southern Oscillation phenomenon and decadal variability over the main ocean basins, it is increasingly important to understand the effects on terrestrial ecosystems of the time-dependent structure of climate variability. The promise also exists that the model might eventually be used directly in paleoclimate reconstruction, through the process of inverse modeling. Whether for paleoclimate reconstruction or for better understanding of the formation of tree rings, there is no doubt that improvement of the model and our understanding of its results would be useful. The eco-physiological model described in Chap. 9 illustrates one approach to this.

We do not offer this modeling approach as a substitute for the established, primarily empirical–statistical techniques of dendroclimatology, but as a complement to them. We are with John Harte (2002) when he argues that "particularity and contingency, which characterize the ecological sciences, and generality and simplicity, which characterize the physical sciences, are miscible, and indeed necessary, ingredients in the quest to understand humankind's home in the universe". Empirical–statistical tools are of great value in dealing with "particularity and contingency", just as process-based modeling may be a way of approaching generality and simplicity.

The task of developing improved models of climate control of tree-ring variability must be seen in the context of challenges from two directions – their scientific basis and the problems to which they should be applied. The type and quantity of information that may be extracted from tree rings has recently expanded considerably, adding, as it were, further color and texture to the image of past environment that is contained in the tree ring. This presents the possibility of constraining the interpretation of tree-ring structure in terms of climatic variability. New analytical instruments, for example, have increased the quantity and variety of chemical measurements (including but not limited to stable isotopic ratios) that may be made and have radically decreased the amount of material needed for such analyses. The measurement of cell dimensions in large numbers of samples has become less onerous with the development of image-processing technology. Thus it becomes more and more possible to treat the tree ring as a natural archive containing many proxy records of past environment. In order to extract these records efficiently and without distortion, it is essential to understand these new measurements in physiological and ecological terms and to develop process-based concepts to explain their variability. These should then be incorporated in new or improved process models. Similarly, new experimental and observational techniques are improving knowledge of the process of ring formation and its relationship to the environment. The recent results pointing to direct local environmental control of cambial activity and identifying the specific molecular agents of control in the cambial zone are particularly apposite examples of this.

These developments in the understanding of the formation of the tree-ring natural archive take place against a background of pressing new scientific questions. As the focus on high-resolution paleoclimatology has increased in recent years, and as its results have assumed a central role in the study of current and future environmental change, new demands are placed on dendroclimatology. For example, it is no longer enough to trace a possible history of winter precipitation in a particular region or the pattern of summer temperature anomalies across the northern hemisphere. A clear understanding of the errors and biases of these histories, or reconstructions, is needed. Just how much of the actual climate control of tree-ring variability is captured by the existing models? We know what fraction is captured relative to some a priori selection of climatic forcing factors, but we do not, of course, know whether we are taking into account all the climate variables that should be included. Put another way, does the part of tree-ring variability that we have not captured with models such as the VS model have a climatic nature? Other than climate, what environmental factors are responsible for tree-ring variability and how do they work?

We have discussed the apparent changes in tree-ring response observed, primarily at high northern latitudes, in recent decades. Why is this happening? Are its causes entirely climatic or are other factors involved? What does this mean for the reconstruction of past climate using tree rings? Might there have been such changes in the past, and if so, how might we detect them? There are other examples of greatly increased growth rates in the last century and in just the last two or three decades, when compared to several prior centuries or even millennia, at sites as scattered as Tasmania, New Mexico, and the Great Basin of the western United States. Is it possible to reproduce these changes using the VS or eco-physiological models described here? If not, does the structure of the residuals between the simulated and observed values point to another possible cause? The answers to these questions may have significance for modeling future changes in the interaction between plants, forest, and the atmospheric environment. Is it perhaps necessary to incorporate the potential effects of varying atmospheric carbon dioxide in the VS model? A number of empirical results suggest that it may be valuable to differentiate the effects of minimum and maximum temperatures. One of the most robust observations of climate change over the last century has been the increase in night-time (minimum) temperatures and the decrease in the diurnal temperature range. How might this affect the development of tree rings?

One of the strongest arguments for developing process-based forward models of environmental control of tree-ring variability comes from the wish to anticipate future changes and understand the remote past. These may well be outside the range of the currently available empirical data on which any statistical model must be based. Our choice of processes to include in a process-based model may also be limited in an analogous way by the range of factors known to be relevant at the time of development of the model. In the use and development of the models described in Chaps. 7, 9, it will be necessary to continue to attempt to strike the right balance between keeping the model simple and including strongly relevant features. However this is to be done, we strongly recommend a firm focus on the cells that the wood comes from – the vascular cambium.

References

Abaimov AP, Bondarev AI, Zyryanova OA, Shitova SA (1997) Forests above the polar circle in Krasnoyarsk region (in Russian). Nauka, Novosibirsk, 208 pp

Ajmal S, Iqbal M (1987a) Annual rhythm of cambial activity in *Streblus asper*. IAWA Bull 8:275–283

Ajmal S, Iqbal M (1987b) Seasonal rhythms of structure and behavior of vascular cambium in *Ficus rumphii*. Ann Bot 60:649–656

Alexeev VA (1975) The light regime in forests (in Russian). Nauka, Leningrad, 227 pp

Alfieri FJ, Evert RF (1968) Seasonal development of the secondary phloem in *Pinus*. Am J Bot 55:518–528

Alfieri FJ, Kemp RI (1983) The seasonal cycle of phloem development in *Juniperus californica*. Am J Bot 70:891–896

Alisov BP (1961) The Climate of USSR (in Russian). Moscow State University, Moscow, 126 pp

Aloni R (1987) Differentiation of vascular tissues. Annu Rev Plant Physiol 38:179–204

Amthor JS (1989) Respiration and crop productivity. Springer, Berlin Heidelberg New York, 215 pp

Anchukaitis KJ, Evans MN, Vaganov EA, Grissino-Mayer H, Hughes MK, Kaplan A, Cane MA (2005) Nonlinear climate–tree growth relationships in the southeastern United States (in press)

Andreev SG, Vaganov EA, Naurzbaev MM, Tulokhonov AK (1999) Registration of long-term variations in atmospheric precipitation, Selenga river runoff, and Lake Baikal level by pine annual tree rings. Dokl Earth Sci 368:1008–1012

Anfodillo T, Di Bisceglie DP, Urso T (2002) Minimum cuticular conductance and cuticle features of *Picea abies* and *Pinus cembra* needles along an altitudinal gradient in the Dolomites (NE Italian alps). Tree Physiol 22:479–487

Antanaitis VV, Zagreev VV (1981) Forest increment (in Russian). Lesnaya Promishlennost, Moscow, 198 pp

Antonova GF, Stasova VV (1993) Effects of environmental factors on wood formation in Scots pine stems. Trees 7:214–219

Antonova GF, Stasova VV (1997) Effects of environmental factors on wood formation in larch (*Larix sibirica* Ldb.) stems. Trees 11:462–468

Antonova GF, Cherkashin VP, Stasova VV, Varaksina TN (1995) Daily dynamics in xylem cell radial growth of Scots pine (*Pinus sylvestris*). Trees 10:24–30

Arbatskaya MK (1998) Long-term variability of climate, tree ring growth and fire frequency in Middle Siberia taiga (in Russian). PhD thesis, Russian Academy of Sciences, Krasnoyarsk, 154 pp

Arbatskaya MK, Vaganov EA (1997) Long-term variation in fire frequency and radial increment in pine from the middle taiga subzone of central Siberia. Russ J Ecol 28:291–297

Arno SF, Hammerly RP (1984) Timberline: mountain and arctic forest frontiers. Mountaineers, Seattle

Arsenault KR (2000) A modeling and observational investigation of North Atlantic SST anomalies and their effects on Eurasian snow cover. MS thesis, Ohio State University, Cleveland

Baillie MGL, Pilcher JR (1973) A simple cross-dating program for tree-ring research. Tree-Ring Bull 33:7–14

Baldocchi D, Collineau S (1994) The physical nature of solar radiation in heterogenous canopies: spatial and temporal attributes. In: Caldwell MM, Pearcy RW (eds) Exploitation of environmental heterogeity in plants. Ecophysiological processes above and below ground. Academic, San Diego, pp 21–71

Ball JT, Woodrow LE, Berry JA (1987) A model predicting stomatal conductance and its contribution to the control of photosynthesis under different environmental conditions. In: Biggins J (ed) Progress in photosynthesis reseach, vol IV. Nijhoff, Dordrecht, pp 221–224

Balodis VA (1968) Some patterns of mitosis distributions in the root apex (in Russian). Tsitologia 10:1374–1381

Baluska F, Barlow PW, Kubica S (1994) Importance of the post-mitotic growth (PIG) region for growth and development of roots. Plant Soil 167:31–42

Baluska F, Volkmann D, Barlow PW (1996) Specialized zones of development in roots: view from the cellular level. Plant Physiol 112:3–4

Baluska F, Volkmann D, Barlow PW (2001) A polarity crossroad in the transition growth zone of maize root apices: cytoskeletal and developmental implications. J Plant Growth Regul 20:170–181

Bannan MW (1954) Ring-width, tracheid size, and ray volume in stem wood of *Thuja occidentalis* L. Can J Bot 32:875–884

Bannan MW (1955) The vascular cambium and radial growth in *Thuja occidentalis* L. Can J Bot 33:113–138

Bannan MW (1957) The relative frequency of the different types of anticlinal divisions. Can J Bot 35:875–884

Bannan MW (1962a) Cambial behavior with reference to cell length and ring width in *Pinus strobus* L. Can J Bot 40:1057–1062

Bannan MW (1962b) The vascular cambium and tree-ring development. In: Kozlovski TT (ed) Tree growth. Ronald, New York, pp 3–21

Bannan MW, Bindra M (1970) Variations in cell length and frequency of anticlinal division in the vascular cambium throughout a white spruce tree. Can J Bot 48:1363–1371

Barber V, Juday GP, Finney B (2000) Reduced growth of Alaskan white spruce in the twentieth century from temperature-induced drought stress. Nature 405:668–673

Bard E, Raisbeck G, Yiou F, Jouzel J (2000) Solar irradiance during the last 1200 years based on cosmogenic nuclides. Tellus 52B:985–992

Barlow PW (1976) Towards an understanding of the behavior of root meristems. J Theor Biol 57:433–451

Barnes BV, Zak DR, Denton SR, Spurr SH (1998) Forest ecology, 4th edn. Wiley, New York, 774 pp

Barnes RD, Gibson GL, Bardey MA (1983) Variation and genotype environment interaction in international provenance trials of *Pinus caribaea* var. *hondurensis* and implications for population improvement strategy. Silvicult Sao Pedro 8:35–43

Barnett JR (1973) Seasonal variation in the ultra structure of the cambium in New Zealand grown *Pinus radiata* D. Don. Ann Bot 37:1005–1011

Barnett JR (ed) (1981) Xylem cell development. Castle House, Tunbridge Wells, pp 307
Baskin TI (2000) On the constancy of cell division rate in root meristem. Plant Mol Biol 43:545–554
Beets PN, Gilchrist K, Jeffrey MP (2001) Wood density of radiate pine: effect of nitrogen supply. For Ecol Manage 145:173–180
Benkova AV, Shashkin AV (2003) The photosynthesis of pine and larch and its relation with radial increment. Lesovedenie 5:1–6
Benn DL, Evans DJA (1998) Glaciers and glaciation. Arnold, London
Benthel JS (1964) The developing architecture of secondary xylem in conifers. For Sci 10:89–91
Biondi F (2000) Are climate–tree growth relationships changing in north-central Idaho? Arct Antarct Alp Res 32:111–116
Biondi F, Perkins DL, Cayan DR, Hughes MK (1999) July temperature during the second millennium reconstructed from Idaho tree rings. Geophys Res Lett 26:1445–1448
Birot Y, Nepveu G (1979) Variabilité clonale et liaisons ortets-ramets dans un population d'Epicêa. Silv Genet 28:37–47
Bitvinskas TT (1974) Dendroclimatic studies (in Russian). Gitrometeoizdat, Leningrad, 172 pp
Bolshov LN, Smirnov NV (1983) The tables of mathematical statistics (in Russian). Nauka, Moscow, 416 pp
Bonan GB, Pollard D, Thompson SL (1993) Effect of boreal vegetation on global climate. Nature 359:716–718
Borovikov AM, Ugolev BN (1989) The hand-book of wood (in Russian). Lesnaya Promislennost, Moscow, 296 pp
Boulton GS, Mason P, Ballantyne CK, Karlen W, Matthews JA, Nesje A (1997) Holocene glacier fluctuations in Scandinavia. Paleoclim Res 24:5–33
Box GE, Jenkins GM (1970) Time series analysis forecasting and control. Holden–Day, San Francisco, 428 pp
Boyer JS (1985) Water transport. Annu Rev Plant Physiol 36:473–516
Boyle TJ, Balatinecz JJ, McCaw PM (1987) Genetic control of some wood properties of black spruce. Can Tree Improv Assoc Meet 21, 198 pp
Bradley RS (1999) Paleoclimatology: reconstructing climates of the Quaternary. Academic, San Diego, 610 pp
Bradley RS, Diaz HF, Eischeid JK, Jones PD, Kelly PM, Goodess CM (1987) Precipitation fluctuation over Northern Hemisphere land areas since the mid-19th century. Science 237:171–175
Bradley RS, Hughes MK, Diaz H F (2003) Climate in medieval times. Science 302:404–405
Braeker OU (1981) Der Alterstrend bei Jahrringdichten und Jahrringbreiten von Nadelhölzern und sein Ausgleich. Mitt Forstl Bundes-Versuchsanst Wien 142:75–102
Bräuning A (1999) Dendroclimatological potential of drought-sensitive tree stands in Southern Tibet for the reconstruction of monsoon activity. IAWA J 20:325–338
Brett C, Waldron K (1996) Physiology and biochemistry of plant cell wall, 2nd edn. Chapman and Hall, London, 255 pp
Briffa KR, Jones PD (1990) Basic chronology statistics and assessment. In: Cook ER, Kairiukstis LA (eds) Methods of dendrochronology: applications in the environmental sciences. Kluwer, Dordrecht, pp 137–152
Briffa KR, Jones PD, Schweingruber FH (1988) Summer temperature patterns over Europe: a reconstruction from 1750 AD based on maximum latewood density indices of conifers. Quat Res (NY) 30:36–52
Briffa KR, Bartholin TS, Eckstein D, Jones PD, Karlen W, Schweingruber FH, Zetterberg PA (1990) A 1,400-year tree ring record of summer temperatures in Fennoscandia. Nature 346:434–439

Briffa KR, Jones PD, Bartholin TS, Eckstein D (1992a) Fennoscandian summers from AD 500: temperature changes on short and long timescales. Clim Dyn 7:111–119

Briffa KR, Jones PD, Schweingruber FH (1992b) Tree-ring density reconstructions of summer temperature patterns across western North America since 1600. J Clim 5:735–754

Briffa KR, Jones PD, Schweingruber FH, Shiyatov SG, Cook ER (1995) Unusual twentieth-century summer warmth in a 1000-year temperature record from Siberia. Nature 376:156–159

Briffa KR, Schweingruber FH, Jones PD, Osborn TJ, Shiyatov SG, Vaganov EA (1998) Reduced sensitivity of recent tree growth to temperatures at high northern latitudes. Nature 391:678–682

Briffa KR, Osborn TJ, Schweingruber FH, Harris IC, Jones PD, Shiyatov SG, Vaganov EA (2001) Low-frequency temperature variations from a northern tree ring density network. J Geophys Res 106:2929–2941

Bristow KL (1992) Prediction of daily mean vapor density from daily minimum air temperature. Agric For Meteorol 59:309–317

Bristow KL, Campbell GS (1984) On the relationship between incoming solar radiation and daily maximum and minimum temperature. Agric For Meteorol 31:159–166

Brown CL, Sax K (1962) The influence of pressure on the differentiation of secondary tissues. Am J Bot 49:693

Brown PM (1994) OLDLIST: a database of maximum tree ages. In: Dean JS, Meko DM, Swetnam TW (eds) Tree rings, environment and humanity. (Proceedings of the international conference.) Radiocarbon, Tucson, pp 727–733

Brown RD, Goodison BE (1996) Interannual variability in reconstructed Canadian snow cover, 1915–1992. J Clim 9:1299–1318

Buckley TN, Mott KA, Farquhar GD (2003) A hydromechanical and biochemical model of stomatal conductance. Plant Cell Environ 26:1767–1785

Bunn AG (2004) Temporal and spatial patterns at alpine treeline in the Sierra Nevada USA: implications for global change. Montana State University, Bozeman

Bunn AG, Lawrence RL, Bellante GJ, Waggoner LA, Graumlich LJ (2003) Spatial variation in distribution and growth patterns of old growth strip-bark pines. Arct Antarct Alp Res 35:323–330

Bunn AG, Graumlich LG, Urban DL (2005a) Interpreting the climatic significance of trends in twentieth-century tree growth at high elevations. Holocene (in press)

Bunn AG, Waggoner LA, Graumlich LJ (2005b) Topographic mediation of growth of subalpine trees in the Sierra Nevada, USA. Glob Ecol Biogeogr (in press)

Burholt DR, Van't Hof J (1971) Quantitative thermal-induced changes in growth and cell population kinetics of Helianthus roots. Am J Bot 58:386–397

Buwalda JG, Smith GS (1990) Effects of partial defoliation at various stages of the growing season on fruit yields, root growth and return bloom of kiwifruit vines. Sci Horticult 42:29–44

Buwalda JH (1991) A mathematical model of carbon acquisition and utilization by kiwi fruit vines. Ecol Model 57:43–64

Buzykin AI (ed) (1977) The forests of the Middle Angara (in Russian). Nauka, Novosibirsk, 263 pp

Callaway RM, Delucia EH, Schlesinger WH (1994) Biomass allocation of montane and desert ponderosa pine: an analog for response to climate change. Oecologia 75:1474–1481

Cannell MGR, Smith RI (1986) Climatic warming, spring bud burst, and frost damage of trees. J Appl Ecol 23:177–191

Canny MJ (1973) Phloem translocation. Cambridge University, London, 301 pp

Carlquist S (1988a) Comparative wood anatomy. Systematic, ecological and evolutionary aspects of dicotyledon wood. Springer, Berlin Heidelberg New York, 460 pp
Carlquist S (1988b) Nearvesselessness in *Ephedra* and its significance. Am J Bot 75:598–601
Carlquist S (2001) Comparative wood anatomy. Systematic, ecological and evolutionary aspects of dicotyledon wood, 2nd edn. Springer, Berlin Heidelberg New York, 448 pp
Carrer M, Urbinati C (2004) Age-dependent tree-ring growth responses to climate in *Larix decidua* and *Pinus cembra*. Ecology 85:730–740
Casgrove DJ (1993) How do plant cell walls extend? Plant Physiol 102:1–6
Catesson AM (1964) Origine, fonctionement et variations cytologiques saisonnière du cambium de *Acer pseudoplatanus*. Ann Sci Nat (Bot) Ser 5:229–498
Catesson AM (1980) The vascular cambium. In: Little CHA (ed) Control of shoot growth in trees. (IUFRO workshop proceedings.) Maritime Forest Research Centre, Fredericton, pp 12–40
Catesson AM (1984) La dynamique cambial. Ann Sci Nat Bot Ser 13:23–43
Catesson AM (1990) Cambial cytology and biochemistry. In: Iqbal M (ed) The vascular cambium. Res Studies, Taunton, pp 63–112
Catesson AM (1994) Cambial ultrastructure and biochemistry: changes in relation to vascular tissue differentiation and the seasonal cycle. Int J Plant Sci 155:251–261
Catesson AM, Roland JC (1981) Sequential changes associated with cell wall formation and fusion in the vascular cambium. IAWA Bull 2:151–162
Chaffey N, Cholewa E, Regan S, Sundberg B (2002) Secondary xylem development in *Arabidopsis*: a model for wood formation. Physiol Plant 114:594–600
Chapin FS III, Eugster W, McFadden JP, Lynch AH, Walker DA (2000) Summer differences among arctic ecosystems in regional climate forcing. J Climate 13:2002–2010
Chavchavadze ES (1979) Conifer wood (in Russian). Nauka, Leningrad, 192 pp
Chen W, Chen J, Cihlar J (2000) An integrated terrestrial ecosystem carbon-budget model based on changes in disturbance, climate, and atmospheric chemistry. Ecol Model 135:55–79
Chernavskaya MM, Grissino-Mayer HD, Krenke AN, Pushin AV (1999) *Pinus tropicalis* growth response to seasonal precipitation changes in Western Cuba. In: Wimmer R, Vetter RE (eds) Tree-ring analysis: biological methodological and environmental aspects. CABI, Wallingford, pp 185–190
Cherubini P, Piussi P, Schweingruber FH (1996) Spatiotemporal growth dynamics and disturbances in subalpine spruce forest in the Alps: a dendroecological reconstruction. Can J For Res 26:991–1001
Ciais P, Tans PP, Trolier M, White JWC, Francey RJ (1995) A large northern hemisphere terrestial CO_2 sink indicated by 13C/12C of atmospheric CO_2. Science 269:1098–1101
Clark NE, Blasing TJ, Fritts HC (1975) Influence of interannual climate fluctuations on biological systems. Nature 256:302–304
Clowes FAL (1961a) Apical meristems. Blackwell, Oxford, 217 pp
Clowes FAL (1961b) Duration of the mitotic cycle in a meristem. J Exp Bot 12:283–293
Clowes FAL (1976) The root apex. In: Cell division in higher plants. Academic, London, pp 96–118
Clowes FAL, Juniper BE (1968) Plant cells. Blackwell, Oxford, 546 pp
Cohen J, Entekhabi D (1999) Eurasian snow cover variability and northern hemisphere climate predictability. Geophys Res Lett 26:345–348
Conkey LE (1986) Red spruce tree-ring width and densities in eastern North America as indicator of past climate. Quat Res (NY) 26:232–243
Conner WS, Schowengerdt RA, Munro M, Hughes MK (2000) Engineering design of an image acquisition and analysis system for dendrochronology. Opt Eng 39:453–463

Connor KF, Lanner RM (1990) Effects of tree age on secondary xylem and phloem anatomy in stems of Great Basin bristlecone pine *Pinus longaeva*. Am J Bot 77:1070–1077

Contributors of the International Tree-Ring Data Bank (2002) International Tree-Ring Data Bank. (IGBP PAGES/World data center for paleoclimatology.) NOAA/ NGDC Paleoclimatology Program, Boulder, Colo. (http://www.ngdc.noaa.gov/paleo/ treering.html)

Cook ER (1985) A time series analysis approach to tree-ring standardization. PhD thesis, University of Arizona, Tucson, 171 pp

Cook ER, Kairiukstis LA (eds) (1990) Methods of dendrochronology. Application in the Environmental Sciences. Kluwer, Dordrecht, 394 pp

Cook ER, Briffa K, Shiyatov S, Mazepa V (1990) Tree-ring standardization and growth-trend estimation. In: Cook E, Kairiukstis L (eds) Methods of dendrochronology. Application in the environmental sciences. Kluwer, Dordrecht, pp 104–123

Cook ER, Briffa KR, Jones PD (1994) Spatial regression methods in dendroclimatology: a review and comparison of two techniques. Int J Climatol 14:379–402

Cook ER, Meko DM, Stahle DW, Cleaveland MK (1999) Drought reconstructions for the continental United States. J Climate 12:1145–1162

Cowan IR (1982) Regulation of water use in relation to carbon gain in higher plants. In: Lange OL, Nobel PS, Osmond CB, Ziegler H (eds) Encyclopedia of plant physiology (NS), vol 12b. Springer, Berlin Heidelberg New York, pp 589–613

Creber GT, Chaloner WG (1984) Influence of environmental factors on the wood structure of living and fossil trees. Bot Rev 1984:357–448

Cregg BM (1993) Seed-source variation in water relations, gas-exchange, and needle morphology of mature ponderosa pine trees. Can J For Res 23:749–755

Cullen LE, Stewart GH, Duncan RP, Palmer JG (2001) Disturbance and climate warming influences on New Zealand *Nothofagus* tree-line population dynamics. J Ecol 89:1061–1071

Dahl-Jensen D, Mosegaard K, Gundestrup N, Clow GD, Johnsen SJ, Hansen AW, Balling N (1998) Past temperatures directly from the Greenland ice sheet. Science 282:268–271

D'Arrigo RD, Jacoby GC, Free RM (1992) Tree-ring width and maximum latewood density at the North American tree line: parameters of climatic change. Can J For Res 22:1290–1296

D'Arrigo RD, Kaufmann RK, Davi N, Jacoby GC, Laskowski C, Myneni RB, Cherubini P (2004) Thresholds for warming-induced growth decline at elevational tree line in the Yukon Territory, Canada. Glob Biogeochem Cycles 18:GB3021 (doi:10.1029/ 2004G02249)

De Wit CT, Brouwer R, Penning de Vries FWT (1970) The simulation of photosynthetic systems/ In: Setlik I (ed) Prediction and measurement of photosynthetic productivity. Wageningen, Pudoc, pp 47–70

Deleuze C, Houller F (1995) Prediction of stems profile of *Picea abies* using a process-based tree growth model. Tree Physiol 15:113–120

Deleuze C, Houllier F (1998) A simple process-based xylem growth model for describing wood densitometric profiles. J Theor Biol 193:99–113

Demura T, Fukuda H (1994) Novel vascular cell-specific genes whose expression is regulated temporally and spatially during vascular system development. Plant Cell 6:967–981

Denne MP (1966) Morphological changes in the shoot apex of *Trifolium repens* L.1. Changes in the vegetative apex during the plastochron. NZ J Bot 4:300–314

Denne MP (1971a) Temperature and tracheid development in *Pinus sylvestris* seedlings. J Exp Bot 22:362–370

Denne MP (1971b) Tracheid length in relation to seedling height in conifers. Wood Sci Technol 5:135–146

Denne MP (1977) Some effects of wounding on tracheid differentiation in *Picea sitchensis*. IAWA Bull 3:49–50

Denne MP (1979) Effects of environmental change on wood production and wood structure in *Picea sitchensis* seedlings. Ann Bot 40:1017–1028

Denne MP (1988) Definition of latewood according to Mork (1928). IAWA Bull 10:59–62

Denne MP, Dodd RS (1981) The environmental control of xylem differentiation. In: Barnett JR (ed) Xylem cell development. Castle House, Tunbridge Wells, pp 237–255

Deslauriers A, Morin H, Begin Y (2003) Cellular phenology of annual ring formation of *Abies balsamea* in the Quebec boreal forest (Canada). Can J For Res 33:190–200

Djanseitov YK, Park W-K, Vaganov EA (2000) A source of high-frequency variations of tracheid size in annual tree rings of conifers. Izv Akad Nauk Ser Biol 27:41–49

Dodd RS, Fox P (1990) Kinetics of Tracheid differentiation in Douglas-fir. Ann Bot 65:649–657

Dodds JD (1948) On the shapes of cells in the cambial zone of *Pinus sylvestris L.* Am J Bot 35:666–682

Dodds JH (1981) Relationship of the cell cycle to xylem differentiation: a new model. Plant Cell Environ 4:145–146

D'Odorico P; Revelli R, Ridolfi L (2000) On the use of neural networks for dendroclimatic reconstructions. Geophys Res Lett 27:791–794

Doley D (1967) Water relations of *Eucalyptus marginata Sm.* under natural conditions. J Ecol 55:597–614

Doley D (1981) Tropical and subtropical forests and woodlands. In: Kozlowski TT (ed) Water deficits and plant growth, vol 6. Academic, New York, pp 209–323

Donaldson JR, Tryon PV (1987) User's guide to STARPAC, the standards time series and regression package. NIST, Boulder, Colo.

Dormer KJ (1972) Shoot organization in vascular plants. Syracuse University, New York, 240 pp

Douville H, Royer JF (1996) Sensitivity of the Asian summer monsoon to an anomalous Eurasian snow cover within the Meteo-France GCM. Climate Dyn 12:449–466

Dowdeswell JA, Hagen JO, Bjornsson H, Glazovsky AF, Harrison WD, Holmlund P, Jania J, Koerner RM, Lefauconnier B, Ommanney CSL, Thomas RH (1997) The mass balance of circum-Arctic glaciers and recent climate change. Quat Res (NY) 48:1–14

Downes GM, Turvey ND (1990) The effect of nitrogen and copper on the characteristics of woody tissue in seedling of *Pinus radiata*. Can J For Res 20:369–1377

Duff GH, Nolan NJ (1953) Growth and morphogenesis in the Canadian forest species I. The controls of cambial and apical activity in *Pinus resinosa Ait.* Can J Bot 31:471–513

Dunin-Barkovsky IV, Smirnov NV (1955) Probability theory and mathematical statistics in engineering (in Russian). Nauka, Moscow, 634 pp

Eckstein D, Frisse E, Liese W (1974) Holzanatomische Untersuchungen an umweltgeschädigten Strassenbäumen der Hamburger Innenstadt. Eur J For Pathol 4:232–244

Efremov SP (1987) Pioneer stands of draining swamps (in Russian). Nauka, Novosibirsk, 249 pp

Ellsworth DS, Reich PB (1992) Water relations and gas exchange of *Acer saccharum* seedlings in contrasting natural light and water regimes. Tree Physiol 10:1–20

Erickson RO, Sax KB (1956) Rates of cell division and cell elongation in the growth of primary root of *Zea mais*. Proc Am Phil Soc 100:499–514

Esper J, Cook ER, Schweingruber FH (2002) Low-frequency signals in long tree-ring chronologies for reconstructing past temperature variability. Science 295:2250–2253

Ettl GJ, Peterson DL (1995) Extreme climate and variation in tree growth- individualistic response in sub-alpine fir (*Abies lasiocarpa*). Glob Change Biol 1:231–241
Evans LS (2000) Diversity of cell length in terminal portion of roots: implications to cell proliferation. Environ Exp Bot 43:239–251
Evans LS, Lagrazon K, Pancrudo J (2001) Diversity of cell lengths in terminal portions of roots: location of the proliferative cell population. Environ Exp Bot 45:85–94
Evans MN, Reichert BK, Kaplan A, Vaganov EA, Hughes MK, Cane MA (2005) A forward modeling approach to paleoclimatic interpretation of tree-ring data (submitted)
Evans R (1994) Rapid measurement of the transferse dimensions of tracheids in radial wood sections from *Pinus radiata*. Holzforschung 48:168–172
Evans R, Downes G, Menz D, Stringer S (1995) Rapid measurement of variation in tracheid transverse dimensions in a radiata pine tree. Appita 48:134–138
Evert RF, Deshande BP (1970) An ultrastructural study of cell devision in the cambium. Am J Bot 57:942–961
Fahn A (1982) Plant anatomy, 3rd edn. Pergamon, Oxford, 544 pp
Fahn A (1990) Plant anatomy, 4th edn. Pergamon, Oxford, 611 pp
Fahnestock JT, Jones MH, Welker JM (1999) Wintertime CO_2 efflux from arctic soils: Implications for annual carbon budget. Glob Biogeochem Cycles 13:775–779
Farquhar GD, Sharkey TD (1982) Stomatal conductance and photosynthesis. Annu Rev Plant Physiol 33:317–345
Farquhar GD, Caemmerer S von (1982) Modelling of photosynthetic response to environmental conditions. In: Lange OL, Nobel PS, Osmond CB, Ziegler H (eds) Encyclopedia of plant physiology (NS), vol 12b. Springer, Berlin Heidelberg New York, pp 550–588
Farquhar GD, Caemmerer S von, Berry JA (1980) A biochemical model of photosynthetic CO_2 assimilation in lives of C_3 species. Planta 149:78–90
Farquhar GD, O'Leary MH, Berry JA (1982) On the relationship between carbon isotope discrimination and the intercellular carbon dioxide concentration in leaves. Aust J Plant Physiol 9:121–137
Farquhar GD, Ehleringer JR, Hubick KT (1989) Carbon isotope discrimination and photosynthesis. Annu Rev Physiol Plant Mol Biol 40:503–537
Fedorov VD, Gilmanov TG (1989) Ecology (in Russian). Moscow University, Moscow, 514 pp
Filion L, Cournoyer L (1995) Variation in wood structure of eastern larch sawfly in subarctic. Can J For Res 25:1263–1268
Filion L, Payette S (1987) Spatial distribution of light-ring years in subarctic Quebec. In: Jacoby GC (ed) On ecological aspects of tree-ring analysis. (Proceedings of the international symposium.) US Department of Energy, Washington, D.C., p. 79
Filion L, Payette S, Gauthier L, Boutin Y (1986) Light rings in subarctic conifers as a dendrochronological tool. Quat Res (NY) 26:272–279
Filroze EM, Gladushko GM (1986) The method of contrasting the boundary between tree rings. In: Galazii GI (ed) Dendrochronology and dendroclimatology (in Russian). Nauka, Novosibirsk, pp 68–71
Fink S (1986) Microscopical investigations of wood formation and function in diseased trees. IAWA Bull 7:351–355
Fischer B (1993) Jahrring und Luftschadstoffe; Einfluss von Schwefeldioxidbegasungen auf Schwarzföhren und Weisstannen. PhD thesis, Geographisches Institut, Zürich, 163 pp
Foley JA, Kutzbach JE, Coe MT, Levis S (1994) Feedbacks between climate and boreal forest during the Holocene epoch. Nature 371:52–54
Ford ED (1981) Can we model xylem production by conifers? Stud For Suec 160:19–29
Ford ED, Robards AW (1976) Short term variation in tracheid development in the early

wood of *Picea sitchensis*. In: Baas P, Bolton AJ, Catling DM (eds) Wood structure in biological and technological research. Leiden University, Leiden, pp 212–221
Ford ED, Robards AW, Piney MD (1978) Influence of environmental factors on cell production and differentiation in the earlywood of *Picea sitchensis*. Ann Bot 42:683–692
Francis D, Barlow PB (1988) Temperature and the cell cycle. In: Long SP, Woodward FL (eds) Plants and temperature. Comp Biologists, Cambridge, pp 181–201
Friend A, Kellomaki S, Kruijt B (1998) Modelling leaf, tree and forest responses to increasing atmospheric CO_2 and temperature. European forests and global change. In: Jarvis PG (ed) The likely impacts of rising CO_2 and temperature. Cambridge University, Cambridge, pp 293–334
Friend AD (1995) EN: an integrated model of leaf photosynthesis, transpiration, and conductance. Ecol Model 77:233–255
Friis-Christensen E, Lassen K (1991) Length of the solar cycle: an indicator of solar activity closely associated with climate. Science 254:698–700
Fritts HC (1956) Radial growth of beech and soil moisture in a Central Chio forest during the growing season of 1952. Ohio J Sci 56:17–28
Fritts HC (1962). An approach to dendroclimatology: screening by means of multiple regression techniques. J Geophys Res 67:1413–1420
Fritts HC (1974) Relationships of ring widths in arid-site conifers to variations in monthly temperature and precipitation. Ecol Monogr 44:411–440
Fritts HC (1976) Tree-rings and climate. Academic, London, 576 pp
Fritts HC (1990) Modeling tree-ring and environmental relationship for dendrochronological analysis. In: Dixon RK, Meldahl RS, Ruark GA, Warren WG (eds) Forest growth: process modeling of responses to environmental stress. Timber, Oregon, pp 360–382
Fritts HC (1991) Reconstructing large scale climatic patterns from tree ring data. University of Arizona, Tucson, 286 pp
Fritts HC, Fritts EC (1955) A new dendrograph for recording radial changes of a tree. For Sci 1:271–276
Fritts HC, Shashkin AV (1995) Modeling tree-ring structure as related to temperature, precipitation, and day length. In: Lewis TE (ed) Tree rings as indicators of ecosystem health. CRC, New York, pp 17–58
Fritts HC, Swetnam TW (1989) Dendroecology: a tool for evaluating variations in past and present forest environments. Ecol Res 19:111–189
Fritts HC, Smith DG, Cardis JW, Budelsky CA (1965) Tree-ring characteristics along a vegetation gradient in northern Arizona. Ecology 46:393–401
Fritts HC, Blasing TJ, Hayden BP, Kutzbach JE (1971) Multivariate techniques for specifying tree-growth and climate relationships and for reconstructing anomalies in paleoclimate. J Appl Meteorol 10:845–864
Fritts HC, Lofgren GR, Gordon GA (1980) Past climate recontructed from tree-rings. J Interdiscip Hist 10:773–793
Fritts HC, Vaganov EA, Sviderskaya IV, Shashkin AV (1991) Climatic variation and tree-ring structure in conifers: empirical and mechanistic models of tree-ring width, number of cell, cell size, cell-wall thickness and wood density. Climate Res 1:97–116
Fritts H, Shashkin AV, Downes GM (1999) Simulation model of conifer ring growth and cell structure. In: Wimmer R, Vetter RE (eds) Tree ring analisis: biological, methodological and environmental aspects. CABI, New York, pp 2–25
Fukuda H (1994) Redifferentiation of single mesophyll cells into tracheary elements. Int J Plant Sci 155:262–271
Fukuda H (1996) Xylogenesis: initiation, progression, and cell death. Annu Rev Plant Physiol Plant Mol Biol 47:299–325

Gamaley YuV (1972) Cytological basics of xylem differentiation (in Russian). Nauka, Leningrad, 208 pp
Gandar PW (1983) Growth in root apices. I. The kinematic description of growth. Int J Plant Sci 144:1–10
Gash JHC (1979) An analytical model of interception by forests. Q J R Meteorol Soc 105:43–55
Gates DM (1980) Biophysical ecology. Springer, Berlin Heidelberg New York, 611 pp
Gholz HL (1980) Structure and productivity of *Juniperus occidentalis* in central Oregon. Am Midl Nat 103:251–261
Gifford RM, Evans LT (1981) Photosynthesis, carbon partitioning, and yield. Annu Rev Plant Physiol 32:485–509
Gindl W (1999) Climatic significance of light rings in timberline spruce, *Picea abies*, Austrian Alps. Arct Antarct Alp Res 31:242–246
Glerum C, Farrar JL (1966) Frost rings in the stems of some coniferous species. Can J Bot 44:879–886
Goggans JF (1962) The corellation, variation and inheritance of wood properties in loblolly pine. (School of Forestry Research Technical Report 14) North Carolina State University, Raleigh, 155 pp
Gorchakovsky PL, Shiyatov SG (1985) Phytoindication of environment and natural processes in Alpine mountains (in Russian). Nauka, Moscow, 208 pp
Gordon JC, Larson PR (1968) Seasonal course of photosynthesis, respiration, and distribution of ^{14}C in young *Pinus resinosa* trees as related to wood formation. Plant Physiol 43:1617–1628
Gottinger HW (1998) Global environmental economics. Kluwer, Boston, 326 pp
Goujon T, Minic Z, Amrani AE, Lerouxel O, Aletti E, Lapierre C, Joseleau J-P, Jouanin L (2003) *AtBXL1*, a novel higher plant (*Arabidopsis thaliana*) putative beta-xylosidase gene, is involved in secondary cell wall metabolism and plant development. Plant J 33:677–690
Gower ST, Grier CC, Vogt DJ, Vogt KA (1987) Allometric relations of deciduous (*Larix occidentalis*) and evergreen (*Pinus contorta* and *Pseudotsuga menziesii*) of the Cascade mountains in central Washington. Can J For Res 17:630–634
Gower ST, Vogt KA, Grier CC (1992) Carbon dynamics of Rocky Mountain Douglas-fir: influence of water and nutrient availability. Ecol Monogr 62:43–65
Gower ST, Isebrands JG, Sheriff DW (1995) Carbon allocation and accumulation in conifers. In: Smith WK, Hinckley TM (eds) Resource physiology of conifers: acquisition, allocation, and utilization. Academic, London, pp 217–254
Graham IA (1996) Carbohydrate control of gene expression in higher plants. Res Microbiol 147:572–580
Graumlich LJ (1993) A 1000-year record of temperature and precipitation in the Sierra Nevada. Quat Res (NY) 39:249–255
Graybill DA, Idso SB (1993) Detecting the aerial fertilization effects of atmospheric CO_2 enrichment in tree-ring chronologies. Glob Biogeochem Cycles 7:81–95
Gregory RA (1971) Cambial activity in Alaskan white spruce. Am J Bot 58:160–171
Gregory RA, Wilson BF (1968) A comparison of cambial activity of white spruce in Alaska and New England. Can J Bot 46:733–734
Grier CC, Waring RH (1974) Conifer foliage mass related to sapwood area. For Sci 20:205–206
Grif VG (1981) Mitotic cycle and functional morphology of plant chromosomes under low temperatures (in Russian, abstract). PhD thesis, Botanical Institute RAS, Leningrad, 41 pp
Grif VG, Ivanov VB (1975) The time parameters of the mitotic cycle in flowering (in Russian). Tsitoilogia 17:694–717

Grif VG, Ivanov VB (1980) Data on the time parameters of the mitotic cycle in flowering (in Russian). Tsitologia 22:107–120
Grif VG, Ivanov VB (1995) Parameters of the mitotic cycle in flowering (in Russian). Tsitologia 37:723–743
Grillos SJ, Smith FH (1959) The secondary phloem of Douglas-fir. For Sci 5:377–388
Groisman PY, Easterling DR (1993) Precipitation changes over the northern hemisphere extratropics during the last hundred years. In: Desbois M, Desalmand F (eds) Global precipitation and climate changes. (NATO ASI series, vol 26) Springer, Berlin Heidelberg New York, pp 119–134
Groisman PY, Karl TR, Knight RW (1994) Observed impact of snow cover on the heat balance and the rise of continental spring temperatures. Science 263:198–200
Guiot J, Tessier L, Serre-Bachet F (1982) Application de la modelisation ARMA en dendroclimatologie. C R Seances Acad Sci 294:133–136
Guiot J, Keller T, Tessier L (1995) Relational databases in dendroclimatology and new non-linear methods to analyse the tree response to climate and pollution. In: Ohta S, Fujii T, Okada N, Hughes MK, Eckstein D (eds) Tree rings: from the past to the future. (Proceedings of the international workshop on Asian and Pacific dendrochronology, Tsukuba City.) Forestry and Forest Products Research Institute, Tsukuba, pp 17–23
Guyette RP, Rabeni CF (1995) Climate response among growth increments of fish and trees. Oecologia 104:272–279
Haigh JD (1996) The impact of solar variability on climate. Science 272:981–984
Haigler CH (1994) Commentary from signal transduction to biophysics: tracheary element differentiation as a model system. Int J Plant Sci 155:248–250
Hallet J-N (1976) Les cycle mitotiques dans le point vegetatif de l'*Hookeria lucens* (Hedw.) SM (Mousse pleurocarpe, Hookeriale). J Bryol 9:97–103
Hanninen H (1990) Modeling dormancy release in trees from cool and temperate regions. In: Dixon RK, Meldahl RS, Ruark GA, Warren WG (eds) Process modeling of forest growth responses to environmental stress. Timber, Portland, Ore., pp 159–165
Hanninen H (1991) Does climate warming increase risk of frost damage in northern trees? Plant Cell Environ 14:449–454
Harte J (2002) Toward a synthesis of the Newtonian and Darwinian worldviews. Phys Today 55:29–34
Hejnowicz Z (1959) Growth and cell division in the apical meristem of wheat roots. Phys Plant 12:124–138
Hemming D, Fritts H, Leavitt SW, Wright W, Long A, Shashkin A (2001) Modelling tree-ring $\delta^{13}C$. Dendrochronologia 19:23–38
Henhappl G (1965) Über die Stärkenänderungen der peripheren Stammzone von Waldbäumen im Jahresablauf. PhD thesis, University of Freiburg, Freiburg, 132 pp
Hertzberg M, Aspeborg H, Schrader J, Andersson A, Erlandsson R, Blomqvist K, Bhalerao R, Uhle M, Teeri TT, Lundberg J, et al (2001) A transcriptional roadmap to wood formation. Proc Natl Acad Sci USA 98:14732–14737
Hesketh JD, Jones JW (1975) Some comments on computer simulators for plant growth. Ecol Model 2:235–247
Hinckley TM, Bruckerhoff DN (1975) Effects of drought on water relations and stem shrinkage of *Quercus alba*. Can J Bot 53:62–72
Holmes RL (1983) Computer-assisted quality control in tree-ring dating and measurement. Tree-Ring Bull 44:69–75
Holmlund P (1997) Climatic influence on the size of glaciers in Northern Scandinavia during the last two centuries. In: Frenzel B, Boulton GS, Glaser B, Huckriede U (eds) Glacier fluctuations during the Holocene. Paleoclim Res 24:115–124
Holtzer H (1970) Proliferative and quantal mitosis in differentiation. In: Padykula HA (ed) Control mechanisms in the expression of cellular phenotypes. Academic, New York, pp 69–88

Horacek P, Slezingerova J, Gandelova L (1999) Effects of environment on xylogenesis of Norway spruce (*Picea abies* L. Karst.). In: Wimmer R, Vetter RE (eds) Tree-ring analysis. Biological, methodological and environmental aspects. CABI, Wallingford, pp 33–54

Howard RA, Wilson BF (1972) A stochastic model for cambial activity. Bot Gaz 133:410–414

Howe GT, Hackett WP, Furnier GR, Klevorn RE (1995) Photoperiodic responses of a northern and southern ecotypes of black cottonwood. Physiol Plant 93:695–708

Hsio TS (1973) Plant response to water stress. Annu Rev Plant Physiol 24:519–570

Hughes MK (1992) Dendroclimatic evidence from the Western Himalaya. In: Bradley RS, Jones PD (eds) Climate since AD 1500. Routledge, London, pp 415–431

Hughes MK (2001) An improved reconstruction of summer temperature at Srinagar, Kashmir since 1660 AD, based on tree-ring width and maximum latewood density of *Abies pindrow* [Royle] Spach. Palaeobotanist 50:13–19

Hughes MK (2002) Dendrochronology in climatology – the state of the art. Dendrochronologia 20:95–116

Hughes MK, Diaz HF (1994a) The medieval warm period. Clim Change 26:342

Hughes MK, Diaz HF (1994b) Was there a medieval warm period, and if so, where and when? Clim Change 26:109–142

Hughes MK, Funkhouser G (1998) Extremes of moisture availability reconstructed from tree rings for recent millennia in the Great Basin of western North America. In: Innes M, Beniston JL (eds) The impacts of climate variability on forests. Springer, Berlin Heidelberg New York, pp 99–107

Hughes MK, Funkhouser G (2003) Frequency-dependent climate signal in upper and lower forest border trees in the mountains of the Great Basin. Clim Change 59:233–244

Hughes MK, Graumlich LJ (1996) Multimillennial dendoclimatic records from western North America. In: Jones PD, Jouzel J, Bradley RS (eds) Climatic variations and forcing mechanisms of the last 2000 years. Springer, Berlin Heidelberg New York, pp 109–124

Hughes MK, Kelly PM, Pilcher JR, La Marche VC (eds) (1982) Climate from tree rings. Cambridge University, Cambridge, 223 pp

Hughes MK, Vaganov EA, Shiyatov S, Touchan R, Funkhouser G (1999) Twentieth-century summer warmth in northern Yakutia in a 600-year context. Holocene 9:603–608

Hungerford RD, Nemani RR, Running SW, Coughlan JC (1989) MTCLIM: a mountain microclimate simulation model. (Research Paper Int-414, Intermountain Research Station, Ogden, Utah) US Department of Agriculture Forest Service, Washington, D.C., 52 pp

Hunt FR, Martin FC, Running SW (1991) Simulating the effect of climate variation on stem carbon accumulation of ponderosa pine stand: comparison with annual growth increment data. Tree Physiol 9:161–171

Imagawa H, Ishida S (1970) Study on the wood formation in trees, report 1. Seasonal development of the xylem ring of Japanese larch stem, *Larix leptolepis* Gordon. Res Bull Coll Exp For Hokkaido Univ 27:373–394

Imagawa H, Ishida S (1972) Study on the wood formation in trees, report III. Occurence of the overwintering cells in cambial zone in several ring-porous trees. Res Bull Coll Exp For Hokkaido Univ 29:207–221

Imagawa H, Ishida S (1981a) An observation of the cambial cells in *Larix leptolepis* by semueltrathin sections. Res Bull Coll Exp For Hokkaido Univ 38:45–54

Imagawa H, Ishida S (1981b) New marking method by electrical stimulation for studying xylem formation. Res Bull Coll Exp For Hokkaido Univ, 38:241–248

Innes JL (1990) General aspects in the use of tree rings for environmental studies. In: Cook ER, Kairiukstis LA (eds) Methods of dendrochronology. Application in the environmental sciences. Kluwer, Dordrecht, pp 224–229

Innes JL (1993) Forest health: its assesssment and status. CABI, Wallingford, 677 pp

Innes JL, Cook ER (1989) Tree-ring analysis as an aid to evaluating the effects of pollution on tree growth. Can J For Res 19:1174–1189

IPCC (2001) Climate change 2001: the scientific basis. Cambridge University, Cambridge, 944 pp

Iqbal M (ed) (1990) The vascular cambium. Wiley, New York, 246 pp

Iqbal M, Ghouse AKM (1985a) Cell events of radial growth with special reference to cambium of tropical trees. In: Malik CP (ed) Widening horizons of plant sciences. Cosmo, New Delhi, pp 218–252

Iqbal M, Ghous AKM (1985b) Impact of climatic variation on the structure and activity of vascular cambium in *Prosopis spicigera*. Flora 177:147–156

Ishikawa H, Evans ML (1995) Specialized zones of development in roots. Plant Physiol 109:725–727

Ito J, Fukuda H (2002) ZEN1 is a key enzyme in the degradation of nuclear DNA during programmed cell death of tracheary elements. Plant Cell 14:3201–3211

Itoh T (1971) On the ultrastructure of dormant and active cambium of conifers. Wood Res 51:33–45

Itoh T, Hayashi S, Kishima T (1968) Cambial activity and radial growth in Sugi trees (*Japanese Cryptomeria*). Wood Res Kyoto 45:23–35

Ivanov VB (1974) The cellular basics of plant growth (in Russian). Nauka, Moscow, 224 pp

Ivanov VB (1987) Cell proliferation in plants (in Russian). (The results of science and technology) Ser Cytol 5:3–217

Ivanov VB, Dubrovsky JG (1997) Estimation of the cell-cycle duration in the root apical meristem: a model of linkage between cell-cycle duration, rate of cell production and rate of root growth. Int J Plant Sci 158:757–763

Ivanov VB, Dobrochaev AE, Baskin TI (2002) What the distribution of cell lengths in the root meristem does and does not reveal about cell division. J Plant Growth Regul 21:60–67

Ivshin AP, Shiyatov SG (1995) The assessment of subtundra forests degradation by dendrochronological methods in the Norilsk industrial area. Dendrochronologia 13:113–126

Jacoby GC, d'Arrigo R (1995) Tree ring width and density evidence of climatic and potential forest change in Alaska. Global Biogeochem Cycles 9:227–234

Jagel R, Telewski FW (1990) Image analysis. In: Kairiukstis L, Cook E (eds) Method of dendrochronology: applications in the environmetal sciences. Kluwer, Dordrecht, pp 76–93

Jardon Y, Filion L, Cloutier C (1994) Long term impact of insect defoliation on growth and mortality of eastern larch in subarctic Quebec. EcoScience 1:231–238

Jardon Y, Filion L, Cloutier C (1994) Tree-ring evidence for endemicity of the larch sawfly in Northern America. Can J For Res 24:742–747

Jarvis PG (1985) Transpiration and assimilation of tree and agricultural crops: the omega factor. In: Cornell MGR, Jackson JE (eds) Trees as crop plants. Institute of Terrestrial Ecology, Huntingdon, pp 461–480

Jarvis PG, Stewart JB (1979) Evaporation of water from plantation forest. In: Ford ED, Malcolm DC, Atterson J (eds) The ecology of even-aged forest plantations. Institute of Terrestrial Ecology/NERC, Cambridge, pp 327–350

Jenkins PA (1975) Seasonal trends in translocation of ^{14}C photosynthates and their association with wood formation in Radiata pine seedlings. NZ J For Sci 5:62–73

Johnson IR, Melkonian JJ, Thornley JHM, Rina SJ (1991) A model of water flow through plants incorporating shoot/root 'message' control of stomatal conductance. Plant Cell Environ 14:531–544

Jones HG (1983) Plants and microclimate. Cambridge University, Cambridge, 323 pp

Jones PD, Briffa KR (1995) Growing season temperature over the former Soviet Union. Int J Climatol 15:943–959

Jones PD, Briffa KR, Barnett TP, Tett SFB (1998) High-resolution palaeoclimatic records for the last millennium: interpretation, integration and comparison with general circulation model control-run temperatures. Holocene 8:455–471

Kandelaki AA (1979) The formation of wood in larch stems in Taymir (in Russian). Lesovedenie 6:64–69

Kandelaki AA, Dem'yanov VA (1982) The dynamics of tree ring formation in trees of Putoran Mountains (in Russian). Lesovedenie 5:46–49

Keeling CD, Whorf TP, Wahlen M, Plicht J van der (1995) Interannual extremes in the rate of rise of atmopheric carbon dioxide since 1980. Nature 375:666–670

Kelly PM, Wigley TML (1992) Solar cycle length, greenhouse forcing and global climate. Nature 360:328–330

Kennedy RW (1966) Intra-increment variation and heritability of specific gravity, parallel to grain tensile strength, stiffness and tracheid lendth in clonal Norway spruce. TAPPI 49:292–295

Kennedy RW (1969) Wood formation in jack pine seedlings of various provenances. (11th Commission for Tree Breeding, Canada, part 2) Commission for Tree Breeding, Ottawa, pp 101–109

Kennedy RW (1970) Specific gravity of early- and late-flushing Douglas-fir trees. TAPPI 53:1479–1481

Kennedy RW, Farrar JL (1965) Tracheid development in tilted seedlings. In: Côté WA (ed) Cellular ultrastructure of woody plants. Syracuse University, New York, pp 419–453

Kharuk V, Nilsson S, Samarskaia E (1996a) Antropogenic and technogenic stress factors in Siberia. (WP-96-104) International Institute for Applied System Analysis, Laxenberg, 114 pp

Kharuk VI, Winterberger K, Tsibul'skii GM, Yakhimovich AP, Moroz SN (1996b) Technogenic disturbance of pretundra forests in Noril'sk Valley. Russ J Ecol 27:406–410

Kienholz R (1934) Leader, needle, cambial and root growth of certain conifers and their interrelations. Bot Gaz 96:73–92

Kimmings JP (1997) Forest ecology, 6th edn. Prentice–Hall, Upper Saddle River, N.J., 596 pp

Kipfmueller KF (2003) Fire–climate–vegetation interactions in subalpine forests of the Selway-Bitterroot wilderness area, Idaho and Montana, USA. PhD thesis, University of Arizona, Tucson

Kirdyanov AV (1999) The usage of wood density characteristics in dendroclimatic research (in Russian). Sib J Ecol 6:193–202

Kirdyanov AV, Zarharjewski DV (1996) Dendroclimatological study on tree-ring width and maximum density chronologies from *Picea obovata* and *Larix Sibirica* from the north of Krasnoyarsk region (Russia). Dendrochronologia 14:227–236

Kirdyanov A, Hughes H, Vaganov E, Schweingruber F, Silkin (2003) The importance of early summer temerature and date of snow melt for tree growth in Siberian subarctic. Trees 17:61–69

Kirst M, Johnson AF, Bancom C, Ulrich E, Hubbard K, Straggs R, Paule C, Retzel E, Whetten R, Sederoff R (2003) Apparent homology of expressed genes from wood forming tissues of loblolly pine (*Pinus taeda* L.) with *Arabidopsis thaliana*. Proc Natl Acad Sci USA 100:7383–7388

Klein SA (1977) Calculation of monthly average insolation on tilted surfaces. Sol Energy 19:325–329
Kofman GB (1986) The growth and form of trees (in Russian). Nauka, Novosibirsk, 210 pp
Konstantinov VK (1981) The exploitation and improvement of drainage systems in forestry (in Russian, abstract). PhD thesis, Forestry Academy, Leningrad, 40 pp
Kostin SI, Pokrovskaya TV (1961) Climatology (in Russian). Gidrometeoizdat, Leningrad, 485 pp
Kozlowski TT (ed) (1968) Water deficits and plant growth, vol 1. Academic, New York, 398 pp
Kozlowski TT, Constantinidou HA (1986) Responses of woody plants to environmental pollution. Part I. Sources, types of pollutants, and plant responses. For Abstr 47:5–51
Kozlowski TT, Pallardy SG (1997) Growth control in woody plants. Academic, San Diego, 641 pp
Kozlowski TT, Winget CH (1964) Diurnal and seasonal change in radii of tree stems. Ecology 45:149–154
Kozlowski TT, Kramer PJ, Pallardy SG (1991) The physiological ecology of woody plants. Academic, San Diego, 657 pp
Kozubov GM, Taskaev AI (1994) Radiobiological and radioecological studies of woody plants (in Russian). Nauka, St Peterburg, 256 pp
Kramer PJ, Boyer JS (1995) Water relations of plants and soils. Academic, San Diego, 495 pp
Kramer PD, Kozlowski TT (1983) The physiology of trees (in Russian, translation from English). Lesnaya Promishlennost, Moscow, 464 pp
Kramer PJ, Kozlowski TT (1979) Physiology of woody plants. Academic, New York, 811 pp
Krause C, Eckstein D (1992) Holzzuwachs an Ästen, Stamm und Wurzeln bei normaler und extremer Witterung. In: Michaelis W, Bauch J (eds) Luftverunreinigungen und Waldschäden am Standort "Postturm", pp 215–242
Kuroda K (1986) Wound effects on cytodifferentiation in secondary xylem of woody plants. Wood Res 72:67–118
Kuroda K, Kiyono Y (1997) Seasonal rhythms of xylem growth measured by the wounding method and with a band-dendrometer: an instance of *Chamaecyparis obtuse*. IAWA J 18:291–299
Kutscha NP, Hyland F, Schwarzmann JM (1975) Certain seasonal changes in balsam fir cambium and its derivatives. Wood Sci Technol 9:175–188
Kuzmichev VV (1977) Patterns of stand growth (in Russian). Nauka, Novosibirsk, 157 pp
Kuzmin PP (1961) The snow melting process (in Russian). Gidrometeoizdat, Leningrad, 367 pp
Ladefoged K (1952) The periodicity of wood formation. Det Kongel Danske Vidensk Selsk Skrift Dan Biol 7 :1–98
LaMarche VC (1974a) Frequency-dependent relationships between tree-ring series along an ecological gradient and some dendroclimatic implications. Tree-Ring Bull 34:1–20
LaMarche VC (1974b) Paleoclimatic inferences from long tree-ring records. Science 183:1043–1088
LaMarche VC (1982) Sampling strategies. In: Hughes MK, et al (eds) Climate from tree rings.: Cambridge University, Cambridge, pp 2–6
Landsberg JJ (1974) Apple fruit bud development and growth: analysis and an empirical model. Ann Bot 38:1013–1023
Landsberg JJ (1986) Physiological ecology of forest production. Academic, London, 192 pp

Larcher W (1980) Physiological plant ecology, 2nd edn. Springer, Berlin Heidelberg New York, 271 pp

Larson PR (1962) The indirect effect of photoperiod on tracheid diameter in *Pinus resinosa*. Am J Bot 49:132–137

Larson PR (1963) The indirect effect of drought on tracheid diameter in red pine. For Sci 9:52–62

Larson PR (1964) Some indirect effects of environment on wood formation. In: Zimmermann M (ed) The formation of wood in forest trees. Academic, New York, pp 345–366

Larson PR (1969) Wood formation and the concept of wood quality. (School of Forestry and Environmental Studies, Bulletin 74) Yale University, New Haven, pp 3–17

Larson PR (1994) The vascular cambium. Development and structure. Springer, Berlin Heidelberg New York, 725 pp

Lavoie C, Payette S (1997) Late-Holocene light-ring chronologies from subfossil black spruce stems in mires of subarctic Quebec. Holocene 7:129–137

Lean J, Beer J, Bradley R (1995) Reconstruction of solar irradiance since 1610: implications for climate change. Geophys Res Lett 22:3195–3198

Lebedenko LA (1969) The cambial activity of larch in relation to tree growth energy (in Russian). (Research paper on forestry N3) Leningrad Forestry Science and Technological Institute, Leningrad, pp 149–167

LeBlanc D, Terrell M (2001) Dendroclimatic analyses using Thornthwaite–Mather-type evapotranspiration models: a bridge between dendroecology and forest simulation models. Tree-Ring Res 57:55–66

LeBreton GTO, Beamish FWH (2000) Interannual growth variation in fish and tree rings. Can J Fish Aquat Sci 57:2345–2356

Leikola M (1969) The influence of environmental factors on the diameter growth of young trees. Acta For Fenn 92:1–44

Lenz O, Schär E, Schweingruber FH (1976) Methodische Probleme bei der radiographisch-densitometrischen Bestimmung der Dichte und der Jahrringbreiten von Holz. Holzforschung 30:114–123

Levitt J (1980) Chilling, freezing, and high temperature stresses, 2nd edn. (Responses of plants to environmental stresses, vol 1) Academic, New York

Lev-Yadun S, Aloni R (1995) Differentiation of the ray system in woody plants. Bot Rev 61:45–84

Lewis FT (1935) The shape of the tracheid in the pine. Am J Bot 22:741–762

Liang C, Filion L, Cournoyer L (1997) Wood structure of biologically and climatologically induced light rings in eastern larch *(Larix laricina)*. Can J For Res 27:1538–1547

Liming FG (1957) Homemade dendrometers. J For 55:575–577

Linder S, Axelsson B (1982) Changes in carbon uptake and allocation patterns as a result of irrigation and fertilization of young *Pinus sylvestris* stand. In: Waring RH (ed) Carbon uptake and allocation in sub-alpine ecosystems as a key management. Oregon State University, Corvallis, pp 38–44

Lindsay AA, Newman JE (1956) Uses of official weather data in spring time – temperature analysis of an Indiana phenological record. Ecology 37:812–823

Lloyd AH, Fastie CL (2002) Spatial and temporal variability in the growth and climate response of treeline trees in Alaska. Clim Change 52:481–509

Lloyd J, Shibistova O, Zolotoukhin D, Kolle O, Arneth A, Wirth C, Styles JM, Tschebakova NM, Schulze ED (2002) Seasonal and annual variations in the photosynthetic productivity and carbon balance of central Siberian pine forest. Tellus 54B:590–610

Lobzhanidze EO (1961) The cambium and tree ring formation (in Russian). Izd AN GSSR, Tbilisi, 159 pp

Lopez-Saez JF, Gimenez-Martin G, Gonzalez-Fernandez A (1966) Duration of the cell division cycle and its dependence on temperature. Z Zellforsch 75:591–602

Lovelius NV (1979) The variability of tree growth. Dendroindication of natural processes and anthropogenic impacts (in Russian). Nauka, Leningrad, 230 pp

Lowe JJ, Walker MJC (1997) Reconstructing Quaternary environments. Longman, New York

Luck J, Barlow PW, Luck HB (1994) Cell genealogies in a plant meristem deduced with the aid of a bootstrap L-system. Cell Prolif 27:1–21

Luck J, Barlow PW, Luck HB (1994) Deterministic pattern of cellular growth and division within a meristem. Ann Bot 73:1–11

Luckman BH (1993) Glacier fluctuations and tree-ring records for the last millennium in the Canadian Rockies. Quat Sci Rev (NY) 12:441–450

Lyndon RF (1973) The cell cycle in the shoot apex. In: Balls M, Billett FS (eds) The cell cycle in development and differentiation. Cambridge University, Cambridge, pp 167–183

Lyndon RF (1979) Rates of growth and primordial initiation during flower development in *Silene* at different temperatures. Ann Bot 43:539–551

Lyndon RF (1998) The shoot apical meristem. Cambridge University, Cambridge, 277 pp

Lyr H, Polster G, Fidler GI (1974) The physiology of trees (in Russian, translation from German). Lesnaya Promishlennost, Moscow, 421 pp

Lyr H, Fiedler HJ, Tranquillini W (1992) Physiologie und Ökologie der Gehölze. Fischer, Jena 620 pp

Magda VN (2003) Radial growth of conifers as moisture indicator in southern Siberia. PhD thesis, Krasnoyarsk University, Krasnoyarsk, 188 pp

Magill AN, et al (1997) Biogeochemical response of forest ecosystems to simulated chronic nitrogen deposition. Ecol Appl 7:402–415

Mahmood A (1968) Cell grouping and primary wall generations in the cambial zone, xylem, and phloem in *Pinus*. Aust J Bot 16:177–195

Mahmood A (1971) Number of initial-cell divisions as a measure of activity in the yearly cambial growth pattern in *Pinus*. Pak J For 21:27–42

Mahmood A (1990) The parental cell walls. In: Iqbal M (ed) The vascular cambium. Res Studies, Taunton, pp 113–126

Makela A (1986) Implications of the pipe model theory on dry matter partitioning and height growth in trees. J Theor Biol 123:103–120

Makinen H, Saranpaa P, Linder S (2002) Wood-density variation of Norway spruce in relation to nutrient optimization and fiber dimensions. Can J For Res 32:185–194

Mann ME, Bradley RS, Hughes MK (1998) Global-scale temperature patterns and climate forcing over the past six centuries. Nature 392:779–787

Mann ME, Bradley RS, Hughes MK (1999) Northern hemisphere temperatures during the past millennium: inferences, uncertainties and limitations. Geophys Res Lett 26:759–762

Mann ME, Gille Ed, Bradley RS, Hughes MK, Overpeck J, Keimig FT, Gross W (2000) Global temperature patterns in past centuries: an interactive presentation. Earth Interact 4:1–29

Matyssek R, Reich P, Oren R, Winner WE (1995) Response mechanisms of conifers to air pollutants. In: Smith WK, Hinckley TM (eds) Ecophysiology of conifers forests. Academic, New York, pp 255–309

Mazepa V (1990) Spectral approach and narrow band filtering for assessment of cyclic components and ecological prognoses. In: Cook ER, Kairiukstis LA (eds) Methods of dendrochronology: applications in the environmental sciences. Kluwer, Dordrecht, pp 302–308

McCabe GJ, Fountain AG, Dyurgerov M (2002) Variability in winter mass balance of northern hemisphere glaciers and relations with atmospheric circulation. Arct Antarct Alp Res 32:64–72

McCarroll D, Loader NJ (2004) Stable isotopes in tree rings. Quat Sci Rev (NY) 23:771–801

McKimmy MD (1959) Factors related to variation in specific gravity in young-growth Douglas-fir. Oregon For Prod Res Bull 8, 52 pp

McMurtie RE (1985) Forest productivity in relation of carbon partitioning and nutrient cycling: a mathematical model. In: Cannel MGR, Jackson JE (eds) Trees as crop plants. Institute of Terrestrial Ecology/NERC, Huntington, pp 194–207

McMurtie RE (1990) Modelling the yield of *Pinus radiata* on a site limited by water and nitrogen. For Ecol Manage 30:381–413

McNaughton KG, Jarvis PG (1983) Predicting effects of vegetation changes on transpiration and evaporation. In: Kozlowski TT (ed) Water deficits and plant growth. Academic, New York, pp 1–47

Meeuse ADJ (1942) A study of intercellular relationships among vegetable cells with special reference to "sliding growth" and to cell shape. Rec Trav Bot Neerl 38:18–140

Meko DM (1981) Applications of Box–Jenkins methods of time series analysis to the reconstruction of drought from tree rings. PhD thesis, University of Arizona, Tucson, 149 pp

Melekhov IS (1979) The importance of tree ring structure and its dynamics in forestry and dendroclimatology (in Russian). Izv Vuzov 4:6–14

Melekhov IS, Melekhova TA (1958) The effect of drainage of pine swamps on wood formation (in Russian). Izv Vuzov 4:16–28

Mellerowicz EJ, Riding RT (1992) Does DNA endoreduplication occur during differentiation of secondary xylem and phloem in *Abies balsamea*? Int J Plant Sci 153:26–30

Mellerowicz EJ, Coleman WK, Riding RT, Little CHA (1992) Periodicity of cambial activity in *Abies balsamea*. I. Effects of temperature and photoperiod on cambial dormancy and frost hardiness. Physiol Plant 85:515–525

Mergen F, Burley J, Yeatman CW (1964) Variation in growth characteristics and wood properties of Norway spruce. TAPPI 48:499–504

Mikola P (1962) Temperature and tree growth near the northern timberline. In: Kozlowski TT (ed) Tree growth. Ronald, New York, pp 265–287

Mikola PU (1950) Puiden kasvun vaihteluista ja niiden merkityksestae kasvututkimuksissa. Comm Inst For Fenn 38:1–131

Milsom SJ, Hughes MK (1978) X-ray densitometry as a dendrochronological technique. BAR Int Ser 51:317–324

Mina MV, Klevezal' GA (1970) The principles of studies of register structures (in Russian). Usp Sovr Biol 70:341–352

Misson L (2004) MAIDEN: a model for analyzing ecosystem processes in dendroecology. Can J For Res 34:874–887

Misson L, Rathgeber C, Guiot J (2004) Dendroecological analysis of climatic effects on *Quercus petraea* and *Pinus halepensis* radial growth using the process-based MAIDEN model. Can J For Res 34:888–898

Mitchell KJ (1961) Apical growth in relation to the transition from springwood to summerwood in conifers. (Univ B C For Club 19) University of British Columbia, Vancouver, 71 pp

Mitropolsky AK (1971) The technique of statistical calculations (in Russian). Nauka, Moscow, 576 pp

Monsi M, Marata Y (1970) Development of photosynthetic systems as influenced by distribution of dry matter. In: Setlik I (ed) Prediction and measurement of photosynthetic productivity. Pudoc, Wageningen, pp 115–130

Monteith JL, Unsworth MH (1990) Principles of environmental physics. Arnold, London, 291 pp

Moore ID, Grayson RB, Ladson AR (1991) Digital terrain modelling: a review of hydrological, geomorphological, and biological applications. Hydrol Proc 5:3–30

Mullen JL, Ishikawa H, Evans ML (1998) Analysis of changes in relative elemental growth rate patterns in the elongation zone of *Arabidopsis* roots upon gravistimulation. Planta 206:598–603

Munro MAR, Brown PM, Hughes MK, Garcia EMR (1996) Image analysis of tracheid dimensions for dendrochronological use. In: Dean J, Meko D, Swetnam T (eds) Tree rings, environment and humanity. Radiocarbon, Tucson, pp 843-853

Munting AJ, Willemse MTM (1987) External influence on development of vascular cambium and its derivatives. Phytomorphology 37:261–274

Murin A (1964) Der Mitosezyklus und seine Zeitparameter in den Wurzelspitzen von *Vicia faba* L. Chromosoma 15:457–464

Murin A (1967) Mitosenzyklus und dessen Zeitparameter in den Achsenscheiteln von *Vicia faba* L. Biologia 22:178–185

Murmanis L (1970) Locating the initial in the vascular cambium of *Pinus strobus* L. by electron microscopy. Wood Sci Technol 4:1–14

Murmanis L (1971) Structural changes in the vascular cambium of *Pinus strobus* L. during an annual cycle. Ann Bot 35:133–141

Murray FW (1967) On the computation of saturation vapor pressure. J Appl Meteorol 6:203–204

Musaev EK (1993) The impact of radiation on tree rings of pines in the Chernobyl AES area (in Russian). Lesovedenie 4:41–49

Musaev EK (1996) Seasonal growth and anatomy of tree rings of Scots pine in the region of the Chernobyl disaster (in Russian). Lesovedenie 1:16–28

Myneni RB, Zhou L (2000) Vegetation activity in northerly latitudes during the past two decades. In: Cenlemans R, Bogaert J, Deckmyn G, Nijs I (eds) Topics in ecology: structure and function in plants and ecosystems. University of Antwerp, Antwerp

Myneni RB, Keeling CD, Tucker CJ, Ascrar G, Nemani RR (1997) Increased plant growth in the northern high latitudes from 1981–1991. Nature 337:388–392

Nadelhoffer KJ, Emmett BA, Gundersen P, Kjonaas OJ, Koopmans CJ, Schleppi P, Tietema A, Wright RF (1999) Nitrogen deposition makes a minor contribution to carbon sequestration in temperate forests. Nature 398:145–148

Naurzbaev MM, Vaganov EA (1999) Variability of the radial growth in larch from the north-east of Taimyr and Putoran over the last 2000 years (in Russian). Lesovedenie 6:24–34

Naurzbaev MM, Hughes MK, Vaganov EA (2004) Tree-ring growth curves as sources of climatic information. Quat Res (NY) 62:126–133

Naurzbaev MM, Vaganov EA (2000) Variation of early summer temperature in east Taymir and Putoran (Siberia) over the last two millennia inferred from tree rings. J Geophys Res 105:7317–7326

Nesje A, Dahl SO (2000) Glaciers and environmental change. Arnold, London, 203 pp

Newman IV (1956) Pattern in meristems of vascular plants 1. Cell partition in the living apices and in the cambial zone in relation to the concepts of initial cells and apical cells. Phytomorphology 6:1–19

Newman IV (1961) Pattern in meristems of vascular plants II. A review of shoot apical meristems of gymnosperms, with comments on apical biology and taxonomy, and a statement of some fundamental concepts. Proc Linn Soc N S Wales 86:9–59

Ni F, Cavazos T, Hughes MK, Comrie AC, Funkhouser G (2002) Cool season precipitation in the Southwestern United States since AD 1000: comparison of linear and nonlinear techniques for reconstruction. Int J Climatol 22:1645–1662

Nicholls JM, Dadswell HE, Fielding JM (1964) The heritability of wood characteristics of *Pinus radiata*. Silv Genet 13:68–71

Nicholls JW, Morris JD, Pederick LA (1980) Heritability estimates of density characteristics in juvenile radiata wood. Silv Genet 29:54–61

Nieminen KM, Kauppinen L, Helariutta Y (2004) A weed for wood? *Arabidopsis* as a genetic model for xylem development. Plant Physiol 135:653–659

Nikitin NI (1962) Chemistry of wood and cellulose (in Russian). Izd AN USSR, Moscow, 711 pp

Nikolov NT, Zeller KF (1992) A solar radiation algorithm for ecosystem dynamic models. Ecol Model 61:149–168

Noble PS (1974) Introduction to biophysical plant physiology. Freeman, San Francisco

Nobuchi TY, Ogata, Siripatanadilok S (1995) Seasonal characteristics of wood formation in *Hopea odorata* and *Shorea henryana*. IAWA J 16:361–369

Nougarede A, Michele MN di, Rondet P, Saint-Come R (1990) Plastochrone, cycle cellulaire et teneurs en AND nucleaire du meristeme caulinaire de plants de *Chrysanthemum segetum soumis* a deux conditions lumineuses differentes, sous une photoperiode de 16 heures. Can J Bot 68:2389–2397

Odum EP (1971) Fundamentals of ecology, 3rd edn. Saunders, Philadelphia, 574 pp

Olesen PO (1976) The interrelation between basic density level and tracheid width within the juvenile and mature wood of Norway spruce. For Tree Improv 12:1–21

Olivas-Garcia JM, Cregg DM, Hennessey TC (2000) Genotypic variation in carbon isotope discrimination and gas exchange of ponderosa pine seedlings under two levels of water stress. Can J For Res 30:1581–1590

Olszyk D, Wise C, VanEss E, Tingey D (1998) Elevated temperature but not elevated CO_2 affects long-term patterns of stem diameter and height of Douglas-fir seedlings. Can J For Res 28:1046–1054

O'Neill BC, MacKellar FL, Lutz W (2001) Population and climate change. Cambridge University, Cambridge, 266 pp

Orwig DA, Abrams MD (1997) Variation in radial growth response to drought among species, site, and canopy strata. Trees 11:474–484

Ottestad P (1960) Forecasting the annual yield in sea fisheries. Nature 185:183

Pahlavanian A, Silk WK (1988) Effect of temperature on spatial and temporal aspects of growth in the primary maize root. Plant Physiol 87:529–532

Palmer J, Ogden J (1983) A dendrometer band study of the seasonal pattern of radial increment of Kauri (*Agathis australis* Salisb.). NZ J Bot 21:121–125

Panyushkina IP, Hughes MK, Vaganov EA, Munro MAR (2003) Summer temperature in northeastern Siberia since 1642 reconstructed from tracheid dimension and cell number of *Larix cajanderi*. Can. J For Res 33:1905–1914

Park W-K (1990) Development of anatomical tree-ring chronologies from Southern Arisona conifers using image analysis. PhD thesis, University of Arizona, Tucson, 234 pp

Park W-K (1991) Effects of increasing atmopheric carbon dioxide on the growth and wood quality of Korean conifers-dendrochronological analysis. (Final report of the Korean Institute for Science and Enginering Fund.) Korean Institute for Science and Enginering, Cheonju, 58 pp

Park W-K (1993) Increasing atmospheric dioxide and growth trends of Korea subalpine conifers – dendrochronological analysis. J Kor For Soc 82:17–25

Park W-K, Vaganov EA (1996) Monitoring cambial activity and growth for predicting the response of trees on global warming and air pollution. (First year report.) Chungbuk National University, Cheongjn, 112 pp

Park WK, Vaganov EA, Arbatskaya MK, Seo JW, Kim JS (2000) Global warming effects on the cambial growth of *Larix leptolepis* in Central Korea: prediction from simulation modeling. Kor J Quat Res 14:57–63

Park W-K, Vaganov EA (1997) Monitoring cambial activity and growth for predicting the response of trees on global warming and air pollution. (Second year report.) Chungbuk National Univesity, Cheongju, 97 pp

Parker ML, Henoch WES (1971) The use of Engelmann spruce latewood density for dendrochronological purposes. Can J For Res 1:90–98

Parry ML (1990) Climate change and world agriculture. Earthscan, London

Parry ML (1993) Climate change and the future of agriculture. Int J Environ Pollut 3:13–30

Patefield WM, Austin RB (1971) A model for the simulation of the growth of *Betula vulgaris*. Ann Bot 35:1227–1250

Pellatt MG, Smith MJ, Mathewes RW, Walker IR, Palmer SL (2000) Holocene treeline and climate change in the subalpine zone near Stoyoma Mountain, Cascade Mountains, southwestern British Columbia, Canada. Arct Antarct Alp Res 32:73–83

Peltola H, Kilpelainen A, Kellomaki S (2002) Diameter growth of Scots pine (*Pinus sylvestris*) trees grown at elevated temperature and carbon dioxide concentration under boreal conditions. Tree Physiol 22:963–972

Penman HL, Schofield RK (1951) Some physical aspects of assimilation and transpiration. Symp Soc Exp Biol 5:115–129

Pereira JS (1995) Gas exchange and growth. In: Schulze E-D, Caldwell MM (eds) Ecophysiology of photosynthesis. Springer, Berlin Heidelberg New York, pp 147–181

Pereligin LM (1969) Wood science (in Russian). Lesnaya Promishlennost, Moscow, 316 pp

Peterson DW, Peterson DL, Ettl DJ (2002) Growth response of subalpine fir to climatic variability in the Pacific Northwest. Can J For Res 32:1503–1517

Peterson TC, Vose RS (1997) An overview of the global historical climatology network temperature data base. Bull Am Meteorol Soc 78:2837–2849

Philipson WK, Ward JM, Butterfield BG (1971) The vascular cambium: its development and activity. Chapman and Hall, London, 182 pp

Phillips R (1980) Cytodifferentiation. Perspectives in plant cell and tissue culture. Int Rev Cytol [Suppl 11a]:55–70

Polge H (1966) Etablissement des courbes de variation de la densité du bois par exploration densitometrique de radiographies d'echantillons prelèves a la tarière sur des arbres vivants. Ann Sci For 23:3–115

Polge H (1969) Influence de la fertilisation sur la qualité du bois de Pin maritime. Ann Sci For 26:45–64

Polge H, Keller R (1969) La xylochronologie, perfectionnement logique de la dendrochronologie. Ann Sci For 26:225–256

Poluboyarinov OI (1976) The density of wood (in Russian). Lesnaya Promishlennost, Moscow, 160 pp

Pozdnyakov LK (1986) Permafrost forestry (in Russian). Nauka, Novosibirsk, 192 pp

Preston RD (1974) The physical biology of plant cell walls. Chapman and Hall, London, 250 pp

Pritchard J (1994) The control of cell expansion in roots. New Phytol 127:3–26

Pritchard J, Barlow PW, Adams JS, Tomos AD (1990) Biophysics of the inhibition of the growth of maize roots by lowered temperature. Plant Physiol 93:222–230

Protopopov VV (1975) Environmental effect of dark-needle forest (in Russian). Nauka, Novosibirsk, 327 pp

Rao KS, Dave YS (1981) Seasonal variations in the cambial anatomy of *Tectona grandis* (Verbenaceae). Nord J Bot 1:535–542

Rao KS, Dave YS (1985) Seasonal variations in the vascular cambium of *Holoptelea intergrifolia* (Ulmaceae). Beitr Biol Pflanz 59:321–331

Rauscher HM, Isebrands JG, Host GE, Dickson RE, Dickman DI, Crow TR, Michael DA (1990) ECOPHYS: An ecophysiological growth process model for juvenile poplar. Tree Physiol 7:255–281

Rayle DL, Cleland RE (1992) The acid growth theory of auxin-induced cell elongation is alive and well. Plant Physiol 99:1271–1274

Reid GC (1987) Influence of solar variability on global sea surface temperature. Nature 329:142–143

Reimers NF (1991) The popular biological dictionary (in Russian). Nauka, Moscow, 544 pp

Reynolds JF, Thorley JHM (1982) A shoot-root partitioning model. Ann Bot 49:585–597

Richardson DD (1964) The external environment and tracheid size in conifers. In: Zimmermann M (ed) The formation of wood in forest trees. Academic, New York, pp 367–388

Rinn F (1996) TSAP ver. 3.6. Reference manual: computer program for tree-ring analysis and presentation. TSAP, Heidelberg, 236 pp

Roberts LW (1976) Cytodifferentiation in plants. Xylogenesis as a model system. Cambridge University, Cambridge, 160 pp

Roberts LW (1983) The influence of physical factors on xylem differentiation. In: Dodds JH (ed) Tissue culture of trees. Croom Helm, London, pp 88–102

Roberts LW, Gahan PB, Aloni R (1988) Vascular differentiation and plant growth regulators. Springer, Berlin Heidelberg New York

Robinson WJ, Evans R (1980) A microcomputer-based tree-ring measuring system. Tree-Ring Bull 40:59–64

Roland JC (1978) Early differences between radial walls and tangential walls of actively growing cambial zone. IAWA Bull 1:7–10

Rosenzweig C, Hillel D (1998) Climate change and the global harvest: potential impact of the greenhouse effect on agriculture. Oxford University, New York

Rothrock DA, Yu Y, Maykut GA (1999) Thinning of the Arctic ice cover. Geophys Res Lett 26:3469–3472

Running SW, Coughlan JC (1988) A general model of forest ecosystem processes for regional applications I. Hydrological balance, canopy, gas exchange and primary production processes. Ecol Model 42:125–154

Running SW, Gower ST (1991) FOREST-BGC, a general model of forest ecosystem processes for regional applications II. Dynamic carbon allocation and nitrogen budgets. Tree Physiol 9:147–160

Running SW, Ramakrishna RN, Hungerford RD (1987) Extrapolation of synoptic meteorological data in mountainous terrain and its use for simulating forest evapotranspiration and photosynthesis. Can J For Res 17:472–483

Russell E (1955) Soil conditions and plant growth (in Russian, translation from English). Foreign Literature, Moscow, 623 pp

Ryan MG, Waring RH (1992) Maintenance respiration and stand development in a subalpine lodgepole pine forest. Ecology 73:2100–2108

Ryan MG, Linder S, Vose JM, Hubbard RM (1994) Dark respiration of pines. Ecol Bull 43:50–63

Ryan MG, Hubbard RM, Pongracic D, Raison RJ, McMurtrie RE (1996) Foliage, fine-root, woody-tissue and stand respiration in *Pinus radiata* in relation to nitrogen status. Tree Physiol 16:333–343

Sastry CBR, Wellwood RW (1971) A quartz ultra-micro balance for weighing single tracheids. Wood Sci 3:179–182

Savelieva NI, Semiletov IP, Vasilevskaya LN, Pugach SP (2000) A climate shift in seasonal values of meteorological and hydrological parameters for northeastern Asia. Prog Oceanogr 47:279–297

Savidge RA (1985) Prospects for manipulating vascular cambium productivity and xylem-cell differentiation. In: Cannell MGR, Jackson JE (eds) Attributes of trees as crop plants. Institute for Terrestrial Ecology, Monks Wood, pp 208–227

Savidge RA (1991) Seasonal cambial activity in *Larix laricina* seedlings in relation to endogenous indol-3-ylacetic acid, sucrose , and coniferin. Forest Sci 37:953–958

Savidge RA (1993) Formation of annual rings in trees. In: Rensing L (ed) Oscillations and morphogenesis. Dekker, New York, pp 343–363

Savidge RA (1994) The tracheid-differentiation factor of conifer needles. Int J Plant Sci 155:272–290

Savidge RA (1996) Xylogenesis, genetic and environmental regulation. IAWA J 17:269–310

Savidge RA (2000) Intrinsic regulation of cambial growth. J Plant Growth Regul 20:52–77

Savva YV, Schweigruber F, Milyutin L, Vaganov E (2002a) Genetical and environmental signals in tree rings from different provenances of *Pinus sylvestris* L. planted in the southern taiga, Central Siberia. Trees 16:313–324

Savva YV, Schweingruber FH, Kuzmina NA, Vaganov EA (2002b) Sensitivity of diameter growth to annual weather conditions in Scots pine provenances at a Central Siberian location. Silv Genet 51:48–55

Savva YV, Schweingruber FH, Vaganov EA, Milyutin LI (2003a) Influence of climate changes on tree-ring characteristics of Scots pine provenances in Southern Siberia (forest-steppe). IAWA J 24:371–383

Savva YV, Vaganov EA, Milyutin LI (2003b) Poculiarities in growth response of different provenances of *Pinus sylvestris* on changes in climatic factors (in Russian). Bot Zh 88:68–82

Schlesinger ME, Ramankutty N (1992) Implications for global warming of intercycle solar irradiance variations. Nature 360:330–333

Schmid R (1976) The elusive cambium – another terminological contribution. IAWA Bull 4:51–59

Schrader J, Baba K, May ST, Palmek K, Bennett M, Bhalerao RP, Sandberg G. (2003) Polar auxin transport in the wood forming tissues of hybrid aspen is under simultaneous control of developmental and environmental signals. Proc Natl Acad Sci USA 100:10096–10101

Schrader J, Nilsson J, Mellerowicz E, Berglund A, Nilsson P, Hertzberg M, Sandberg G (2004) A high-resolution transcript profile across the wood-forming meristem of poplar identifies potential regulators of cambial stem cell identity. Plant Cell 16:2278–2292

Schulman E (1958) Bristlecone pine, oldest known living trees. Nat Geogr Mag 113:355–372

Schultness J (1990) Der Einfluss von Entwässerung auf die Bewaldung eines Hochmoors. PhD thesis, Geographical Institute, Zurich, 190 pp

Schulze E-D, Hall AE (1982) Stomatal responses, water loss and CO_2 assimilation rates of plants in contrasting environments. In: Pirson A, Zimmermann MH (eds) Encyclopedia of plant physiology (NS), vol 12b. Springer, Berlin Heidelberg New York, pp 181–230

Schulze E-D, Schulze W, Kelliner FM, et al (1995) Aboveground biomass and nitrogen nutrition in a chronosequence of pristine Dahurian *Larix* stands in eastern Siberia. Can J For Res 25:943–960

Schweingruber FH (1979) Auswirkungen des Lärchenwicklerbefalls auf die Jahrringesstruktur der Lärche; Ergebnisse einer Jahrringanalyse mit röntgendensitometrischen Methoden. Schweiz Z Forstwes 130:1071–1093

Schweingruber FH (1987) Flächenhafte dendroklimatische Temperaturrekonstruktionen für Europa. Naturwissenschaften 74:205–212

Schweingruber FH (1988) Tree ring: basics and applications of dendrochronology. Reidel, Dordrecht, 276 pp
Schweingruber FH (1990) Anatomie europäischer Hölzer. Haupt, Bern, 806 pp
Schweingruber FH (1993) Trees and wood in dendrochronology. Springer, Berlin Heidelberg New York, 386 pp
Schweingruber FH (1996) Tree rings and environment. Dendroecology. Haupt, Berne, 609 pp
Schweingruber FH, Braeker OU, Schaer E (1979) Dendroclimatic studies on conifers from central Europe and Great Britain. Boreas 8:427–452
Schweingruber FH, Voronin V (1996) Eine dendrochronologisch-bodenchemische Studie aus dem Waldschadengebiet Norilsk, Sibirien und die Konsequenzen für die Interpretation großflächiger Kronentaxationsinventuren. Allg Forst Jagdztg 167:53–67
Serreze MC, Walsh JE, Chapin FS, Osterkamp T, Dyurgerov M, Romanovsky V, Oechel WC, Morison J, Zhang T, Barry RG (2000) Observational evidence of recent change in the northern high latitude environment. Clim Change 46:159–207
Shashkin AV, Fritts HC (1995) TREERING 2.0: a model that simulates cambial activity and ring structure in conifers. Tree rings from the past and to the future. Forestry and Forest Products Institute, Tsukuba, pp 24–30
Shashkin AV, Vaganov EA (1993) Simulation model of climatically determined variability of conifers annual increment (on the example of Scots pine in the steppe zone). Russ J Ecol 24:275–280
Shashkin EA, Vaganov EA (2000) Dynamics of the stem cross-area of trees from different regions of Siberia in relation to temperature changes in the northern hemisphere: a comparative analysis along the latitudinal and meridional transects. Lesovedenie 3:3–20
Sheehy JE, Cobby JM, Ryle GJA (1979) The growth of perennial ryegrass: a model. Ann Bot 43:335–354
Sheldrake AR (1971) Auxin in the cambium and its differentiating derivatives. J Exp Bot 22:735–740
Sheppard PR, Graumlich LJ (1996) A reflected-light video imaging system for tree-ring analysis of conifers. In: Dean J, Meko D, Swetnam T (eds) Tree rings, environmental and humanity. Radiocarbon, Tucson, pp 879–889
Shiyatov SG (1969) Snow cover at the upper timberline and its influence on trees (in Russian). Trans Inst Plant Anim Ecol Sverdlovsk 1969:141–157
Shiyatov SG (1970) To the methods of estimation of tree-ring width indices (in Russian). Ecologia 3:85–87
Shiyatov SG (1973a) Dendrochronological scale of Siberian cedar on the northern timberline in River Taz valley (in Russian). Lesovedenie 4:40–45
Shiyatov SG (1973b) Dendrochronology, its principles and methods (in Russian). Trans Sverdlovsk Dep VBO 6:53–81
Shiyatov SG (1986) Dendrochronology of upper timberline in Ural Mountains (in Russian). Nauka, Moscow, 136 pp
Shugart HH (1984) A theory of forest dynamics. The implications of forest succession models. Springer, Berlin Heidelberg New York, 216 pp
Siddiqi TO (1991) Impact of seasonal variation on the structure and activity of vascular cambium in *Ficus religiosa*. IAWA Bull 12:177–185
Silk WK, Erickson RO (1978) Kinematics of hypocotyls curvature. Am J Bot 65:310–319
Silk WK, Erickson RO (1979) Kinematics of plant growth. J Theor Biol 76:481–501
Silk WK, Lord EM, Eckard KJ (1989) Growth patterns inferred from anatomical records. Plant Physiol 90:708–713

Silkin PP, Kirdyanov AV (1999) The cell wall mass of early wood and late wood tracheids in larch tree rings. Lesovedenie 6:54–58

Simachev IV, Vaganov EA, Vysotskaya LG (1992) Dendroclimatic analysis of larch growth under pollution impact of Norilsk metallurgical plant (in Russian). Geogr Prirod Resursi 4:65–73

Sinnot E (1963) Morphogenesis of plants (in Russian, translation from English). Mir, Moscow, 603 pp

Sisson WE (1968) Cambial divisions in *Pseudotsuga menziesii*. Am J Bot 55:923–926

Skene DS (1969) The period of time taken by cambial derivatives to grow and differentiate into tracheids in *Pinus radiata* D. Don. Ann Bot 33:253–262

Skene DS (1972) The kinetics of tracheid development in *Tsuga canadensis* Carr and its relation to tree vigour. Ann Bot 36:179–187

Smirnov VV (1964) Seasonal growth of the main species of trees (in Russian). Nauka, Moscow, 167 pp

Smith DM, Wilsie MC (1961) Some anatomical responses of loblolly pine to soil-water deficiencies. TAPPI 44:179–185

Snowdon P, Benson ML (1992) Effects of combinations of irrigation and fertilization on the growth and above-ground biomass production of *Pinus radiata*. For Ecol Manage 52:87–116

Spiecker H (1987) Düngung, Niederschlag und der jährliche Volumenzuwachs einiger Fichtenbestände Südwestdeutschlands. Ergebnisse einer neuen Methode der Zuwachsvermittlung. Allg Forst Jagdztg 158:70–76

Spiecker H (1991) Liming, nitrogen and phosphorous fertilization and the annual volume increment of Norway spruce stands on long-term permanent plots in south-western Germany. Fertil Res 27:87–93

Spurr SH, Barnes BV (1980) Forest ecology. Wiley, New York, 687 pp

Srivastava LM (1966) On the fine structure of the cambium of *Fraxinus americana* L. J Cell Biol 31:79–93

Stevens DW (1975) A computer program for simulating cambial activity and ring growth. Tree-Ring Bull 35:29–56

Stokes MA, Smiley TL (1968) An introduction to tree-ring dating. University of Chicago, Chicago, 73 pp

Stuiver M (1980) Solar variability and climate change during the current millennium. Nature 286:868–871

Sudachkova NE (1977) Metabolism and wood formation of conifers (in Russian). Nauka, Novosibirsk, 228 pp

Sviderskaya IV (1996) Kinetics of cambial activity in seasonal growth of tree rings in relation to tree vigor. In: Sudachkova NE (ed) Proceedings for the international conference on ecological and physiological aspects of xylogenesis in conifers. IF SB RAS, Krasnoyarsk, pp 96–102

Sviderskaya IV (1999) Histometric analysis of patterns of seasonal formation of conifer wood (in Russian). PhD thesis, IF SB RAS, Krasnoyarsk, 167 pp

Szeicz JM, MacDonald GM (1994) Age-dependent tree-ring growth-responces of subarctic white spruce to climate. Can J For Res 24:120–132

Szeicz JM, MacDonald GM (1995) Dendroclimatic reconstruction of summer temperatures in northwestern Canada since AD 1638 on age-dependent modeling. Quat Res (NY) 44:257–266

Taiz L (1984) Plant cell expansion: regulation of cell wall mechanical properties. Annu Rev Plant Physiol 35:583–657

Taylor EL, Taylor TN, Cuneo NR (2000) Permian and Triassic high latitude paleoclimates: evidence from fossil biotas. In: Huber BT, MacLeod KG, Wing SL (eds) Warm climates in earth history. Cambridge University, Cambridge, pp 321–350

Terskov IA, Vaganov EA (1978) Concerning quantitative patterns of individual tree growth. In: Terskov IA (ed) Analysis of growth dynamics of biological objects (in Russian). Nauka, Moscow, pp 15–27

Terskov IA, Vaganov EA, Zyryanov GI, Filimonov VS (1978) Analysis of tree growth using the variations of cell dimensions within tree rings (in Russian). Izv SO RAN Ser Biol 3:26–36

Terskov IA, Vaganov EA, Sviderskaya IV (1981) Tracheidograms of tree rings as characteristics of seasonal growth rate (in Russian). Izv SO RAN Ser Biol 2:22–30

Telewski FW, Lynch AM (1991) Measuring growth and development of stem. In: Lassoie JP, Hinkley TM (eds) Techniques and approaches in forest tree ecophysiology. CRC, Boca Raton, pp 503–555

Thornley JHM (1970) Respiration, growth and maintenance in plants. Nature 227:304–305

Thornley JHM (1991) A transport resistance model of forest growth and partitioning. Ann Bot 68:211–226

Thornthwaite CW, Mather JR (1955) The water balance. (Publications in climatology, vol 1) Drexel Institute of Technology, Philadelphia, pp 1–104

Till C, Guiot J (1990) Reconstruction of precipitation in Morocco since 1100 AD based on *Cedrus atlantica* tree-ring widths. Quat Res (NY) 33:337–351

Timell TE (1973) Studies on opposite wood in conifers II. Histology and ultrastructure. Wood Sci Technol 7:79–91

Timell TE (1979) Formation of compression wood in Balsam fir (*Abies balsamea*) 1. Ultrastructure of the active cambial zone and its enlarging derivatives. Holzforschung 33:137–143

Timell TE (1986) Compression wood in gymnosperms, vol 3. Springer, Berlin Heidelberg New York, 2,150 pp

Touchan R, Meko DM, Hughes Mm (1999) A 396-year reconstruction of precipitation in southern Jordan. J Am Water Res Assoc 35:45–59

Tranquillini W (1979) Physiological ecology of the Alpine timberline. Springer, Berlin Heielberg New York, 137 pp

Trenberth KE (1984) Signal versus noise in the southern oscillation. Mon Weather Rev 112:326–332

Turner H, Streule A (1983) Wurzelwachstum und Sprossentwicklung junger Koniferen im Klimastress der alpinen Waldgrenze, mit Berücksichtigung von Mikroklima, Photosynthese und Stoffproduktion. In: Bohm W, Kutschera L, Lichtenegger EB (eds) Wurzelökologie und ihre Nutzanwendung. Springer, Berlin Heidelberg New York, 775 pp

Uggla C, Moritz T, Sandberg G, Sundberg B (1996) Auxin as a positional signal in pattern formation in Plants. Proc Natl Acad Sci USA 95:9282–9286

Uggla C, Mellerowicz EJ, Sundberg B (1998) Indole-3-acetic acid controls cambial growth in Scots pine by positional signaling. Plant Physiol 117:113–121

Uggla C, Magel E, Moritz T, Sundberg B (2001) Function and dynamics of auxin and carbohydrates during earlywood/latewood transition in Scots pine. Plant Physiol 125:2029–2039

Urban DL, Miller C, Halpin PN, Stephenson NL (2000) Forest gradient response in Sierran landscapes: the physical template. Landscape Ecol 15:603–620

Usoltsev VA (1988) Growth and phytomass structure of stands (in Russian). Nauka, Novosibirsk, 253 pp

Utkin AI (1982) The methods to study biological primary productivity of forests (in Russian). In: Biological Productivity of near Volga River Forests. Nauka, Moscow, pp 59–71

Utkin AI (1995) Carbon cycle and forestry (in Russian). Lesovedenie 5: 3–20

Utkin AI, Zamolodchikov DG, Gubbe TA, Gubbe YI (1996) Allometric equations for tree phytomass of pine, spruce, birch and aspen growing in European part of Russia (in Russian). Lesovedenie 6:36–46

Vaganov EA (1981) The layer structures in analysis of growth dynamics of biological organisms. In: Terskov IA (ed) The studies of organisms growth dynamics (in Russian). Nauka, Novosibirsk, pp 55–67

Vaganov EA (1989) Method for forecasting grain yields with the aid of dendrochronological data. Sov J Ecol 20:139–146

Vaganov EA (1990) The tracheidogram method in tree-ring analysis and its application. In: Cook E, Kairiukstis L (eds) Methods of dendrochronology. Kluwer, Dordrecht, pp 63–75

Vaganov EA (1996a) Mechanisms and simulation of tree ring formation in conifer wood (in Russian). Lesovedenie 1:3–15

Vaganov EA (1996b) Recording of warming in current century by tracheids of the annual tree rings. Dokl Biol Sci 351:281–283

Vaganov EA, Djanseitov YK (2000) Phenomenological approach to the analysis of the kinetic characteristics in the cambial zone of stems in conifers (in Russian). Usp Sov Biol 120:190–200

Vaganov EA, Kachaev AV (1992) Dendroclimatic study of pine growth in forest-peat phytocoenoces of the Tomsk region (in Russian). Lesovedenie 6:3–10

Vaganov EA, Park W-K (1995) Seasonal growth dynamics of tree rings of red and pitch pines in Korean Peninsula (in Russian). Lesovedenie 2:31–41

Vaganov EA, Terskov IA (1977) Analysis of tree growth by tree ring structure (in Russian). Nauka, Novosibirsk, 98 pp

Vaganov EA, Smirnov VV, Terskov IA (1975) Concerning the possibility to determine seasonal growth rate of tree stem and variations in hydrological regime using the photometric curves of annual rings (in Russian). Russ J Ecol 2:45–53

Vaganov EA, Petrenko ES, Dryanikh NM (1979) Response of young pine trees to defoliation: In: Pleshanov AS (ed) Physiology of plant stability to natural and anthropogenic factors (in Russian). Nauka, Irkutsk, pp 45–48

Vaganov EA, Krasovsky KF, Sviderskaya IV, Shashkin AV (1983) The semi-automated system of measurement and processing of tree ring structure data (in Russian). Lesovedenie 2:73–81

Vaganov EA, Sviderskaya IV, Shashkin AV (1984) Well-ordered cell structures in analysis of growth and tissue differentiation. In: Terskov IA (ed) Analysis of population growth by biophysical methods (in Russian). Nauka, Novosibirsk, pp 51–61

Vaganov EA, Shashkin AV, Sviderskaya IV, Vysotskaya LG (1985) Histometric analysis of woody plant growth (in Russian). Nauka, Novosibirsk, 102 pp

Vaganov EA, Sviderskaya IV, Kondrat'eva EN (1990) Weather conditions and tree ring structure: the simulation model of the tracheidogram (in Russian). Lesovedenie 2:37–45

Vaganov EA, Shashkin AV, Sviderskaya IV (1992) Seasonal growth and formation of tree rings: the kinetics approach and simulation modeling. In: Giterson II (ed) Biophysics of cell populations and organisms' systems (in Russian). Nauka, Novosibirsk, pp 140–150

Vaganov EA, Vysotskaya LG, Shashkin AV (1994) Seasonal growth and tree ring structure of larch growing near the polar timberline (in Russian). Lesovedenie 5:3–15

Vaganov EA, Naurzbaev MM, Schweingruber FH, Briffa KR (1996a) A 840-year tree-ring chronology for Taimyr as an indicator of summer temperature changes. Dendrochronologia 14:193–205

Vaganov EA, Vysotskaya LG, Shashkin AV (1996b) Using cell chronologies in seasonal tree growth analysis and dendro climatology. In: Dean JS, Meko DM, Swetnam TW (eds) Tree rings, environment and humanity. Radiocarbon, Tucson, pp 95–106

Vaganov EA, Hughes MK, Shashkin EA, Arbatskaya MK (1996 c) Dendrochronological methods in evalution of carbon cycle of forest ecosystems (in Russian). In: Zavarzin GA (ed) Carbon cycle in the territory of Russia. Nauka, Moscow, pp 96–123

Vaganov EA, Shiyatov SG, Mazepa VS (1996d) Dendroclimatic study in the Ural-Siberian Subarctic (in Russian). Nauka, Novosibirsk, 244 pp

Vaganov EA, Hughes MK, Kirdyanov AV, Schweingruber FH, Silkin PP (1999) Influence of snowfall and melt timing on tree growth in subarctic Eurasia. Nature 400:149–151

Valentine HT (1983) Bud break and leaf growth functions for modelling herbivory in some gypsy moth hosts. For Sci 29:607–617

Valentine HT (1985) Tree growth models: derivations employing the pipe model theory. J Theor Biol 117:579–585

Van Buijtenen JP (1958) Experimental control of environmental factors and their effect upon some aspects of wood anatomy in Lolbolly pine. TAPPI 41:175–178

Van Buijtenen JP (1965) Inheritance of fiber properties in North American conifers. (IUFRO section 41, vol 2) IUFRO, Melbourne, 12 pp

Vargas-Hernandez J (1990) Genetic variation of wood density components in coastal Douglas-fir and their relationship to growth rhythm. PhD thesis, Oregon State University, Corvallis, 113 pp

Vargas-Hernandez J, Adams WT (1994) Genetic relationships between wood density components and cambial growth rhythm in young coastal Douglas-fir. (Forest research laboratory paper 2978) Oregon State University, Corvallis, 20 pp

Verstraete MM, Martin P (1994) Biosphere modeling for climate studies. NATO ASI Ser I 22:153–173

Villalba R, Veblen TT, Ogden J (1994) Climatic influences on the growth of subalpine trees in the Colorado front range. Ecology 75:1450–1462

Visser H, Molenaar J (1992) Estimating trends and stochastic response functions in dendroecology with application to fir decline. For Sci 38:221–234

Visser H, Molenaar J (1988) Kalman filter analysis in dendroclimatology. Biometrics 44:929–940

Vitousek PM, Howarth RW (1991) Nitrogen limitation on land and in the sea: how can it occur? Biogeochemistry 13:87–115

Volney WJA, Mallett KI (1992) Light rings and the age of jack pine trees. Can J For Res 22:2011–2013

Von Pechman H (1960) Haben Mineraldüngung und Lupinenanbau einen Einfluss auf die Eigenschaften von Fichten- und Kieferholz. Forstw Cbl 79:91–105

Vowinckel T, Oechel WC, Boll WG (1975) The effect of climate on the photosynthesis of *Picea mariana* at the subarctic tree-line I. Field experiments. Can J Bot 53:604–620

Vysotskaya LG, Vaganov EA (1989) Components of the variability of radial cell size in tree rings of conifers. IAWA Bull 10:417–428

Vysotskaya LG, Shashkin AV, Vaganov EA (1985) Analysis of the size distribution of tracheids in the annual rings of pines growing under various moisture conditions. Russ J Ecol 16:29–34

Wachter H (1985) Zur Lebensdauer von Fichtennadeln in einigen Waldgebieten Nordrhein-Westfalens. Forst- Holzwirt 40:420–425

Wager DJ, Baker FA (2003) Potential effects of ozone, climate, and spruce budworm on Douglas-fir growth in the Wasatch Mountains. Can J For Res 5:910–921

Waisel Y, Fahn A (1956) The effect of environment on wood formation and cambial activity in *Robinia pseudoacacia* L. New Phytol 64:436–442

Wang L, Payette S, Begin Y (2000) A quantitative definition of light rings in black spruce (*Picea mariana*) at the arctic tree line in Northern Quebec, Canada. Arct Antarct Alp Res 32:324–330

Wang YP, Jarvis PG (1990) Description and validation of an array model – MAESTRO. Agric For Meteorol 1990:257-280

Wareing PF, Roberts DL (1956) Photoperiodic control of cambial activity in *Robinia pseudoacacia* L. New Phytol 55:356–366

Waring RH, Pitman GB (1985) Modifying lodgepole pine stands to change susceptibility to mountain pine beetle attack. Ecology 66:889–897

Webster PL (1979) Variation in sister-cell cycle durations and loss of synchrony in cell lineages in root apical meristems. Plant Sci Lett 14:13–22

Webster PL, MacLeod RD (1980) Characteristics of root apical meristem cell population kinetics: a review of analysis and concepts. Environ Exp Bot 20:335–358

Weinstein DA, Beloin R (1991) TREGRO, a model of carbon, water, and nutrient balances to predict response Kf plants to interacting stresses. In: Kiester AR (ed) Development and use of tree and forest response models, vol 17. (State of science and technology report), Washington, D.C., pp 129–151

Whitehead D, Jarvis PG (1981) Coniferous forests and plantations. In: Kozlowski TT (ed) Water deficits and plant growth, vol 6. Academic, New York, pp 49–133

Whitmore FW, Zahner R (1966) Development of the xylem ring in stems of young red pine trees. For Sci 12:198–210

Wilcox H (1962) Cambial growth characteristics. In: Kozlowski TT (ed) Tree growth. Ronald, New York, pp 57–88

Wilks D S (1995) Statistical methods in the atmospheric sciences. Academic, San Diego, 467 pp

Wilpert K (1990) Die Jahrringstruktur von Fichten in Abhängigkeit vom Bodenwasserhaushalt auf Pseudogley und Parabraunerde. Freiburg Bodenkund, 115 pp

Wilpert K (1991) Intraannual variation of radial tracheid diameters as monitor of site specific water stress. Dendrochronologia 9:95–113

Wilson BF (1963) Increase in cell wall surface area during enlargement of cambial derivatives in *Abies concolor*. Am J Bot 50:95–102

Wilson BF (1964) A model for cell production by the cambium of conifers. In: Zimmermann MH (ed) The formation of wood in forest trees. Academic, New York, pp 19–34

Wilson BF (1966) Mitotic activity in the cambial zone of *Pinus strobus*. Am J Bot 53:364–372

Wilson BF (1973) A diffusion model for tracheid production and enlargement in conifers. Bot Gaz 134:189–196

Wilson BF, Howard RA (1968) A computer model for cambial activity. For Sci 14:77–90

Wilson BF, Wodzicki TJ, Zahner R (1966) Differentiation of cambial derivatives: proposed terminology. For Sci 12:438–440

Wilson R, Elling W (2003) Temporal instability in tree-growth/climate response in the Lower Bavarian Forest region: implications for dendroclimatic reconstruction. Trees 18:19–28

Wilson BF (1968) Effect of girdling on cambial activity in white pine. Can J Bot 46:141–146

Wimmer R (1994) Structural, chemical and mechanical trends within coniferous trees. In: Spiecker, Kahle P (eds) Modelling tree-ring development cell structure and environment. (Workshop proceedings), Freiburg, pp 2–11

Wimmer R (1995) Intra-annual cellular characteristics and their implications for modeling softwood density. Wood Fiber Sci 27:413–420

Wodzicki TJ (1962) A contribution to the rate of xylem cell formation and differention of tracheids in *Pinus sylvestris* L. Acta Soc Bot Pol 31:219–227

Wodzicki TJ (1971) Mechanism of xylem differentiation in *Pinus sylvestris* L. Exp Bot 22:670–687

Wodzicki TJ, Peda T (1963) Investigaton on the annual ring of wood formation in European silver fir (*Abies pectinata* D.C.). Acta Soc Bot Pol 32:609–618

Wolter KE (1968) A new method for marking xylem growth. Forest Sci 14:102–104

Wong SC, Cowan IR, Farquhar GD (1979) Stomatal conductance correlates with photosynthetic capacity. Nature 282:424–426

Woo KS, Fins L, McDonald GI, Wiese MV (2001) Differences in needle morphology between blister rust resistant and susceptible western white pine stocks. Can J For Res 31:1880–1886

Woodhouse CA (1999) Artificial neural networks and dendroclimatic reconstructions: an example from the Front Range, Colorado, USA. Holocene 9:521–529

Woodmansee RG, Wallach LS (1981) Effects of fire regimes on biogeochemical cycles. US For Serv Gen Tech Rep WO 26:379–400

World Data Center-A for Paleoclimatology (2003) WebMapper display of tree-ring data sampling sites. National Geophysical Data Center, Washington, D.C. (http://www.ngdc.noaa.gov/paleo)

Worral J (1970) Interrelationships among some phenological and wood property variables in Norway spruce. TAPPI 53:58–63

Yasue K, Funada R, Noda M, Fukuzawa K (1994) Dendroclimatological study of *Picea glehnii* growing in the Teshio experimental forest of Hokkaido University. Res Bull Hokkaido Univ For 51:243–266

Yasue K, Funada K, Kobayashi O, Ohtani J (2000) A diffusion model for tracheid production and enlargement in conifers. Trees 14:223–229

Yatsenko-Khmelevsky AA (1954) The basics and methods of anatomical study of wood (in Russian). Izd AN USSR, Leningrad, 338 pp

Yoshizawa N (1987) Cambial responses to the stimulus of inclination and structural variation of compression wood tracheids in gymnosperms. Bull Utsunomiya Univ For 23:23–141

Yoshizawa N, Itoh T, Shimaji K (1985a) Helical thickening in normal and compression wood of some softwoods. IAWA Bull 6:131–138

Yoshizawa N, Koike S, Idei T (1985b) Formation and structure of compression wood tracheids induced by repeated inclination in *Taxus aespidata*. Mokusai Gakkaishi 31:325–333

Yoshizawa N, Matsumoto S, Idei T (1985 c) Morphological features of tracheid tips associated with compression wood formation in *Larix leptolepis*. IAWA Bull 6:245–253

Yoshizawa N, Okamoto Y, Idei T (1986a) Righting movement and xylem development in tilted young conifer trees. Wood Fiber Sci 18:579–589

Yoshizawa N, Tanaka Y, Idei T (1986b) Development of vascular cambium and compression wood formation in the shoot of young Spruce (*Picea jezoensis* var. *hondoensis*). IAWA Bull 7:21–30

Yoshizawa N, Kiyomiya M, Idei T (1987) Variations in tracheid length and morphological changes of tracheid tips associated with development of compression wood. Wood Sci Technol 21:1–10

Zahner R (1955) Soil water depletion by pine and hardwood stands during a dry season. For Sci 17:466–469

Zahner R (1968a) Site quality and wood quality in upland hardwoods: theoretical consideration of wood density. Forest Soils Conference, Raleigh, 43 pp

Zahner R (1968b) Water deficits and growth of trees. In: Kozlowski TT (ed) Water deficits and plant growth II. Academic, New York, pp 191–254

Zahner R, Oliver WW (1962) The influence of thinning and pruning on the date of summerwood initiation in red and jack pines. For Sci 8:51–63

Zarnovican R (2000) Climate and volume growth of young yellow birch (*Betula alleghaniensis* Britton) at three sites in the sugar maple-yellow birch forest region in Quebec. EcoScience 7:222–227

Zhang JW, Feug Z, Gregg BM, Schulmann CM (1997) Carbon isotopic composition, gas exchange, and growth of three populations of ponderosa pine differing in drought tolerance. Tree Physiol 17:461–466

Zhang T, Heginbottom JA, Barry RG, Brown J (2000) Further statistics on the distribution of frozen ground and permafrost. Polar Geogr 24:126–131

Zhang T, Barry RG, Gilichinsky D, Bykhovets SS, Sorikov VA, Ye J-P (2001) An amplified signal of climate change in soil temperatures during the last century at Irkutsk, Russia. Clim Change 49:41–76

Zimmermann MH (1964) The formation of wood in forest trees. Academic, New York, 562 pp

Zimmermann MH (1983) Xylem structure and the ascent of sap. Springer, Berlin Heidelberg New York, 143 pp

Zimmermann MH, Brown CL (1971) Trees: structure and function. Springer, Berlin Heidelberg New York, 336 pp

Zobell BJ, Jett JB (1995) Genetics of wood production. Springer, Berlin Heidelberg New York, 337 pp

Zobel BJ, Buijtenen JP van (1989) Wood variation its causes and control. Springer, Berlin Heidelberg New York, 363 pp

Subject Index

Taxonomic Index

Common plant names can be found in the Subject Index.

Ecological Studies

Volumes published since 2001

Volume 144
Coastal Marine Ecosystems of Latin America (2001)
U. Seeliger and B. Kjerfve (Eds.)

Volume 145
Ecology and Evolution of the Freshwater Mussels Unionoida (2001)
G. Bauer and K. Wächtler (Eds.)

Volume 146
Inselbergs: Biotic Diversity of Isolated Rock Outcrops in Tropical and Temperate Regions (2000)
S. Porembski and W. Barthlott (Eds.)

Volume 147
Ecosystem Approaches to Landscape Management in Central Europe (2001)
J.D. Tenhunen, R. Lenz, and R. Hantschel (Eds.)

Volume 148
A Systems Analysis of the Baltic Sea (2001)
F.V. Wulff, L.A. Rahm, and P. Larsson (Eds.)

Volume 149
Banded Vegetation Patterning in Arid and Semiarid Environments (2001)
D. Tongway and J. Seghieri (Eds.)

Volume 150
Biological Soil Crusts: Structure, Function, and Management (2001, 2003)
J. Belnap and O.L. Lange (Eds.)

Volume 151
Ecological Comparisons of Sedimentary Shores (2001)
K. Reise (Ed.)

Volume 152
Global Biodiversity in a Changing Environment: Scenarios for the 21st Century (2001)
F.S. Chapin, O. Sala, and E. Huber-Sannwald (Eds.)

Volume 153
UV Radiation and Arctic Ecosystems (2002)
D.O. Hessen (Ed.)

Volume 154
Geoecology of Antarctic Ice-Free Coastal Landscapes (2002)
L. Beyer and M. Bölter (Eds.)

Volume 155
Conserving Biological Diversity in East African Forests: A Study of the Eastern Arc Mountains (2002)
W.D. Newmark

Volume 156
Urban Air Pollution and Forests: Resources at Risk in the Mexico City Air Basin (2002)
M.E. Fenn, L. I. de Bauer, and T. Hernández-Tejeda (Eds.)

Volume 157
Mycorrhizal Ecology (2002, 2003)
M.G.A. van der Heijden and I.R. Sanders (Eds.)

Volume 158
Diversity and Interaction in a Temperate Forest Community: Ogawa Forest Reserve of Japan (2002)
T. Nakashizuka and Y. Matsumoto (Eds.)

Volume 159
Big-Leaf Mahogany: Genetic Resources, Ecology and Management (2003)
A. E. Lugo, J. C. Figueroa Colón, and M. Alayón (Eds.)

Volume 160
Fire and Climatic Change in Temperate Ecosystems of the Western Americas (2003)
T. T. Veblen et al. (Eds.)

Volume 161
Competition and Coexistence (2002)
U. Sommer and B. Worm (Eds.)

Volume 162
How Landscapes Change: Human Disturbance and Ecosystem Fragmentation in the Americas (2003)
G.A. Bradshaw and P.A. Marquet (Eds.)

Volume 163
Fluxes of Carbon, Water and Energy of European Forests (2003)
R. Valentini (Ed.)

Volume 164
Herbivory of Leaf-Cutting Ants: A Case Study on *Atta colombica* in the Tropical Rainforest of Panama (2003)
R. Wirth, H. Herz, R.J. Ryel, W. Beyschlag, B. Hölldobler

Volume 165
Population Viability in Plants: Conservation, Management, and Modeling of Rare Plants (2003)
C.A Brigham, M.W. Schwartz (Eds.)

Volume 166
North American Temperate Deciduous Forest Responses to Changing Precipitation Regimes (2003)
P. Hanson and S.D. Wullschleger (Eds.)

Volume 167
Alpine Biodiversity in Europe (2003)
L. Nagy, G. Grabherr, Ch. Körner, D. Thompson (Eds.)

Volume 168
Root Ecology (2003)
H. de Kroon and E.J.W. Visser (Eds.)

Volume 169
Fire in Tropical Savannas: The Kapalga Experiment (2003)
A.N. Andersen, G.D. Cook, and R.J. Williams (Eds.)

Volume 170
Molecular Ecotoxicology of Plants (2004)
H. Sandermann (Ed.)

Volume 171
Coastal Dunes: Ecology and Conservation (2004)
M.L. Martínez and N. Psuty (Eds.)

Volume 172
Biogeochemistry of Forested Catchments in a Changing Environment: A German Case Study (2004)
E. Matzner (Ed.)

Volume 173
Insects and Ecosystem Function (2004)
W. W. Weisser and E. Siemann (Eds.)

Volume 174
Pollination Ecology and the Rain Forest: Sarawak Studies (2005)
D. Roubik, S. Sakai, and A.A. Hamid (Eds.)

Volume 175
Antarctic Ecosystems: Environmental Contamination, Climate Change, and Human Impact (2005)
R. Bargagli

Volume 176
Forest Diversity and Function: Temperate and Boreal Systems (2005)
M. Scherer-Lorenzen, Ch. Körner, and E.-D. Schulze (Eds.)

Volume 177
A History of Atmospheric CO_2 and its Effects on Plants, Animals, and Ecosystems (2005)
J.R. Ehleringer, T.E. Cerling, and M.D. Dearing (Eds.)

Volume 178
Photosynthetic Adaptation: Chloroplast to Landscape (2005)
W.K. Smith, T.C. Vogelmann, and C. Chritchley (Eds.)

Volume 179
Lamto: Structure, Functioning, and Dynamics of a Savanna Ecosystem (2005)
L. Abbadie et al. (Eds.)

Volume 180
Plant Ecology, Herbivory, and Human Impact in Nordic Mountain Birch Forests (2005)
F.E. Wielgolaski (Ed.) and P.S. Karlsson, S. Neuvonen, D. Thannheiser (Ed. Board)

Volume 181
Nutrient Acquisition by Plants: An Ecological Perspective (2005)
H. BassiriRad (Ed.)

Volume 182
Human Ecology: Biocultural Adaptations in Human Cummunities (2006)
H. Schutkowski

Volume 183
Growth Dynamics of Conifer Tree Rings: Images of Past and Future Environments (2006)
E.A. Vaganov, M.K. Hughes, and A.V. Shashkin

MIX
Papier aus verantwortungsvollen Quellen
Paper from responsible sources
FSC® C105338
FSC
www.fsc.org

If you have any concerns about our products,
you can contact us on
ProductSafety@springernature.com

In case Publisher is established outside the EU,
the EU authorized representative is:
Springer Nature Customer Service Center GmbH
Europaplatz 3, 69115 Heidelberg, Germany

Printed by Libri Plureos GmbH
in Hamburg, Germany